Opportunities in Protection Materials Science and Technology for Future Army Applications

Committee on Opportunities in Protection Materials Science and Technology for Future Army Applications

National Materials Advisory Board
and
Board on Army Science and Technology

Division on Engineering and Physical Sciences

NATIONAL RESEARCH COUNCIL
OF THE NATIONAL ACADEMIES

THE NATIONAL ACADEMIES PRESS
Washington, D.C.
www.nap.edu

THE NATIONAL ACADEMIES PRESS 500 Fifth Street, N.W. Washington, DC 20001

NOTICE: The project that is the subject of this report was approved by the Governing Board of the National Research Council, whose members are drawn from the councils of the National Academy of Sciences, the National Academy of Engineering, and the Institute of Medicine. The members of the committee responsible for the report were chosen for their special competences and with regard for appropriate balance.

This study was supported by Contract No. W911NF-09-C-0164 between the National Academy of Sciences and the Department of Defense. Any opinions, findings, conclusions, or recommendations expressed in this publication are those of the authors and do not necessarily reflect the views of the organizations or agencies that provided support for the project.

International Standard Book Number-13: 978-0-309-21285-4
International Standard Book Number-10: 0-309-21285-5

This report is available in limited quantities from

National Materials and Manufacturing Board
500 Fifth Street, N.W.
Washington, DC 20001
nmab@nas.edu
http://www.nationalacademies.edu/nmab

Additional copies of this report are available from the National Academies Press, 500 Fifth Street, N.W., Lockbox 285, Washington, DC 20055; (800) 624-6242 or (202) 334-3313 (in the Washington metropolitan area); Internet: http://www.nap.edu.

Cover: A soldier wearing protective equipment (left); up-armored high-mobility multipurpose wheeled vehicle (HMMWV) (center); drawing showing penetration of target (right, upper) and interface defeat—the goal of protective material (right, lower). The lower border serves as a reminder of the continued increase in threat that drives the need for advances in protective materials.

Printed in the United States of America

THE NATIONAL ACADEMIES

Advisers to the Nation on Science, Engineering, and Medicine

The **National Academy of Sciences** is a private, nonprofit, self-perpetuating society of distinguished scholars engaged in scientific and engineering research, dedicated to the furtherance of science and technology and to their use for the general welfare. Upon the authority of the charter granted to it by the Congress in 1863, the Academy has a mandate that requires it to advise the federal government on scientific and technical matters. Dr. Ralph J. Cicerone is president of the National Academy of Sciences.

The **National Academy of Engineering** was established in 1964, under the charter of the National Academy of Sciences, as a parallel organization of outstanding engineers. It is autonomous in its administration and in the selection of its members, sharing with the National Academy of Sciences the responsibility for advising the federal government. The National Academy of Engineering also sponsors engineering programs aimed at meeting national needs, encourages education and research, and recognizes the superior achievements of engineers. Dr. Charles M. Vest is president of the National Academy of Engineering.

The **Institute of Medicine** was established in 1970 by the National Academy of Sciences to secure the services of eminent members of appropriate professions in the examination of policy matters pertaining to the health of the public. The Institute acts under the responsibility given to the National Academy of Sciences by its congressional charter to be an adviser to the federal government and, upon its own initiative, to identify issues of medical care, research, and education. Dr. Harvey V. Fineberg is president of the Institute of Medicine.

The **National Research Council** was organized by the National Academy of Sciences in 1916 to associate the broad community of science and technology with the Academy's purposes of furthering knowledge and advising the federal government. Functioning in accordance with general policies determined by the Academy, the Council has become the principal operating agency of both the National Academy of Sciences and the National Academy of Engineering in providing services to the government, the public, and the scientific and engineering communities. The Council is administered jointly by both Academies and the Institute of Medicine. Dr. Ralph J. Cicerone and Dr. Charles M. Vest are chair and vice chair, respectively, of the National Research Council.

www.national-academies.org

COMMITTEE ON OPPORTUNITIES IN PROTECTION MATERIALS SCIENCE AND TECHNOLOGY FOR FUTURE ARMY APPLICATIONS

EDWIN L. THOMAS, *Chair*, Massachusetts Institute of Technology
MICHAEL F. McGRATH, *Vice Chair*, Analytic Services Inc. (ANSER)
RELVA C. BUCHANAN, University of Cincinnati
BHANUMATHI CHELLURI, IAP Research, Inc.
RICHARD A. HABER, Rutgers University
JOHN WOODSIDE HUTCHINSON, Harvard University
GORDON R. JOHNSON, Southwest Research Institute
SATISH KUMAR, Georgia Institute of Technology
ROBERT M. McMEEKING, University of California, Santa Barbara
NINA A. ORLOVSKAYA, University of Central Florida
MICHAEL ORTIZ, California Institute of Technology
RAÚL A. RADOVITZKY, Massachusetts Institute of Technology
KALIAT T. RAMESH, Johns Hopkins University
DONALD A. SHOCKEY, SRI International
SAMUEL ROBERT SKAGGS, Los Alamos National Laboratory (retired), Consultant
STEVEN G. WAX, Defense Applied Research Projects Agency (retired), Consultant

Staff

ERIK SVEDBERG, NMAB Senior Program Officer
ROBERT LOVE, BAST Senior Program Officer
NANCY T. SCHULTE, BAST Senior Program Officer
HARRISON T. PANNELLA, BAST Senior Program Officer
JAMES C. MYSKA, BAST Senior Research Associate
NIA D. JOHNSON, BAST Senior Research Associate
LAURA TOTH, NMAB Senior Program Assistant
RICKY D. WASHINGTON, NMAB Administrative Coordinator
ANN F. LARROW, BAST Research Assistant

NATIONAL MATERIALS ADVISORY BOARD

ROBERT H. LATIFF, *Chair*, R. Latiff Associates
LYLE H. SCHWARTZ, *Vice Chair*, University of Maryland
PETER R. BRIDENBAUGH, Alcoa, Inc. (retired)
L. CATHERINE BRINSON, Northwestern University
VALERIE BROWNING, ValTech Solutions, LLC
YET MING CHIANG, Massachusetts Institute of Technology
GEORGE T. GRAY III, Los Alamos National Laboratory
SOSSINA M. HAILE, California Institute of Technology
CAROL A. HANDWERKER, Purdue University
ELIZABETH HOLM, Sandia National Laboratories
DAVID W. JOHNSON, JR., Stevens Institute of Technology
TOM KING, Oak Ridge National Laboratory
KENNETH H. SANDHAGE, Georgia Institute of Technology
ROBERT E. SCHAFRIK, GE Aircraft Engines
STEVEN G. WAX, Strategic Analysis, Inc.

Staff

DENNIS CHAMOT, Acting Director
ERIK SVEDBERG, Senior Program Officer
RICKY D. WASHINGTON, Administrative Coordinator
HEATHER LOZOWSKI, Financial Associate
LAURA TOTH, Senior Program Assistant

NOTE: In January 2011 the National Materials Advisory Board (NMAB) and the Board on Manufacturing and Engineering Design combined to form the National Materials and Manufacturing Board. Listed here are the members of the NMAB who were involved in this study.

Preface

Armor materials are remarkable: Able to stop multiple hits and save lives, they are essential to our military capability in the current conflicts. But as threats have increased, armor systems have become heavier, creating a huge burden for the warfighter and even for combat vehicles. This study of lightweight protection materials is the product of a committee created jointly by two boards of the National Research Council, the National Materials Advisory Board (NMAB)[1] and the Board on Army Science and Technology (BAST), in response to a joint request from the Assistant Secretary of the Army for Acquisition, Logistics, and Technology and the Army Research Laboratory. The committee examined the fundamental nature of material deformation behavior at the very high rates characteristic of ballistic and blast events. Our goal was to uncover opportunities for development of advanced materials that are custom designed for use in armor systems, which in turn are designed to make optimal use of the new materials. Such advances could shorten the time for material development and qualification, greatly speed engineering implementation, drive down the areal density of armor, and thereby offer significant advantages for the U.S. military. We hope this report will have a revolutionary effect on the materials and armor systems of the future—an effect that will meet mission needs and save even more lives.

Coincidentally, six weeks after the final committee meeting, the Army announced a draft program calling for establishment of a collaborative research alliance for materials in extreme dynamic environments.[2] Since the committee did not review the Army's preliminary request for proposal, it is not discussed in the study.

The committee was composed of a wide range of experts whose backgrounds in processing and characterization of ceramics, metals, polymers, and composites, as well as theory and modeling and high-rate testing of protection materials, combined wonderfully to make this report possible. I want to thank each and every one of the committee members for their hard work, camaraderie, and dedicated efforts over the past year and in particular, Mike McGrath, the vice chair, and chapter leads Richard Haber, John Hutchinson, Nina Orlovskaya, Don Shockey, Bob Skaggs, Raúl Radovitzky, and Steve Wax. Staff of the NMAB and the BAST did a great job supporting the study and in bringing the report to fruition.

Edwin L. Thomas, NAE, *Chair*
Committee on Opportunities in
Protection Materials
Science and Technology for
Future Army Applications

[1]In January 2011 the National Materials Advisory Board (NMAB) and the Board on Manufacturing and Engineering Design combined to form the National Materials and Manufacturing Board. The move underscored the importance of materials science to innovations in engineering and manufacturing.

[2]U.S. Army. 2010. A Collaborative Research Alliance (CRA) for Materials in Extreme Dynamic Environments (MEDE), Solicitation Number W911NF-11-R-0001, October 28. Available online at https://www.fbo.gov/index?s=opportunity&mode=form&id=48a13a80653b1fabe3f83ede9ddc641b&tab=core&tabmode=list&=. Last accessed March 31, 2011.

Acknowledgment of Reviewers

This report has been reviewed in draft form by individuals chosen for their diverse perspectives and technical expertise, in accordance with procedures approved by the National Research Council's (NRC's) Report Review Committee. The purpose of this independent review is to provide candid and critical comments that will assist the institution in making its published report as sound as possible and to ensure that the report meets institutional standards for objectivity, evidence, and responsiveness to the study charge. The review comments and draft manuscript remain confidential to protect the integrity of the deliberative process. We wish to thank the following individuals for their review of this report:

Charles E. Anderson, Jr., Southwest Research Institute,
Diran Apelian, Worcester Polytechnic Institute,
Morris E. Fine, Technological Institute Professor Emeritus, Northwestern University
Peter F. Green, University of Michigan,
Julia R. Greer, California Institute of Technology,
Wayne E. Marsh, DuPont Central Research and Development,
R. Byron Pipes, Purdue University,
Bhakta B. Rath, Naval Research Laboratory,
Susan Sinnott, University of Florida, and
Edgar Arlin Starke, Jr., University of Virginia

Although the reviewers listed above have provided many constructive comments and suggestions, they were not asked to endorse the conclusions or recommendations nor did they see the final draft of the report before its release. The review of this report was overseen by Elisabeth M. Drake, NAE, Massachusetts Institute of Technology Laboratory of Energy and the Environment. Appointed by the National Research Council, she was responsible for making certain that an independent examination of this report was carried out in accordance with institutional procedures and that all review comments were carefully considered. Responsibility for the final content of this report rests entirely with the authoring committee and the institution.

Contents

Tables, Figures, and Boxes

TABLES

FIGURES

BOXES

Acronyms and Abbreviations

AlON	aluminum oxynitride
ARL	Army Research Laboratory
ARO	Army Research Office
ATC	Aberdeen Test Center (Maryland)
ATH	aluminum trihydroxide
ATPD	Army Tank Purchase Description
BAST	Board on Army Science and Technology
CIP	cold isostatic pressing
CNT	carbon nanotubes
CTE	coefficient of thermal expansion
CZM	cohesive zone models
DARPA	Defense Advanced Projects Research Agency
DMC	dynamic magnetic compaction
DoD	Department of Defense
DoE	Department of Energy
ERDC	Engineer Research and Development Center (U.S. Army)
ESAPI	enhanced small arms protective insert
FGAC	functionally graded armor composites
FGM	functionally gradient material
FSP	fragment simulating projectiles
GHz	gigahertz
GPa	gigapascals
HEL	Hugoniot elastic limit
HMMWV	high-mobility multipurpose wheeled vehicle (Humvee)
HP	hot pressing
IBA	Interceptor body armor
ICME	Integrated Computational Materials Engineering (an NRC report)
ITAR	International Traffic in Arms Regulations
JHB	Johnson, Holmquist, and Beissel
M&S	modeling and simulation
MMC	metal matrix composites
MPa	megapascal
MZ	Mescall zone
NDE	nondestructive evaluation
NIJ	National Institute of Justice
NMAB	National Materials Advisory Board
NRC	National Research Council
NSF	National Science Foundation
NVI	normal velocity interferometer
OHPC	Omnipresent High-Performance Computing program
PAN	polyacrylonitrile
PBO	polybenzoxazole
PBZT	poly(benzobisthiazole)
PC	polycarbonate
PE	polyethylene
PMC	polymer matrix composite
PMD	protection materials-by-design
PMMA	polymethyl methacrylate
PPTA	polyparaphenylene terephthalamide
PU	polyurethane
PVB	polyvinyl butyral
QMU	quantification of margins and uncertainties
RHA	rolled homogeneous armor

SAN poly(styrene-co-acrylonitrile)
SAPI small arms protective insert
SCS shear compression (test)
SEM scanning electron microscope
SiC silicon carbide
SiSiC siliconized silicon carbide
SPS spark plasma sintering

TDI transverse displacement interferometer
TEM transmission electron microscopy
TPU thermoplastic polyurethanes

UHMWPE ultrahigh molecular weight polyethylene
UQ uncertainty quantification
UV ultraviolet

VISAR velocity interferometry system for any reflector
V&V verification and validation

XCT x-ray computed tomography

Summary

This report responds to a request by the Assistant Secretary of the Army (Acquisition, Logistics, and Technology) to the National Research Council (NRC) to examine the current theoretical and experimental understanding of the key issues surrounding protection materials, identify the major challenges and technical gaps for developing the future generation of lightweight protection materials, and recommend a path forward for their development. While underscoring the paramount need for lightweight materials, the charge included requirements to consider multiscale shockwave energy transfer mechanisms and experimental approaches for their characterization over short timescales, as well as multiscale modeling techniques to predict mechanisms for dissipating energy.

Accordingly, two NRC boards—the National Materials Advisory Board[1] and the Board on Army Science and Technology—established the Committee on Opportunities in Protection Materials Science and Technology for Future Army Applications to investigate opportunities in protection materials science and technology for the Army. What follows is the evaluation developed by that committee.

The report considers exemplary threats and design philosophy for the three key applications of armor systems: (1) personnel protection, including body armor and helmets, (2) vehicle armor, and (3) transparent armor. For each of these applications, specific constraints drive the armor design and thus the ultimate choice of protection materials.

In developing its recommendations, the committee assessed current knowledge and gaps in that knowledge as it sought to prioritize the various types of lightweight protective materials and armor systems for future research. Key areas and research challenges for protection materials discussed in these pages include the following:

- Penetration mechanisms in metals and alloys, ceramics and glasses, and polymeric materials (Chapter 3).
- Failure mechanisms in cellular-sandwich materials due to blast (Chapter 3).
- Current capabilities for modeling and simulation of protection materials and material systems on scales ranging from the atomic to the macroscopic, including a discussion of state-of-the-art modeling and simulation tools (Chapter 4).
- The state of the art in experimental methods, including defining the length and timescales of interest, evaluating material behavior at the relevant high-strain rates, and investigating shock physics, dynamic failure processes, and impact phenomenology (Chapter 4).
- Ceramic armor materials, including crystalline and amorphous ceramics, ceramic powders, processing and fabrication techniques, and transparent crystalline ceramics (Chapter 5).
- Fibers, including the effect of fiber diameter on strength in high-performance fibers, microstructural advances to approach the theoretical maximum tensile strength and modulus, and the need for mechanical tests at high strain rates and pressures (Chapter 5).
- Ballistic fabrics, including ballistic testing, failure mechanisms, and interactions among fibers and among yarns during loading (Chapter 5).
- Metals and metal-matrix composites and their desirable attributes, especially those of low-density metals such as magnesium alloys (Chapter 5).
- Fabrication and assembly of armor systems, with an emphasis on adhesives for armor and transparent armor, including (1) general considerations for selecting an adhesive interlayer and (2) testing, simulation, and modeling of adhesives and armor systems (Chapter 5).

[1]In January 2011 the National Materials Advisory Board (NMAB) and the Board on Manuacturing and Engineering Design combined to form the National Materials and Manufacturing Board. The move underscored the importance of materials science to innovations in engineering and manufacturing.

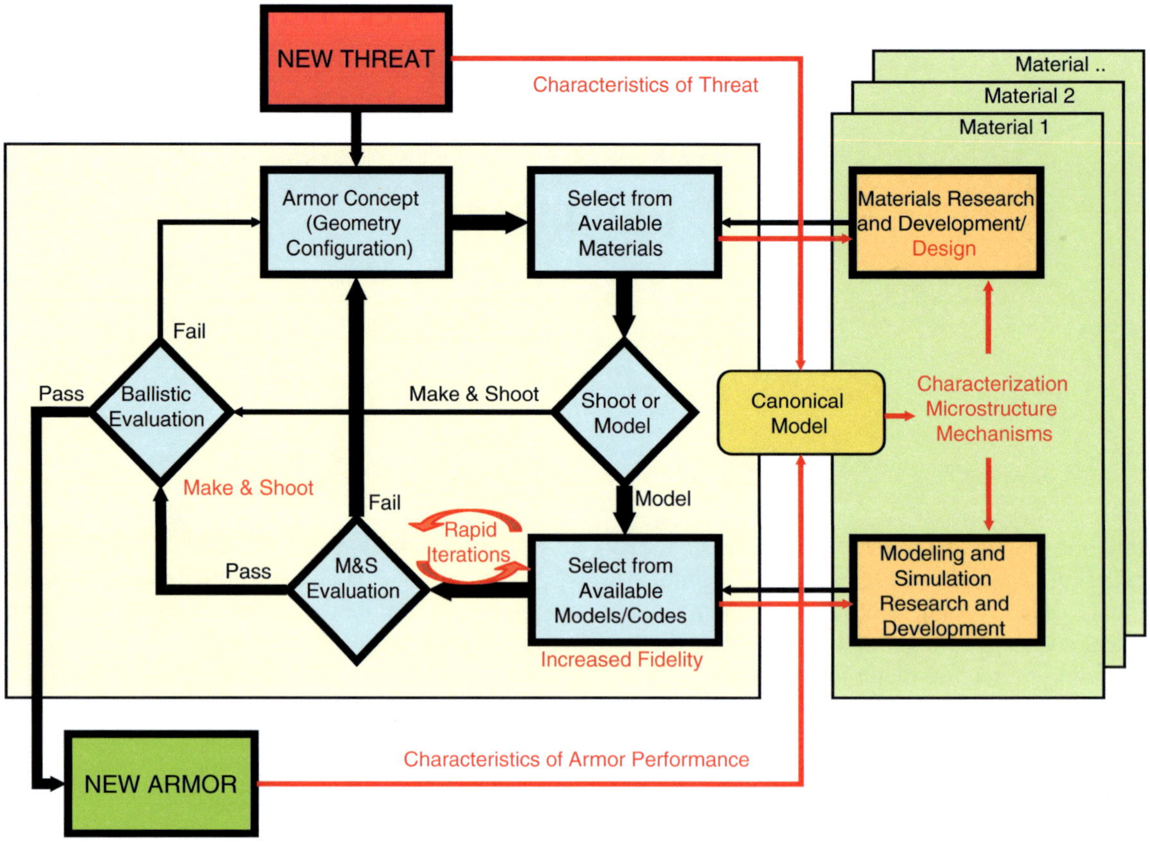

FIGURE S-1 New paradigm for armor development. The new design path for armor provides enhanced and closer coupling of the materials research and development community and the modeling and simulation community, resulting in significantly reduced time for development of new armor. This new approach connects the armor design process to the materials research and development community through canonical models to deal with the restricted information problem. The elements of armor system design are not themselves new, but the emphasis shifts from design-make-shoot-redesign to rapid simulation iterations, and from designing with off-the-shelf materials to designing that exploits materials for their protective properties. The feedback loop between armor system design and material design contrasts with current practice, in which a one-way flow puts new materials on the shelf to be tried in the make-shoot-look process.

Findings and recommendations pertaining to these areas and research challenges appear in Chapters 3 through 5. The single overarching recommendation is repeated here in the summary, along with the four key recommendations in the main text.

OVERARCHING RECOMMENDATION

The conclusion of this study is that the ability to design and optimize protection material systems can be accelerated and made more cost effective by operating in a new paradigm of lightweight protection material development (Figure S-1). In this new paradigm, the current armor system design practice, which relies heavily on a design-make-shoot iterative process, is replaced by rapid iterations of modeling and simulation, with ballistic evaluation used selectively to verify satisfactory designs. Strong coupling with the materials research and development community is accomplished through canonical models that translate armor system requirements (often data with restricted access) into characterizations, microstructures, behaviors, and deformation mechanisms that an open research community can use in designing new lightweight protection materials. The principal objective of this new paradigm is to enable the design of superior protection materials and to accelerate their implementation in armor systems. This new paradigm will build upon the multidisciplinary collaboration concepts and lessons from other applications documented in the report *Integrated Computational Materials Engineering*.[2] It can be focused on the most promising opportunities in lightweight protection materials, bringing such current products as ceramic plates and polymer fiber materials well beyond their

[2]NRC. 2008. Integrated Computational Systems Engineering: A Transformational Discipline for Improved Competitiveness and National Security. Washington, D.C.: The National Academies Press.

present state of performance and opening the possibility for radically new armor system solutions to be explored and optimized in tens of months rather than tens of years.

Overarching Recommendation. Given the long-term importance of lightweight protection materials to the Department of Defense (DoD) mission, DoD should establish a defense initiative for protection materials by design (PMD), with associated funding lines for basic and applied research. Responsibility for this new initiative should be assigned to one of the Services, with participation by other DoD components whose missions also require advances in protection materials. The PMD initiative should include a combination of computational, experimental, and materials testing, characterization, and processing research conducted by government, industry, and academia. The program director of the initiative should be given the authority and resources to collaborate with the national laboratories and other institutions in the use of unique facilities and capabilities and to invest in DoD infrastructure where needed.

This overarching recommendation requires actions in four important elements of the PMD initiative.

RECOMMENDATIONS

Element 1—Fundamental Understanding of Mechanisms of Deformation and Failure Due to Ballistic and Blast Threats

The first element of the PMD initiative would be to develop better fundamental understanding of the mechanisms of high-rate[3] material deformation and failure in various protection materials, discussed in Chapter 3. As part of the new paradigm, armor development should be considered not from the viewpoint of conventional bulk material properties but from the viewpoint of mechanisms. The deeper fundamental understanding could lead to the development of more failure-resistant material compositions, crystal structures, and microstructures and to protective materials with better performance. Moreover, by identifying the operative mechanisms and quantifying their activity, mathematical damage models can be written that may allow computational armor design. Chapter 3 discusses failure mechanisms for the several classes of materials.

Recommendation S-1/6-1. The Department of Defense should establish a program of sustained investment in basic and applied research that would facilitate a fundamental understanding of the mechanisms of deformation and failure due to ballistic and blast events. This program should be established under a director for protection materials by design, with particular emphasis on the following:

- Relating material performance to deformation and failure mechanisms. Developing models and data for choosing materials based on their ability to inhibit or avoid failure mechanisms as opposed to choosing them based on bulk properties as measured in quasistatic and dynamic tests.
- Developing superior armor materials by identifying compositions, crystalline structures, and microstructures that counteract observed failure mechanisms and by establishing processing routes to the synthesis of these materials.
- Reducing the cost of production of protection materials by improving the processes and yields and by enhancing the ability to manufacture small lots.

Element 2—Advanced Computational and Experimental Methods

The second element of the PMD initiative would be to advance and exploit the capabilities of the emerging computational and experimental methods discussed in Chapter 4. The first objective is to predict the ballistic and blast performance of candidate materials and materials systems as a prelude to the armor design process. The second objective is to define requirements that will guide the synthesis, processing, fabrication, and evaluation of protection materials. The PMD initiative would develop the next generation of

- DoD advanced protection codes that incorporate experimentally validated, high-fidelity, physics-based models of material deformation and failure, as well as the necessary high-performance computing infrastructure;
- Experimental facilities and capabilities to assess and certify the performance of new protection materials and system designs, as well as provide insight into fundamental material behaviors under relevant conditions with unprecedented simultaneous high spatial and temporal resolution; and
- Collaborative infrastructure for encouraging direct communication and improved cooperation between modelers and experimenters, through both (1) the establishment of collaborative environments and (2) requirements in proposals when the specific research topic is well served by such collaboration.

The high-priority opportunities identified in Chapter 4 will need sustained investment and program direction to advance computational and experimental capabilities. The envisioned computational capabilities must be developed in partnership with a strong experimental effort that identifies the dynamic mechanisms of material behavior. These mechanisms must be understood and modeled for the activity to be successful, the material characteristics and properties must be known for the simulations to be carried out, and the outcomes of the computational modeling must be validated.

[3]Ballistic velocities typically range from several hundred to several thousand meters per second and can lead to strain rates of up to $10^5\ s^{-1}$.

Recommendation S-2/6-2. The Department of Defense should establish a program of sustained investment in basic and applied research in advanced computational and experimental methods under the director of the protection materials by design (PMD) initiative, with particular emphasis on the following:

- *Dynamic mechanism characterization.* Identify and characterize (1) the failure mechanisms underlying damage to a material caused by projectiles from weapons and detonations and (2) the compositional and microstructural features of each constituent of the material, as well as the material's overall structure. An enhanced experimental infrastructure will be needed to make progress in high-resolution (time and space) experiments on material deformation and failure characterization.
- *Code validation and verification.* Focus on multiscale, multiphysics material models, integrated simulation/experimental protocols, prediction with quantified uncertainties, and simulation-based qualification to help advance the predictive science for protection systems.
- *Challenges and canonical models.* Periodically propose open challenges comprising design, simulation, and experimental validation that will convincingly demonstrate the PMD. Each challenge problem must address the corresponding canonical model and must result in quantifiable improvements in performance within that framework.

Element 3—Development of New Materials and Material Systems

The third element of the PMD initiative is the development and production of new materials and material systems whose characteristics and performance can achieve the behavior validated in modeling and simulation of the new armor system. The recommendations in this element target the most promising opportunities identified in Chapter 5.

Recommendation S-3/6-3. The Department of Defense should establish a program of sustained investment in basic and applied research in advanced materials and processing, under the director of the PMD initiative program, with particular emphasis on the following:

- *A sustained effort to develop a database of high-strain-rate materials for armor.* Material behavior and dynamic properties must be measured and characterized over the range of strains, strain rates, and stress states in the context of penetration and blast events. Develop a comprehensive database of materials that exhibit high-strain-rate behavior and consider them as materials of interest. The PMD director should designate a custodian for this database and arrange for experimental results of the PMD program to be provided to the database and shared with the research community. The database should include ceramics, polymers, metals, glasses, and composite materials in use today and should be expanded as new materials are developed.
  —*Opaque and transparent ceramics and ceramic powders.* The intrinsic properties of opaque and transparent ceramics and ceramic powders are not yet fully realized in armor systems. There is need for understanding at the atomic, nano-, and micron levels of how powders and processing can be designed and manipulated to maximize the intrinsic benefits of dense ceramic armor and reduce production costs.
  —*Polymeric, carbon, glass, and ceramic fibers.* There is an opportunity to develop finer diameter and more ideally microstructured polymeric and carbon fibers with potentially a two- to fivefold improvement in specific tensile strength over the current state of the art. Such improvements would significantly reduce the weight of body armor.
  —*Polymers.* In addition to polymer fibers, thermoplastic and thermoset polymers are used as monolithic components and also serve as matrixes in various composites. Improved measurements of and models for the deformation mechanisms and failure processes are needed for thermoplastic- and thermoset-based protection materials.
  —*Magnesium alloys.* The very low density of magnesium provides potential for the development of very lightweight alternatives to traditional metallic materials in protection material systems. The basic understanding of strengthening mechanisms in magnesium should be advanced, especially the development of ultra-fine-grained magnesium alloys through severe plastic deformation. Magnesium-based fibers are also worthy of exploration.
- *Adhesives and active brazing/soldering materials.* Development of adhesives and active brazing/soldering materials and their processing methods to match the elastic impedance of current materials while minimizing the thermal stresses will improve the ballistic and blast performance of panels made of bonded armor, including transparent armor.
- *Test methods.* Advances are needed in test methods for determining the high strain rates (10^3 to 10^6 s^{-1}) and dynamic failure processes of (especially) fibers, polymers, and ceramics. Results should be passed on to the designated database of materials with high-strain-rate behavior.
- *Material characterization.* The characterization of, composition, crystalline structure, and microstruc-

ture at appropriate length scales is a key task that will need more attention to take advantage of the improved experimental tools for quantifying initial and deformed microstructures.

- *Cost reduction.* Advances are needed to reduce the cost of producing protection materials by improving their processing and yield and by improving small-lot manufacturing capability.
- *Processing science and intelligent manufacturing.* Advances are needed in basic understanding of and ability to model the consequences of material processing for performance and other characteristics of interest. Intelligent manufacturing sensing and control capabilities are needed that can maintain low variance and produce affordable protection materials, even in relatively low volumes.

Element 4—Organizational Approach

The fourth element of the PMD initiative is an organizational construct for multidisciplinary collaboration among academic researchers, government laboratories, and industry, in both restricted-access and open settings. The PMD initiative will need strong top-level leadership with insight into both the open and restricted research environments and the authority to direct funding and set PMD priorities. The program will require committed funding to ensure long-term success and should be subject to periodic external reviews to ensure that high standards of achievement are established and maintained. To meet these requirements, the committee recommends the notional DoD organizational approach depicted in Figure S-2.

Recommendation S-4/6-4. In order to make the major advances needed for the development of protection materials, the Department of Defense should appoint a PMD program director, with authority and resources to accomplish the following:

- Plan and execute the PMD initiative and coordinate PMD activities across the DoD.
- Select an existing facility to be the DoD center for PMD and fund a research director and the staff, equipment, and programs needed by the PMD initiative;
- Award a competitive contract for an open access PMD center whose mission would be to host and foster open collaboration in research and development of protection materials;

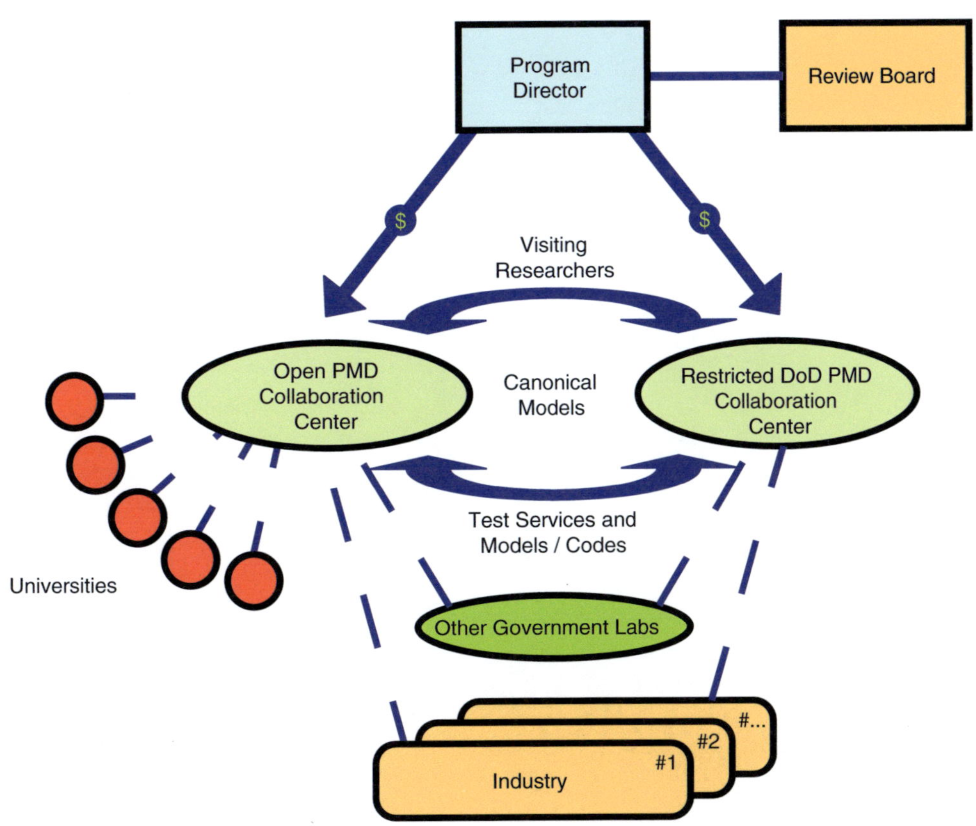

FIGURE S-2 PMD initiative organizational structure involving academic researchers, government laboratories, and industry.

- Establish an external review board to conduct periodic reviews of programs in both centers; and
- Provide liaison with the Department of Energy, the National Institute of Standards and Technology, and other government laboratories on matters related to PMD.

The sponsor asked that the committee suggest an organizational structure for the path forward and a teaming approach for it. In considering the sponsor's request that the study report not include restricted material, which would have precluded wide dissemination to the research and development communities, the committee recognized the broader issue of the role restricted information plays in impeding research collaborations.[4] Such limitations are prudent and necessary but require periodic review to ensure they are consistent with the current state of open knowledge and do not unnecessarily restrict the exchange of information with an open research community when such an exchange would be beneficial to national security.

The chapters that follow develop the rationale and conclusions that underpin the detailed recommendations in Chapter 6 and identify needed actions in the four elements of the initiative.

[4]A detailed discussion of the effects on research of classification guidelines, security, and export control is beyond the scope of this study.

1

Overview

INTRODUCTION

Since the beginning of armed conflict, armor has played a significant role in the protection of warriors. In present-day conflicts, armor has inarguably saved countless lives. Over the course of history—and especially in modern times—the introduction of new materials and improvements in the materials already used to construct armor have led to better protection and a reduction in the weight of the armor. Body armor, for example, has progressed from the leather skins of antiquity, through the flak jackets of World War II to today's highly sophisticated designs that exploit ceramic plates and polymeric fibers to protect a person against direct strikes from armor-piercing projectiles (Figure 1-1). The advances in vehicle armor capabilities have similarly been driven by new materials, as shown in Figures 1-2 and 1-3.

But even with such advances in materials, the weight of the armor required to manage threats of ever-increasing destructive capability presents a huge challenge. For example, body armor, which presently constitutes almost 30 percent of a soldier's fighting load,[1] is the single largest weight carried by an Army rifle squad. For vehicles, up-armored Humvees have reached the limit beyond which armor cannot be added without "compromising essential vehicle capabilities."[2]

FIGURE 1-1 A soldier wearing protective equipment. SOURCE: Adapted from Gaston Bathalon, Commander, U.S. Army Research Institute of Environmental Medicine, "The Soldier as a Decisive Weapon: USAMRMC soldier focused research," presentation to the Board on Army Science and Technology on February 15, 2011.

The Challenge

The challenge for protective material developers, made clear by current military engagements, is twofold: (1) to ensure the rapid (re)design and manufacture of armor systems optimized against specific threats and (2) at the same time, ensure that these systems are as lightweight as possible. As described above, many of the advances in the performance of lightweight armor have historically come from the introduction of new or improved materials. However, it has become increasingly difficult to produce new materials with properties that allow the design of complex new armor systems or the rapid iterations of such designs. Not only must a material be quickly identified, but it must also be reliably produced,

[1]Dean, C. 2008. The modern warrior's combat load: Dismounted operations in Afghanistan. 2003. Medicine and Science in Sports & Exercise 40(5): 60.

[2]Inspector General, U.S. Department of Defense. 2009. Procurement and delivery of joint service armor protected vehicles. Report No. D-2009-046. Available online at http://www.dodig.mil/audit/reports/fy09/09-046.pdf. Accessed April 7, 2001.

FIGURE 1-2 Up-armored high-mobility multipurpose wheeled vehicle (HMMWV, or Humvee). SOURCE: Available at http://www.militaryfactory.com/armor/imgs/hmmwv-m1114uah.jpg. Courtesy of U.S. DoD.

which is not currently possible with the extensive, costly, and time-consuming practice that is perhaps best described as "build it, shoot it, and then look at it." This problem, including specific recommendations for areas of investigation, will be addressed further at the end of Chapter 3.

This seeming technological inability to keep up with evolving needs is not exclusive to protection materials. A recent National Research Council (NRC) study, *Integrated Computational Materials Engineering: A Transformational Discipline for Improved Competitiveness and National Security*,[3] describes how, like advances in armor, the "vast majority of disruptive technologies since the start of the industrial revolution" have been due to materials innovations, but that "the insertion of new materials technologies has become much more difficult and less frequent" as materials development fails to keep pace with the rapid design process. This describes exactly the problems experienced with development of the new protection materials that are the focus of this study. The *Integrated Computational Materials Engineering* (ICME) report cites many advances and several examples of successful implementation. It advocates pushing the large body of existing computational materials science to the next step. Unfortunately, while "the optimization of the materials, manufacturing processes, and component design" is well described in the ICME report, the path forward for protection materials is far more complicated, since designs must deal with highly nonlinear and large deformations typically not encountered in commercial products, where applied stresses are kept well below the elastic limit in the linear regime. Simply put, the key materials properties—for example, tensile strength and toughness—that inform the design of commercial structures and devices are well established and extensively measured. Such is not the case for armor.

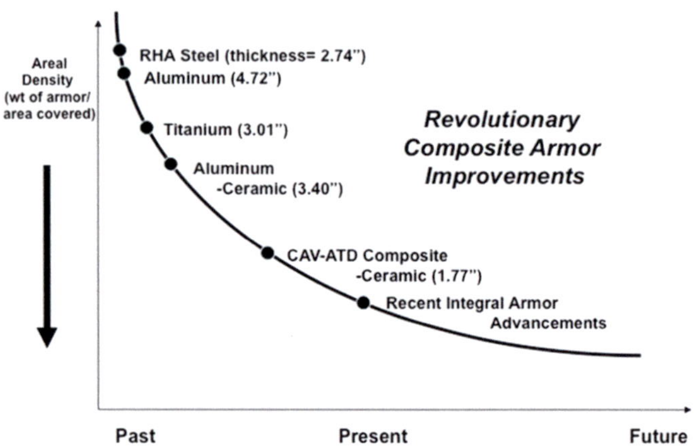

FIGURE 1-3 Areal density of armor versus time, demonstrating that new lightweight materials such as titanium, aluminum, and ceramics have provided increased protection at a lower weight per unit area over time. The flattening curve illustrates that the challenge for the future is to be able to continue to decrease the areal density of the armor despite increased threats. SOURCE: Adapted from Fink, B.K. 2000. Performance Metrics for Composite Integral Armor. ARL-RP-8. Aberdeen Proving Ground, Md.: Army Research Laboratory.

The armor that protects U.S. fighting forces is seldom a single, homogeneous material. More often than not, what is called "armor" is actually a complex system constructed of several, often quite different, materials arranged in a very specific configuration designed to protect against a particular threat. As will be discussed extensively in this study, the properties and behavior of a protection material must be considered in the specific context of how it will be used in the construction of a particular armor system. Further, there is often little understanding of how to link specific material properties to the actual behavior of the materials and armor systems during the many types of ballistic and blast events. It is often the case that new protection materials have not been well characterized with respect to strain rates, pressures, and the like under appropriate conditions, either alone or as part of an armor system, and databases for materials' performance and constitutive relationships are often not available. This is especially true at the high strains and very high strain rates relevant to ballistic and blast threats. This gap in knowledge greatly limits the ability of simulation codes to play a significant role in guiding the development of new materials. Moreover, the design philosophy is completely dependent on how the armor system is to be used.

In this study, the committee was guided by military applications that necessitate lightweight armor, with particular emphasis on (1) personnel protection, which includes body armor and helmets, (2) vehicle armor, and (3) transparent

[3]NRC. 2008. Integrated computational systems engineering: A transformational discipline for improved competitiveness and national security. Washington, D.C.: The National Academies Press.

armor for face shields, vehicle windows, and other applications requiring transparency. For each of these applications, system-level constraints affect armor design and, ultimately, the design and choice of protection materials. The committee viewed the need for strong coupling between armor system designers and protection materials developers as the most difficult challenge to be addressed.

SCOPE OF THE STUDY

The Assistant Secretary of the Army (Acquisition, Logistics, and Technology) requested that the NRC's Board on Army Science and Technology and its National Materials Advisory Board collaborate to form an ad hoc study committee to investigate opportunities in protection materials science and technology for the Army.

The committee was given the following statement of task:

Statement of Task

> An ad hoc committee will conduct a study and prepare a report on protection materials for the Army to explore the possibility of a path forward for these materials. Specifically, the committee will:
>
> 1. Review and assess the current theoretical and experimental understanding of the major issues surrounding protection materials.
> 2. Determine the major challenges and technical gaps for developing the future generation of light weight protection materials for the Army, with the goal of valid multi-scale predictive simulation tools for performance and, conversely, protection materials by design
> 3. Suggest a path forward, including approach, organizational structure and teaming, including processing, material characterization (composition and microstructure), quasi-static and dynamic mechanical testing and model development and simulation and likely timeframes for the Army to deliver the next generation protection materials.

The sponsor requested that in considering the questions posed by the task statement, the committee should consider the following:

- Shock wave energy dissipative (elastic, inelastic and failure) and management mechanisms throughout the full materials properties spectrum (nano through macro).
- Experimental approaches and facilities to visualize and characterize the response at nano and mesoscales over short time scales.
- Multi-scale modeling techniques to predict energy dissipative mechanisms (twinning, stacking faults, etc.) from the atomic scales and bulk material response.
- Materials and material systems issues including processing and characterization techniques focusing on intrinsic (single crystal) properties and processing controlled extrinsic characteristics (phases, microstructure, interfaces).

The sponsor further requested that the study not include restricted material so as to permit wide dissemination of study results to the research and development communities.

STUDY METHODOLOGY

The study consisted of six full two- or three-day committee meetings held mostly in Washington, D.C., but also included a three-day meeting held near Aberdeen Proving Ground, in Maryland, one day of which was devoted to visiting the U.S. Army Research Laboratory and observing some of the relevant experimental testing facilities. The committee received briefings from academic, industrial, military, and government presenters covering lightweight materials for warfighter protection as well as vehicle protection. Topics ranged from ballistic threats to blast threats and from very hard to relatively soft armor materials and included a brief from the National Aeronautics and Space Administration on protection of space vehicles against hypervelocity impacts from meteors. The committee met in closed sessions to develop conclusions and recommendations responsive to the study task, drawing upon the materials presented in open sessions and additional published materials cited throughout the report.

Report Organization

The report contains 6 chapters and 10 appendixes. This first chapter provides the introduction and background to the study and defines the overall perspective of the report. Chapter 2 introduces the reader to some armor systems and gives examples relating to the key concept of reducing the areal density of the protection materials while improving the performance of armor against ever-increasing threats. Chapter 2 also makes the important distinction between armor systems and material systems.

Chapters 3, 4, and 5 provide the technical details of the committee's assessment of current knowledge and discuss the gaps and opportunities meriting high priority in future research. In order to appreciate the task for designing materials for armor, Chapter 3 covers the complex interacting mechanisms and processes that take place during deformation and failure when a material is impacted by a high-velocity penetrator.

Chapter 4 addresses the computational and experimental approaches to armor material design and the need to better couple and integrate these activities to create materials by design and armor systems by design. Multiscale modeling and simulation are reviewed for a few key scenarios for threat-protection materials, illustrating the considerable challenge of accurately capturing the extreme deformations involved in penetration. The goal is to enable much more rapid advances in both materials and systems and, accordingly, a much faster and better response to changing threats.

Chapter 5 provides a broad perspective on the structure and composition of exemplary protection materials including ceramics, polymers, metals, and composites. It highlights the most exciting opportunities in materials research—opportunities that may lead to revolutionary advances in protection and a significant reduction in areal density. This chapter is extensively appended with descriptions and processing for specific materials.

Chapter 6 suggests a path forward and recommends future research tied to the conclusions of the earlier chapters. To realize all the potential gains for protection materials noted in the report, an important new paradigm is proposed, along with an organizational plan for its implementation.

Collectively, these chapters provide technical recommendations and a proposed way forward for long-term research directed at the development of the following:

- A fundamental understanding of how a ballistic object or a blast interacts with a material—in other words, the material's performance. This would include an understanding of which time and length scales are important and how controlling the material's composition and microstructure, and hence its mechanical behavior, contributes to altering the deformation mechanisms and improving performance;
- Experimental approaches to identify and quantitatively characterize the mechanisms and processes that lead to damage during these dynamic events;
- Quantitative relationships for the evolution of the damage during a high-deformation event and extending these relationships to account for multiple events, termed multi-hit relationships;
- Computational approaches—coupled with synergistic experiments that inform and validate—to predict the performance of specific protection materials in an integrated armor configuration;
- Model-driven methods to design new materials or improve existing ones to meet the behavior criteria for successful protection;
- Model-driven synthesis, processing, and manufacturing capabilities to produce affordable materials in quantities needed for defense applications; and
- An environment that allows successful interplay and collaboration among the DoD, government labs, industry, and academe while at the same time addressing security and organizational matters.

Appendix A includes the Statement of Task, Appendix B provides biographical sketches of the committee members, and Appendix C lists committee meetings and speaker topics. Appendixes D through J contain much additional detailed information on protection materials, supplementing the points made in Chapters 3, 4, and 5.

Other Issues

The sponsor asked that the committee suggest both an organizational structure and a teaming approach as part of the path forward. In considering the sponsor's request that the study report not include restricted material so as to enable wide dissemination to the research and development communities, the committee recognized a broader issue—namely, that restricted information is a barrier to research collaborations.[4] Chapter 2 addresses armor system design at the unrestricted level but closes with a comment on the extensive regime of security and export control restrictions that affects research on protection materials. Several speakers from industry, government, and academic organizations told the committee that these restrictions make it extremely difficult for fundamental research in protection materials to be usefully communicated among the various organizations and to be connected to the development of armor systems, which entails restricted information. It notes that a review of classification guidelines and export control restrictions would facilitate clearer, more up-to-date boundaries for the necessary control of information. Chapter 6 proposes an organizational structure to bridge this gap.

Overarching Recommendation

The committee's key recommendations are presented in Chapter 6, with ancillary recommendations found in Chapters 3 and 4. The overall thrust of this report, however, is evident in the following overarching recommendation:

Overarching Recommendation. Given the long-term importance of lightweight protection materials to the Department of Defense (DoD) mission, DoD should establish the defense initiative protection materials by design (PMD), with associated funding lines for basic and applied research. Responsibility for this new initiative should be assigned to one of the Services, with participation by other DoD components whose missions also require advances in protection materials. The PMD initiative should include a combination of computational, experimental, and materials testing, characterization, and processing research conducted by government, industry, and academia. The program director should be given the authority and resources to collaborate with the national laboratories and other institutions in the use of unique facilities and capabilities and to invest in DoD infrastructure where needed.

This overarching recommendation requires actions in four important elements of the PMD initiative:

[4]A detailed discussion of the effects on research of classification guidelines, security, and export controls is beyond the scope of this study.

- *Element 1.* Fundamental understanding of mechanisms of deformation and failure due to ballistic and blast threats.
- *Element 2.* Advanced computational and experimental methods.
- *Element 3.* Development of new materials and material systems.
- *Element 4.* Organizational approach.

The chapters that follow develop the rationale and conclusions that underpin the detailed recommendations in Chapter 6 and identify actions that are needed to address the four elements of the initiative. The committee is unanimous in its support of these recommendations.

2

Fundamentals of Lightweight Armor Systems

As described in Chapter 1, the path forward for development of protection materials must consider the armor systems that form the context in which those protection materials are used. This chapter presents a brief overview of a few armor systems, including the threats to them and the designs for them, to give the reader enough information to inform the discussion.

The first section of this chapter discusses how armor systems are characterized and tested. However, while a general discussion such as this is valid for all classes of armor systems, the threats and the design philosophy are completely dependent on how the armor system is used. Accordingly, the following discussion covers the three applications of armor systems considered in this study: (1) personnel protection, which includes body armor and helmets, (2) vehicle armor, and (3) transparent armor.[1] For each of these applications, very specific constraints drive the armor design and thus the ultimate choice of protection materials. This chapter provides, within the security guidelines discussed in the final section, a general description of the threats and the armor designs against those threats as well as a brief description of some systems fielded as of 2011.

ARMOR SYSTEM PERFORMANCE AND TESTING IN GENERAL

Definition of Armor Performance

The complexities of armor systems make even the assessment of weight situationally dependent: What is lightweight for vehicles is extremely heavy for personnel. Thus, in assessing whether an armor system is sufficiently lightweight, one cannot look at the absolute weight of the system. Rather, because armor is used to protect a particular area, its practical weight is best described by its areal density, A_d:

$$A_d = \text{Weight of the armor system/Area being protected}$$

The units are kilograms per square meter (kg/m^2) or, more commonly in the United States, pounds per square foot. Note that areal density is a physical characteristic of the armor and does not indicate if that armor is effective. The effectiveness of two armor systems can only be assessed by comparing their performance against the same threat. The effectiveness of a given armor system is called its mass effectiveness, E_m, a dimensionless quantity that is simply the ratio of the areal density of rolled homogeneous armor (RHA), a common steel for tank armor (see Box 2-1 for its composition) that will stop a particular threat, to the areal density of the given armor that will stop that same threat:

$$E_m\ (\text{Armor}) = A_d(\text{RHA})/A_d(\text{Armor})$$

The mass effectiveness of an armor system does indeed indicate how effective it is against a specific threat and generally suggests whether the system may be considered lightweight—that is, the higher the E_m value, the lighter the weight of the armor system. However, one of the complications of armor is that E_m does not translate from one threat to another; it is even possible that two armor systems will reverse their relative effectiveness against different threats.

BOX 2-1
Composition of Rolled Homogeneous Armor [L] (MIL-DTL-12560)

- Low-alloy (Ni-Cr-Mo), high-strength steel (0.26-0.28 percent C).
- Quenched and tempered (Stage III, 500°C-600°C) material, cementite strengthening precipitate:/tempered martensite structure.

[1] Transparent armor is the technical term for protective transparent material systems commonly called ballistic-resistant windows.

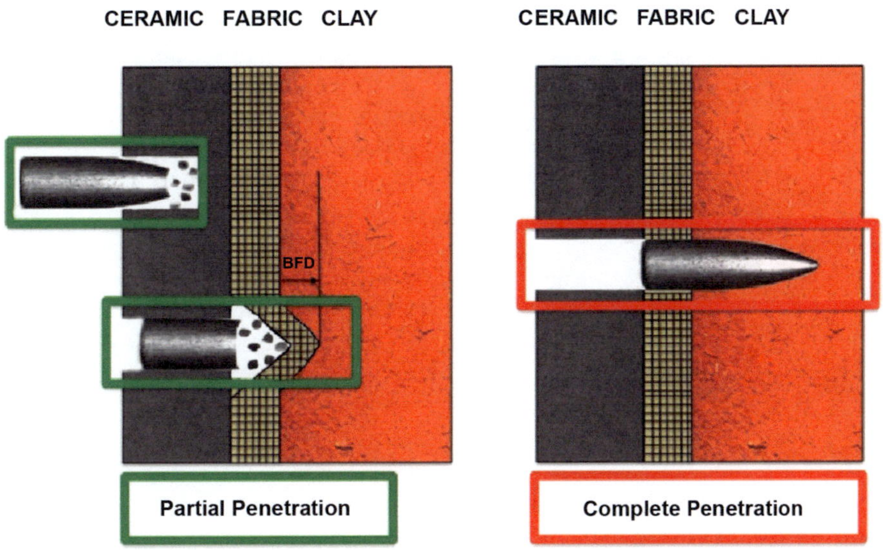

FIGURE 2-1 Partial and complete ballistic penetration. In a partial penetration the projectile stops within the armor structure, whereas in a complete penetration, it exits the armor structure. Note that the clay is not part of the armor structure but is placed behind the armor to record its deformation. BFD, back-face deformation.

Testing of Armor Systems

This section describes the testing and analysis of complete armor systems. The experimental approaches used to understand the behavior and measure the properties of individual materials are discussed in Chapters 3 through 5.

Measurement of both partial and complete penetration by threats of the separate material composing the system and of the full armor system is key to understanding how materials are selected for use in armor systems to protect against ballistics. In the case of body armor, in addition to the ability of the armor to stop the projectile, there is another requirement—namely, that the deflection of the backside of the armor toward the wearer be small.

The specifics of the tests used to qualify armor systems for field use are well documented and will not be described at length here. As an example, the very elaborate requirements for the testing of body armor are described in great detail in the National Institute of Justice (NIJ) standard.[2] In addition, a recent National Research Council (NRC) report examined specific aspects of the techniques used to evaluate body armor.[3] Yet another recent report by the Department of Defense (DoD) Office of the Inspector General[4] described the Army's testing to certify armor. Although the purchase specification for body armor might seem insensitive, it allows for an "acceptable number of complete and partial penetrations," as shown in Figure 2-1. An additional parameter for body armor certification is the maximum depth of the back-face deformation for partial penetrations. (Back-face deformation is the depth of the crater left by each partial penetration in the clay placed behind the armor during testing with threats. It represents the blunt force trauma inflicted on the wearer, which can contribute to injury or even death.) The accepted deformation of the back face of an armor system is currently 44 mm (1.73 in.) or less[5] (see Figure 2-1).

To assess the different threats against a particular armor system, two key measurements, V_0 and V_{50}, are made. V_0, the ballistic limit, is "the maximum velocity at which a particular projectile is expected to consistently fail to penetrate armor of given thickness and physical properties at a specified angle of obliquity."[6] If the measured V_0 exceeds the maximum velocity for a particular threat (see Table 2-1) the armor system is said to defeat that threat. Essentially, the

[2]Department of Justice. 2008. Ballistic Resistance of Body Armor, NIJ Standard–0101.06. Available online at http://www.ncjrs.gov/pdffiles1/nij/223054.pdf. Last accessed April 15, 2011.

[3]NRC. 2009. Phase I Report on Review of the Testing of Body Armor Materials for Use by the U.S. Army: Letter Report. Washington, D.C.: The National Academies Press. Available online at http://www.nap.edu/catalog.php?record_id=12873. Accessed April 7, 2011.

[4]Inspector General, Department of Defense. 2009. DoD Testing Requirements for Body Armor, Report No. D-2009-047. Available online at http://www.dodig.mil/audit/reports/fy09/09-047.pdf. Last accessed April 15, 2011.

[5]Department of Justice. 2008. Ballistic Resistance of Body Armor, NIJ Standard–0101.06. Available online at http://www.ncjrs.gov/pdffiles1/nij/223054.pdf. Last accessed April 15, 2011.

[6]Department of Defense. 1997. Department of Defense Test Method Standard: V50 Ballistic Test for Armor, MIL-STD-662F, December 18. Aberdeen Proving Ground, Md.: U.S. Army Research Laboratory.

TABLE 2-1 National Institute of Justice (NIJ) Ballistic Threat Standards

Level	Projectile	Weight (g)	Velocity (m/s)	Kinetic Energy (Relative to Type IIA)
Type IIA	9 mm full-metal-jacketed round nose (FMJ RN)	8.0	373± 9.1	1.0
	.40 S&W FMJ	11.7	352 ± 9.1	1.3
Type II	9 mm FMJ RN	8.0	398 ± 9.1	1.1
	.357 magnum jacketed soft point (JSP)	10.2	436 ± 9.1	1.7
Type IIIA	.357 SIG FMJ flat nose (FN),	8.1	448 ± 9.1	1.5
	.44 magnum semijacketed hollow point (SJHP)	15.6	436 ± 9.1	2.7
Type III (rifles)	7.62 mm FMJ, steel-jacketed bullets (U.S. military designation M80)	9.6	847 ± 9.1	6.2
Type IV (armor-piercing rifle)	.30 caliber armor-piercing (AP) bullets (U.S. military designation M2 AP)	10.8	878 ± 9.1	7.5

qualification tests described above ensure that V_0 exceeds the performance specification.

However, the expense of firing and the inability to control projectile velocity exactly makes the determination of 0 percent penetration statistically problematic during the experimental phase of armor development. The determination of V_0 is therefore generally reserved for the final stages of development and qualification.

For research and development purposes, the use of V_{50}, "the velocity at which complete penetration and partial penetration are equally likely to occur," is much more prevalent. These tests are done with a configuration similar to that in Figure 2-1 but without the clay, which is replaced by a "witness plate" placed at a distance behind the armor configuration. A complete penetration event takes place when a thin witness plate is fully penetrated, or perforated, by the projectile; partial (or no) penetration takes place when no perforation of the witness plate is observed. To calculate V_{50}, the highest partial/no penetration velocities and the lowest complete penetration velocities are used, generally with at least 4 and often as many as 10 shots—enough to make sure there are at least two partial/no and at least two complete penetrations.

During the development of armor systems, it is much more important to understand what is actually occurring during the penetration event than it is to simply measure V_0 or V_{50}. To this end, ballistic ranges are often equipped with an array of sophisticated diagnostic tools. For example, at Aberdeen Test Center (ATC), projectile velocity is measured with optical screens and electronic counters before, inside, and after passing the target.[7] The ATC range also has high-speed cameras that can capture 6,688 frames per second at full resolution and up to 100,000 frames per second at lower resolutions. In addition, flash x-rays can provide a three-dimensional reconstruction of a material's deformation and failure during a ballistic event. [8] It is clear that researchers wish for additional real-time measurements on ballistic time scales both locally and globally in relation to the point of impact. The ability to make quantitative measurements across many properties would necessitate approaches and methods wholly beyond those that are currently known.

Figure 2-2, taken from an earlier NRC study,[9] shows a typical range at ATC as well as one at New Lenox Machine Co.

Exemplary Threats and Armor Designs

Although the testing and definitions described above hold for all classes of armor systems, the threats and the design philosophy are completely dependent on how the armor is used. Thus, each of the three applications focused on in this report (personnel, vehicle, and transparent armors) are treated separately. It should be noted that military armor systems are currently purchased according to performance specifications that are classified. Descriptions of threats and designs in this study are taken from the open literature and documents approved for public release. As such, they are only illustrative of current threats and designs.

PERSONNEL PROTECTION

Threat

Modern armor for personnel protection includes both body armor and combat helmets. The threats for which personnel armor is designed are small-caliber projectiles, including both bullets and fragments. The level of ballistic protection of personnel armor is taken as the total kinetic energy of a single round that the armor can stop.[10] The stan-

[7]Rooney, J.P. 2008. Army Aberdeen Test Center Light Armor Range Complex. ITEA Journal 29: 347-350.

[8]An example of using flash x-rays to observe the sample and projectile changes during a penetration event is shown in Figure 2-6, which is discussed later in this chapter.

[9]NRC. 2009. Phase I Report on Review of the Testing of Body Armor Materials for Use by the U.S. Army: Letter Report. Washington, D.C. : The National Academies Press. Available online at http://www.nap.edu/catalog.php?record_id=12873. Accessed April 7, 2011.

[10]Montgomery, J.S., and E.S. Chin. 2004. Protecting the future force: A new generation of metallic armors leads the way. AMPTIAC Quarterly 8(4): 15-20.

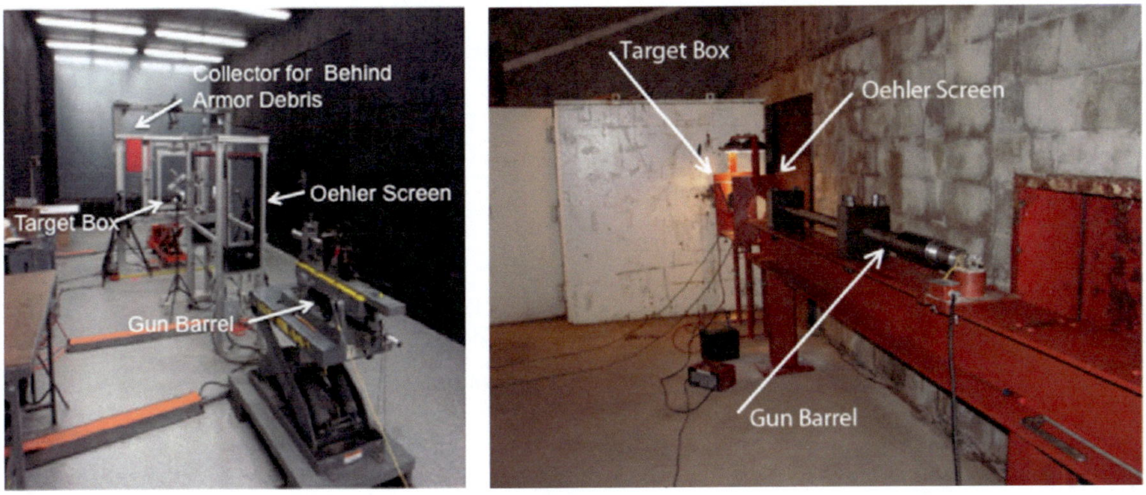

FIGURE 2-2 Indoor firing ranges. Depicted are (left) the gun barrel (foreground) and Oehler screens at the light armor range complex, which measure velocity midway between the barrel and target. The target box contains the target being shot at and debris. The red panel collects behind-armor debris. Depicted at right is an alternative setup for a commercial indoor firing range at New Lenox Machine Co. SOURCE: Adapted from John Wallace, Technical Director, ATC, "Body armor test capabilities," presentation to the Committee to Review the Testing of Body Armor Materials for Use by the U.S. Army, on March 10, 2010.

FIGURE 2-3 Examples of 7.62 mm (.30 cal) small arms projectiles. SOURCE: Courtesy of Robert Skaggs.

dards set by the NIJ shown in Table 2-1[11] are for typical ballistic threats, although not specifically those for military body armor, which are classified. Note that a Type IV projectile has more than 7.5 times the energy of a Type IIA projectile.

In addition to surviving the impact of specific projectiles (see Figure 2-3), there is generally a requirement to withstand multiple hits on the same armor panel. For armor meeting NIJ Type IIA and Type III standards, panels must demonstrate the ability to survive six hits without failure. Only Type IV has no multi-hit requirements.[12] Personnel protection armor is also often designed against fragments.

Finally, for body armor, as previously mentioned, stopping penetration is not the only issue. It is also important that when stopping the projectile, the armor itself does not deflect to an extent that would severely injure the wearer. This puts an additional constraint on body armor systems. (See the preceding discussion on back-face deflection.)

Design Considerations for Fielded Systems

The design of armor for personnel protection depends on the specific threat. For fragments and lower velocity penetrators, vests are typically made from polymer fibers (see Chapter 5). Advances in fibers for personnel armor began with the use of fiberglass and nylon. These were followed in the late 1960s by polyaramid fibers (DuPont PRD 29 and PRD 49), now called Kevlar. Later, high molecular weight polyethylene fibers, made of Spectrashield and Dyneema, were also used as backing in vests. Zylon, made of polybenzobisoxazole (PBO), has also been considered. Figure 2-4 depicts how the evolution of fibers has steadily improved the performance of polymer vests. Thus, the primary factor in the design of armor for vests is the selection of the fiber.

When the threat increases to rifle rounds, including

[11]Department of Justice. 2008. Ballistic Resistance of Body Armor, NIJ Standard–0101.06. Available online http://www.ncjrs.gov/pdffiles1/nij/223054.pdf. Last accessed April 15, 2011.

[12]Ibid.

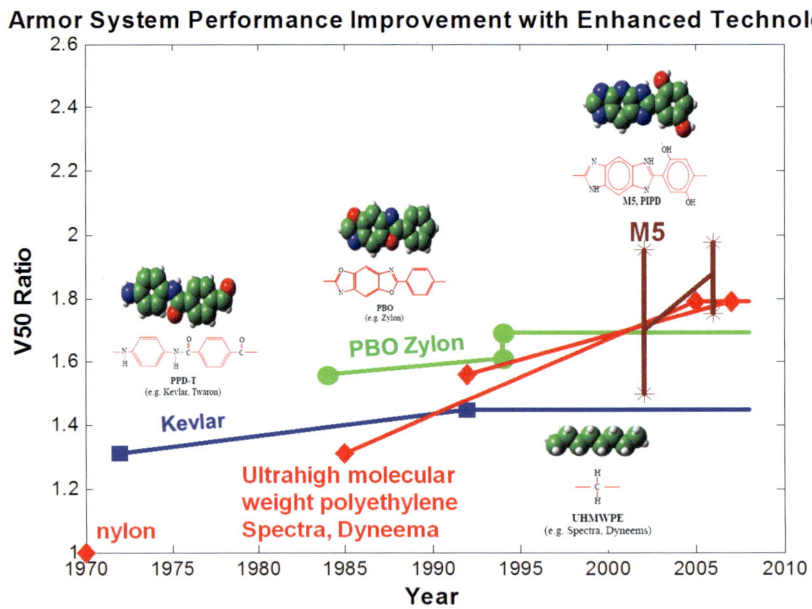

FIGURE 2-4 Increase in ballistic performance as a function of improved fibers. This figure depicts how the V_{50} of fiber-based vests has increased as new fibers have been introduced over the years. SOURCE: Philip Cunniff, U.S. Army Natick Soldier Research, Development and Engineering Center, "Fiber research for soldier protection," presentation to the committee, March 10, 2010.

armor-piercing projectiles (see Table 2-1, Types III and IV), ballistic fabric alone is insufficient. Stopping these threats requires adding a ceramic plate to the outside of the vest. The hard ceramic blunts and/or erodes the projectile nose, which increases the projected area of the projectile and spreads the load across more of the fabric.[13] It is the combination of two independently developed materials—a ceramic faceplate and a fiber fabric—that constitutes the armor system and provides overall protection. The combination creates a complex system where the performance of the ceramic and the polymer backing (vest) are intimately connected. An extended discussion of ceramics and polymer protection materials can be found in Chapter 5.

The currently fielded body armor, the Interceptor body armor (IBA), makes use of the combination of ceramic and fiber described above and shown in Figure 2-5.[14] The main component of this armor is the improved outer tactical vest, which provides protection against fragments and 9-mm rounds.[15] Enhanced small-arms protective insert (ESAPI) ballistic plates and enhanced side ballistic insert plates are inserted into plate carrier pockets in the polymeric vest. These plates can withstand multiple small-arms hits, including armor-piercing rounds.[16]

IBA can stop small-arms ballistic threats and fragments, thus reducing the number and severity of wounds. An improvement, the X small-arms protective insert, is designed for "potential emerging small arms ballistic threats."[17]

The deltoid and axillary protectors, an integral component of the improved outer tactical vest, extend protection against fragments and 9-mm rounds to the upper arm areas (see Figure 1-1).[18]

The combination of ceramic inserts and polymeric fibers in the IBA vest is an example of how particular arrangements of specific materials make up a typical armor system. The complexity goes even further: A change in threat can drastically change the performance of a given armor system. Figure 2-6 shows how the Nammo 7.62-mm M993 tungsten carbide projectile, with a velocity of 970 m/sec, more easily defeats a B_4C ceramic plate than does the Type IV APM2 threat. This indicates how armor systems solutions are intertwined with the specific threat they are intended to defeat.

Because helmets and vests demand similar levels of pro-

[13]Montgomery, J.S., and E.S. Chin. 2004. Protecting the future force: A new generation of metallic armors leads the way. AMPTIAC Quarterly 8(4): 15-20.

[14]Inspector General, Department of Defense. 2009. DoD Testing Requirements for Body Armor. Report No. D-2009-047, January 29. Available online at http://www.dtic.mil/cgi-bin/GetTRDoc?AD=ADA499208&Location=U2&doc=GetTRDoc.pdf. Last accessed April 29, 2011.

[15]Figure 2-5 shows the version of tactical vest before the improved outer tactical vest was introduced.

[16]U.S. Army. 2010. Interceptor Body Armor (IBA) brochure, October. Available online at https://peosoldier.army.mil/FactSheets/PMSPIE/SPIE_SPE_IBA.pdf. Last accessed April 29, 2011.

[17]Ibid.

[18]Ibid.

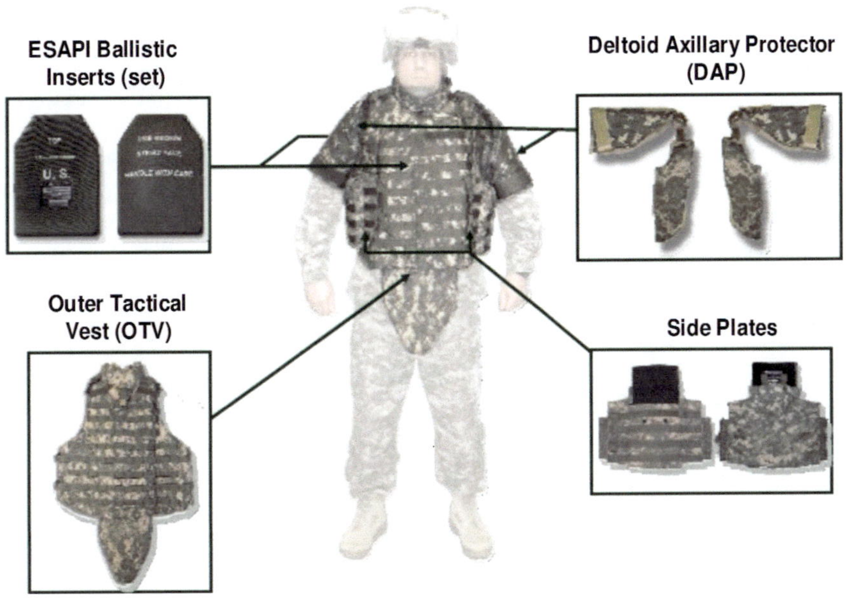

FIGURE 2-5 Interceptor body armor. Shown are the various components that make up the Interceptor body armor system (see DoD Inspector General's Report No. D-2009-047, January 29, 2009). The outer tactical vest, the deltoid axillary protectors, and the carrier for the ESAPI inserts (not shown) are made of Cordura, Kevlar, and/or Twaron fabric. The ESAPI ballistic inserts are composite ceramic plates with ballistic fiber backing (see the Interceptor body armor [IBA] brochure of the Program Executive Office, Soldier, October 2010). SOURCE: DoD Inspector General. 2009. DoD Testing Requirements for Body Armor. Report No. D-2009-047, January 29. Available online at http://www.dtic.mil/cgi-bin/GetTRDoc?AD=ADA499208&Location=U2&doc=GetTRDoc.pdf. Last accessed April 29, 2011.

tection, primary ballistic protection is also based on the performance of the fiber. However, the currently fielded helmet, the advanced combat helmet (see Box 2-2 for materials of construction), must not only provide ballistic protection, but it must also protect against blunt forces. Equally important, the helmet must provide comfort and thermal management without degrading vision or hearing and be able to interface with other equipment, including night vision goggles and

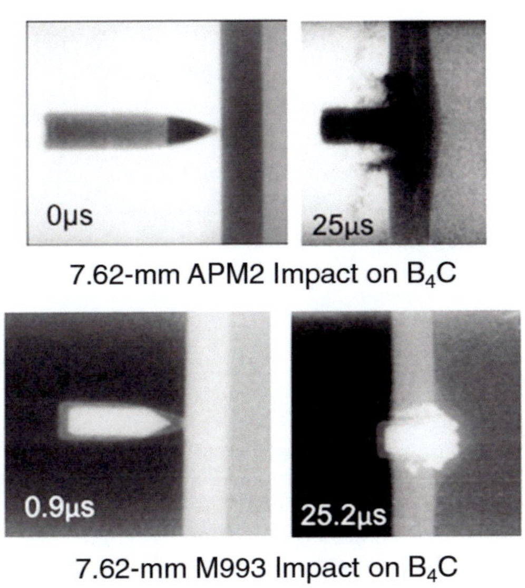

FIGURE 2-6 Effect of a ballistic threat on performance. This figure shows X-ray exposures during two impacts on boron carbide plates, each with a different type of projectile. In the top set, a 7.62-mm Type IV APM2 has not yet fully penetrated the ceramic after 25 microseconds. In the bottom set, in the same time frame, the 7.62-mm M993 projectile has begun to exit the ceramic. This is striking evidence of the effect of different threats on the performance of ballistic armor. SOURCE: Adapted from William Gooch, Jr., U.S. Army Research Laboratory, "Overview of the development of ceramic armor technology—Past, present and the future," presentation at the 30th International Conference on Advanced Ceramics and Composites, Cocoa Beach, Florida, January 24, 2006.

BOX 2-2
Construction of the Advanced Combat Helmet

Component materials:

- Helmet shell: aramid fabric + resin.
- Chin strap: Cotton/polyester webbing and foam nape pad, or nylon webbing and leather nape pad; foam pads are made of polyurethane.

SOURCE: U.S. Army. 2010. Advanced Combat Helmet (ACH) brochure, October. Available online at https://peosoldier.army.mil/Factsheets/PMSPIE/SPIE_SPE_ACH.pdf. Last accessed April 29, 2011.

weapons.[19,20] Ultimately, the weight of the helmet is limited by the ability of the neck to bear weight, especially over long periods of time.

VEHICLE ARMOR

While vehicle armor is generally understood to encompass armor systems to protect all classes of vehicles, this study will focus on armor protection for land vehicles such as the M1A1/M1A2 Abrams main battle tank, the Bradley fighting vehicle, the Stryker combat vehicle, and the high-mobility multipurpose wheeled vehicle (HMMWV, or Humvee) (see Figure 2-7).

Threat

Like personnel armor, vehicle armor is also typically required to protect against small-caliber projectiles and fragments. In addition, however, it is required to stop a host of other threats. These include medium- and large-caliber ballistic threats (20-140 mm);[21] shaped charge munitions, as depicted in Box 2-3; and chemical energy munitions. Rocket-propelled grenades are ubiquitous in the world of terrorists owing to the efforts of the countries that manufacture them to market them to developing countries. Because little effort was made to destroy ammunition dumps during the invasion of Iraq, the artillery projectiles left behind have since been used to fashion improvised explosive devices. Countries such as Iran have taken it upon themselves to manufacture many sizes of projectiles that are nominally concave metal disks propelled by large cylindrical high-explosive charges.

Specific requirements for the multithreat environment to which truck and tactical wheel systems are exposed are defined by the Army's long-term armor strategy specifications, which are classified.

Design Considerations for Fielded Systems

The design of armor systems for vehicles depends on the size of the vehicle, the threat or threats the vehicle is likely to encounter, and, equally important, the weight of the armor that the vehicle can handle. Since the early days of tanks in World War I, metal has been the primary armor material used for large combat vehicles. Table 2-2 gives selected examples of such materials and their applications.

Figure 2-8 depicts the various classes of armor that are in use or under consideration for combat vehicles. This study considers only the passive armor systems; electromagnetic, energetic, and smart armor are beyond its scope, as are reactive armor systems.

As with personnel protection, passive vehicle protection is generally a complicated arrangement of material layers, each serving a different role in the overall protection schedule. Figure 2-9 schematically depicts one such arrangement that comprises six layers of various materials, including ceramics, metals, and polymers.[22] Note that the entire system serves many more functions than just protection against projectiles.

Unlike designs for protecting personnel, armor designs for vehicles are less constrained in thickness. This allows for a concept known as "spaced armor," another option for the arrangement of armor. In spaced armor, a thin armor plate is separated from the main armor system with the goal of breaking up or disrupting the projectile, thus making it easier for the remainder of the armor to stop it. This concept was used by the Germans in World War II[23] and in various armor configurations since. It should also be noted that, even if the threat does not completely exit the armor, pieces of the back face can be accelerated by the shock wave, creating spall, which can have sufficient velocity to considerably damage people and equipment inside the vehicle. Thus, armor design must minimize behind-the-armor damage, which can ad-

[19]U.S. Army. 2010. Advanced Combat Helmet (ACH) brochure, October. Available online at https://peosoldier.army.mil/Factsheets/PMSPIE/SPIE_SPE_ACH.pdf. Last accessed April 29, 2011.

[20]Walsh, S.M., B.R. Scott, T.L. Jones, K. Cho, and J. Wolbert. 2008. A materials approach in the development of multi-threat warfighter head protection, December. Available online at http://www.dtic.mil/cgi-bin/GetTRDoc?AD=ADA504397&Location=U2&doc=GetTRDoc.pdf. Last accessed April 29, 2011.

[21]Normandia, M.J., J.C. LaSalvia, W.A. Gooch Jr., J.W. McCauley, and A.M. Rajendran. 2004. Protecting the future force: Ceramics research leads to improved armor performance. AMPTIAC Quarterly 8(4): 21-27.

[22]William Gooch, Jr., U.S. Army Research Laboratory, "Overview of the development of ceramic armor technology—Past, present and the future," presentation at the 30th International Conference on Advanced Ceramics and Composites, Cocoa Beach, Fla., January 24, 2006.

[23]A. Hurlich. 1950. Spaced Armor. Available online at http://www.dtic.mil/cgi-bin/GetTRDoc?AD=ADA954865&Location=U2&doc=GetTRDoc.pdf. Last accessed April 29, 2011.

FIGURE 2-7 Examples of Army combat vehicles. This figure portrays a subset of combat vehicles for which ballistic and/or blast protection is a critical consideration. SOURCE: Photo courtesy of the U.S. Army.

BOX 2-3
Shaped Charge Characteristics

Shaped charges are made by inverting a soft metal cone (typically copper) that will be propelled by an explosive charge to velocities near 10,000 m/s. The copper slug is hydrodynamically driven into the plastic regime and stretches continuously as it is propelled forward toward the target. At some point it starts to separate into a string of liquidlike particles, but before this occurs it is just like a long rod penetrator that is traveling at supersonic speed and will penetrate great thicknesses. The optimum standoff distance for a chemical energy penetrator is between 2.4 and 4 cone diameters. At this distance, the penetrator will not have started to fragment before it hits the target. The example below shows (1) copper liner, (2) charge, (3) body, (4) booster, and (5) initiation charge.

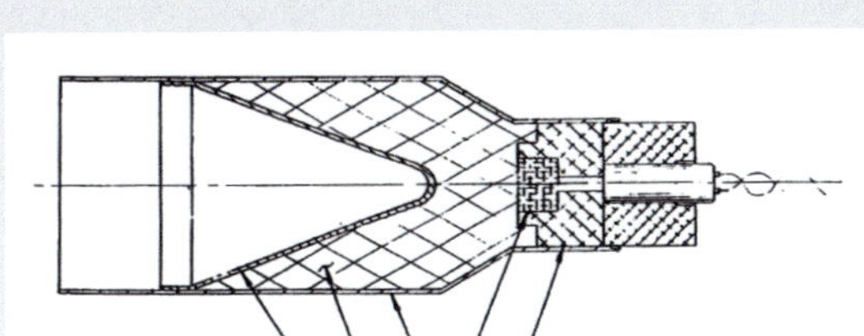

TABLE 2-2 Metallic Armor Materials

Metal	Military Specification	Application
Rolled homogeneous armor	MIL-DTL-12560	M1A1/M1A2 Abrams light armored vehicle, above beltline
High-hardness steel armor	MIL-DTL-46100	M1A1/M1A2 Abrams light armored vehicle, below beltline
Aluminum alloy 5083-H131	MIL-DTL-46027	M113 armored personnel carrier M109 Paladin self-propelled howitzer Bradley fighting vehicle, lower half
Aluminum alloy 7039-T64	MIL-DTL-46063	Bradley fighting vehicle, upper half

SOURCE: Montgomery, J.S., and E.S. Chin. 2004. Protecting the future force: A new generation of metallic armors leads the way. AMPTIAC Quarterly 8(4): 15-20.

versely affect the survival of the crew even if the projectile is stopped.[24]

Before the start of the current conflicts, light vehicles (e.g., Humvees and light trucks) were lightly armored if at all. However, unanticipated threats began to be seen—for example, rocket-propelled grenades and improvised explosive devices—causing a rethinking of that approach. Programs to quickly up-armor the Humvees and other vehicles were

[24]Prakash, A. 2004. Virtual Experiments to Determine Behind-Armor Debris for Survivability Analysis, December. Available online at http://www.dtic.mil/cgi-bin/GetTRDoc?AD=ADA433014&Location=U2&doc=GetTRDoc.pdf. Last accessed April 29, 2011.

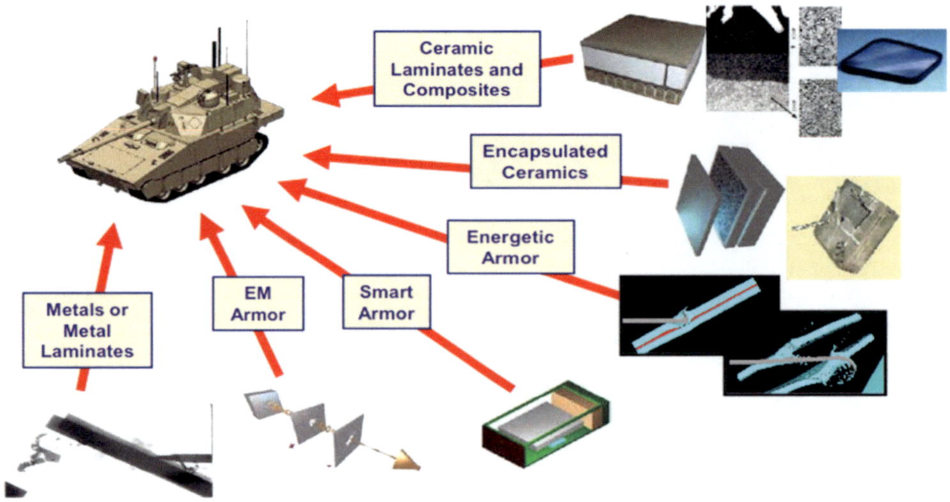

FIGURE 2-8 Examples of vehicle protection. This figure shows the many types of protection systems that are used or under consideration for Army combat vehicles. This study looks at only those materials that passively protect the vehicle from ballistics and blast threats. SOURCE: Christopher Hoppel, Chief, High Rate Mechanics and Failure Branch, Army Research Laboratory, "Multi-scale modeling of armor materials," presentation to the committee, March 10, 2010.

established. Since August 2004, all Marine Corps vehicles operating outside the forward operating bases have had their armor protection upgraded.[25] Consequently, there is now a very large array of armor combinations, often with one kit laid on top of the other, making the scheme shown in Figure 2-9 simple by comparison.

Since a bomb blast severely damaged the U.S.S. *Cole* on October 12, 2000, taking 19 lives, the Navy has also shown more interest in developing structures that can survive a blast. A Navy multidisciplinary research program known as Integrated Cellular Materials Approach to Force Protection is developing complex, topologically designed sandwich panels for this application. Like vehicle armor, these panels must protect against both ballistic and other threats.

TRANSPARENT ARMOR

Threat

The windshields and side windows of vehicles such as Humvees and trucks are an important application for transparent armor. Currently, such windows are designed to protect against armor-piercing threats as well as high-velocity fragments. In addition, they must be able to withstand multiple hits and to fracture in a way that maintains their structural integrity and transparency. Advanced applications of transparent armor often demand additional protection against electromagnetic fields or lasers. This study, however, will cover only the ballistic requirements of transparent armor.

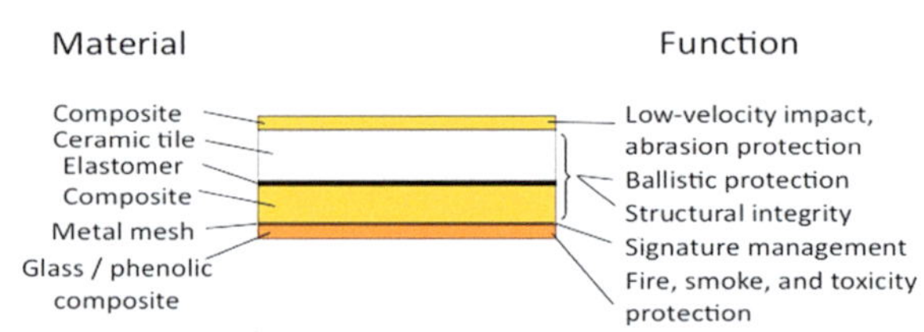

FIGURE 2-9 Schematic of vehicle armor protection system. The armor is made of many layers, each with a different overall function. In this construct, ballistic protection is obtained primarily through the ceramic tile and composite backing. The composite faceplate also contributes to the protective properties of the vehicle armor, while the ballistic components contribute to the structural integrity of the armor. Other configurations (not shown) might include a structure designed primarily for blast resistance. SOURCE: William Gooch, Jr., U.S. Army Research Laboratory, "Overview of the development of ceramic armor technology—Past, present and the future," presentation at the 30th International Conference on Advanced Ceramics and Composites, Cocoa Beach, Fla., January 24, 2006.

The specifications for transparent armor are called out in Army Tank Purchase Description (ATPD) 2352P, July 7, 2008,[26] which describes the general characteristics that transparent armor must possess to qualify for purchase. These

[25]Gen. William L. Nyland, Assistant Commandant of the Marine Corps, and Major General (Select) William D. Catto, Commanding General Marine Corps Systems Command, Statement before the House Armed Services Committee on Marine Corps vehicle armoring and improvised explosive device countermeasures, June 21, 2005.

[26]ATPD 2352P, July 7, 2008, supersedes ATPD 2352N, January 3, 2008. ATPD 2352 defines a standardized four-shot pattern and is used throughout the Army to provide consistent criteria for evaluating multiple impacts on transparent armor. ATPD 2352P is available at https://aais.ria.army.mil/AAIS/award_web_09/W52H0909A00030000/Award_attach/Attach1.pdf.

FIGURE 2-10 Example of transparent armor for a vehicle window. SOURCE: Stephan Bless, Institute for Advanced Technology, University of Texas at Austin, "Transparent armor research issues," presentation to the committee, March 10, 2010.

specs also set forth criteria for the environmental effects of transparent armor. As with other documents on applications of armor, DTA184044, the document that describes the threats that must be defeated, is contained in a classified appendix and so cannot be elaborated on here.

Design Considerations for Fielded Systems

In contrast to conventional opaque ceramic armors, the design of transparent armor is often driven by the multi-hit requirement, a requirement mostly achieved by layering (see Figures 2-10 and 5-14). A typical transparent armor uses a layer of glass or glass ceramic followed by a layer of polycarbonate and then other similar layers until seven or more have been stacked and bonded with polyvinyl butyral adhesive layers. While the backing of transparent armor is primarily polycarbonate, other polymeric materials, such as polyurethane, are showing some potential.

Most current armor windows are laminates of glass and plastic.[27] The three main transparent ceramic candidates are currently aluminum oxynitride (AlON), magnesium aluminate spinel ($MgAl_2O_4$), commonly referred to as spinel, and single-crystal aluminum oxide (Al_2O_3-sapphire).[28] These materials are described further in Chapter 5.

[27]Patel, P.J., G.A. Gilde, P.G. Dehmer, and J.W. McCauley. 2000. Transparent armor. The AMPTIAC Newsletter 4(3): 1, 2-5, 13.

[28]Sands, J.M., P.J. Patel, P.G. Dehmer, A.J. Hsieh, and M.C. Boyce. 2004. Protecting the force: Transparent materials safeguard the Army's vision. AMPTIAC Quarterly 8(4): 28-36.

FROM ARMOR SYSTEMS TO PROTECTION MATERIALS

The goal of armor system development is (1) to continually decrease the weight—that is, to increase E_m—required to protect against a given threat or (2) to not increase the weight required to protect against a greater threat. According to the Army, new armor systems can in fact be delivered to the field relatively quickly. However, this is generally because new armor configurations and materials are not radically different from those that have already been demonstrated to be effective. This is best illustrated by looking at how one presently designs an armor system in response to a new threat.

While there is no unique way to design an armor system, Figure 2-11 reflects what the study committee heard from several presenters representing the Army Research Laboratory and from other invited speakers. Figure 2-11 also demonstrates several major limitations that impact protection materials research. These will be addressed in subsequent chapters.

Existing Paradigm

In response to a new threat against which current armor systems fail, a new armor system concept—including geometry, configuration, and materials—is chosen that, from experience, designers hope will defeat the new threat. Changes in geometry can be as simple as adding thicknesses to various layers in an existing configuration; possible, but less likely, is an entirely new design. Materials are chosen from a set of available materials whose ballistic and blast performance have already been proven both as individual

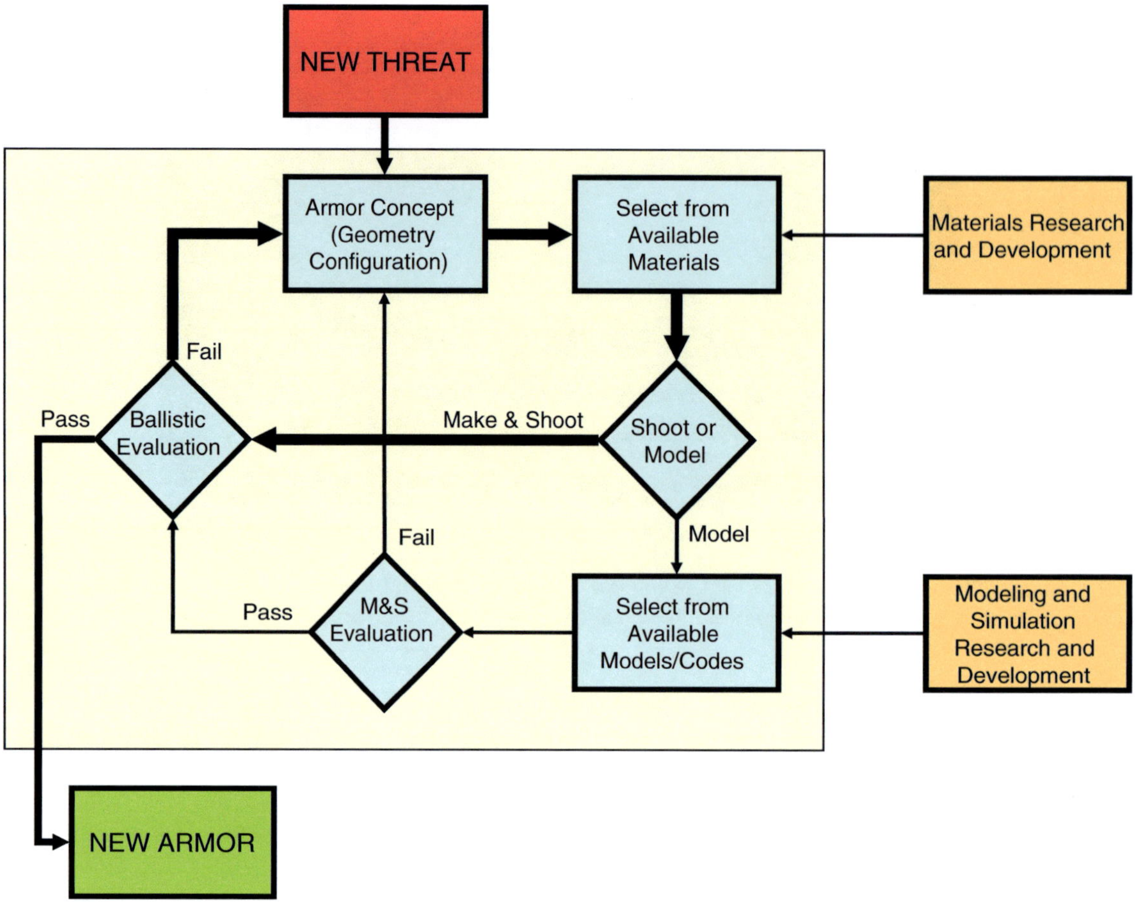

FIGURE 2-11 Current paradigm for armor design. In response to a new threat, a new concept and materials are chosen and then tested or modeled depending on several factors. Armor that fails ballistic testing is redesigned and the (costly) process begins again. The goal is to reduce repetitive looping by making better use of modeling and simulation. It should be noted that, while computations are sometimes used, the shoot-and-look mode is much more common. In addition, the materials research and development community and the modeling and simulation community are not particularly well connected.

materials and combined with other materials. While much excellent materials research is under way, emerging research materials are seldom, if ever, chosen for new armor because there is no way to directly tie how they perform in a research environment to how they will perform in the actual armor configuration. Moreover, most nonarmor applications materials are chosen according to their bulk quasi-static properties, such as hardness, strength, and toughness, even though such properties do not always predict the materials' ballistic or blast performance. This issue will be discussed extensively in Chapter 3.

The next step is deciding how to evaluate the candidate armor system. Although it might seem intuitive to run simulations before expensive testing, the decision on how to test the new configuration actually depends on several factors. If the armor varies only slightly from existing armor, then the most expedient method might well be to build the armor and go straight to a ballistic evaluation. However, if the armor design is significantly different from current armor systems, there are limitations to the effectiveness of modeling and simulation. If the properties of the materials—that is, the constitutive relations needed to run computational material models for them—are not known, then the modeling would have to use information from the most similar existing material, making the result uncertain. This is another reason why armor designers do not consider using research materials that have not yet been sufficiently characterized under appropriate dynamic conditions (see Chapter 4). Consequently, modeling and simulation are often used more as a guide to identify trends due to design changes than as a source of absolute results. Thus, even configurations that survive the modeling and simulation step may fail ballistic testing. Chapter 4 elaborates on the limitations of how well the material can be modeled and addresses shortcomings in the models themselves.

Armor that fails ballistic testing is redesigned and the process begins again. Once there is a successful ballistic test, the armor will be constructed in sufficient quantities to qualify the new configuration for fielding. It can only be hoped that a new armor design comes about without too many loops (as shown in the Figure 2-11 diagram), which are costly in terms of effort. More important, repetitive looping delays the fielding of new armors, which in turn adversely impacts the safety of troops. At the same time, speed must be offset by the need to make sure the armor will perform as expected in the field.

As described, the current armor design paradigm clearly makes it difficult to incorporate a new material into existing armor systems or to use it in an entirely new design. Any path forward for future generations of lightweight armor materials must alter this paradigm, although it will not be an easy undertaking. As has been shown in this chapter, the ultimate performance of an armor system depends on the materials used and on the geometric arrangement of those materials, both of which vary according to threat and the application. The challenge is to represent the complications inherent in materials in a way that allows designers to focus on the materials independently of the specific armor system design or threat that the armor is intended to thwart. As will be shown in Chapter 3, meeting this challenge will require an understanding of how to relate the behavior of a material—especially its failure behavior—during ballistic or blast events to its initial structure and composition. Chapter 4 describes approaches for successfully predicting the theoretical and experimental failure behavior of a protection material for the benefit of the materials research community. Finally, it will be important to convey information about armor performance to those developing armor materials.

SECURITY AND EXPORT CONTROLS

It is important to acknowledge the security restrictions that surround protection materials. Such limitations are prudent and necessary but require periodic review to ensure they are consistent with the current state of open knowledge and do not unnecessarily restrict the exchange of information with an open research community when such an exchange would be beneficial to national security.

The information content of this study, which deals with armor systems and armor performance, is bound by both security regulations and export control law. The security limitations generally imposed by the Army in this area restrict the discussion of performance of certain armor system designs against specific threats. These limitations extend to the ability to test armor systems with militarily relevant threats. The details of specific threats and design are generally not published in the open literature. Even the availability of information on armor systems that is not proprietary or classified is often restricted by DoD to DoD and contractors to DoD. The underlying technical basis for these restrictions may be available to researchers working on armor under contract to DoD, but it is not generally available to researchers outside of that context.

In addition, there are export control restrictions that generally limit the distribution of information to U.S. citizens and lawful permanent residents. The restrictions that apply to armor materials cover almost all of the relevant protection materials, including ceramics near theoretical density—among them B_4C (boron carbide), SiC (silicon carbide), and Al_2O_3 (aluminum oxide, or alumina), discussed in Chapter 5—composite materials, arrays of woven cloth, metals, and ceramics, and layers of metals.

Information in the public domain as defined in 22 CFR 120.11 is generally not subject to International Traffic in Arms Regulations (ITAR). The definition of "technical data" that is subject to the export control regulations does not include "information concerning general scientific, mathematical or engineering principles commonly taught in colleges and universities or information in the public domain."[29]

The combination of security regulations and ITAR makes it extremely difficult for fundamental research in protection materials to connect to the development of restricted armor systems. Ultimately much of work on armor is restricted. For example, a quick search of the Defense Technical Information Center database for "vehicle armor" indicates that only about 30 percent of the technical documents from 2005 through 2010 are approved for public release. Clear, up-to-date boundaries need to be specified between restricted and unrestricted information and related research. Such a review, however, is beyond the scope of this report.

This report is a public document, and its content is limited to general descriptions of threats, performance, and design that may be discussed without restriction. While the restrictions discussed above suffice for the needs of this study, it is important to note that they can significantly complicate the use of available information in basic research.

[29] 22 CFR 120.10(a)(5).

3

Mechanisms of Penetration in Protective Materials

In designing armor, materials high in hardness, strength, and toughness have traditionally been sought, since common sense would dictate that such materials should be most resistant to attack by a projectile. However, according to Shockey et al., ballistic tests often show that the best-performing armor material is not necessarily the strongest, the toughest, or the hardest. Are there other properties that reliably offer guidance in choosing and developing armor materials, if such conventional bulk properties do not? Does ballistic behavior depend on some vague or unknown property or combination of properties, and, if so, how can they be identified, measured, and even enhanced? Can the chemistry and processing of materials be manipulated to achieve microstructures that exhibit nonconventional mechanical properties once they have been identified? Can such manipulation improve penetration resistance?

To answer these questions, armor development should be looked at not from the perspective of conventional bulk material properties but from that of micromechanical mechanisms.[1] An understanding of the mechanisms operating in a target during a penetration event can suggest microstructures—including those that characterize the chemical and phase composition of the building blocks—that are more resistant to penetration and that will lead to protective materials with better performance (see Box 3-1). Moreover, by identifying penetration-induced failure mechanisms and quantifying their activity, mathematical damage models can be developed that may allow what is termed computational armor design.

Penetration mechanisms are perhaps best revealed by post-test examination of penetrated targets. Ejected or otherwise separated target material contains telltale signs of the failure modes that operated during penetration, as does the material in the vicinity of the penetration cavity. The collection of loose material and the sectioning of penetrated material, followed by unaided visual inspection and inspection under a microscope, show the damage features, helping to uncover the mechanisms of material failure. In situ,

BOX 3-1
Microstructural Options for Influencing Failure Mechanisms in Metals, Ceramics, and Polymers

The nucleation, growth, and coalescence of cracks and shear instabilities in metals and ceramics could be suppressed by manipulating the grain structure or by adding second phase particles. The size, shape, and orientation of the grains could be configured to disrupt failure mechanisms. The mechanical properties of the grain boundaries can, moreover, dictate a transgranular or intergranular failure mode. And the chemical and phase composition of the grains themselves and their crystalline structure can be specified to affect deformability, mode of deformation (dislocation activity, twinning, phase changes), and propensity to rupture.

Likewise, the size, shape, orientation, crystal structure, spatial distribution, and mechanical properties of second-phase particles as well as the strength of particle and matrix interfaces can be manipulated to deter failure mechanisms. Second-phase particles such as coherent nanocrystallites have been shown to improve the ballistic performance of glasses, although there is not yet a detailed understanding of their effect on failure mechanisms. Pores can also inhibit cracks, and judicious open-architecture geometries may provide a lightweight solution to a penetration or blast problem.

Microstructural variables in polymers include chemical makeup, length and degree of branching of molecular chains, degree of alignment and entanglement, and extent of cross-linking. The types and strengths of bonds in the chains and between chains affect polymer strength and deformability (for instance, in thermosets versus thermoplastics) and can be expected to affect failure mechanisms.

[1] Shockey, D.A., J.W. Simons, and D.R. Curran. 2010. The damage mechanism route to better armor materials. International Journal of Applied Ceramic Technology 7(5): 566-573.

real-time, high-speed dynamic observations can in principle provide even better indications of failure modes. However, it is difficult to simultaneously achieve both high spatial and high temporal resolution. Future advances in instrumentation will bring new insights to the complex interplay of deformation and failure mechanisms during penetration.

Partially penetrated targets are particularly useful for determining failure mechanisms. A close examination of areas where the damaged material remains in place and of polished cross sections taken on a plane containing the shot line demonstrates how damage varies with distance from the side and distance ahead of the penetrating object. Such observations also suggest how damage evolves, thereby providing notions for equations describing damage development. The next section illustrates the failure mechanisms invoked by a penetrator by presenting damage observations in penetrated and partially penetrated targets of metals and alloys, ceramics and glasses, and polymeric materials. This is followed by a short discussion on the damage mechanisms in cellular materials invoked by blast loads.

PENETRATION MECHANISMS IN METALS AND ALLOYS

Consider the case of a rod impacting a steel plate (Figure 3-1). If the plate is relatively soft compared to the rod, perforation may occur by homogeneous plastic flow of the plate, with little or no damage to the rod. A hardened plate, on the other hand, may fail by shear banding and consequent liberation of a plug of material pushed out by the projectile (Figure 3-1). Reflected stress waves from the rear surface of the plate may produce tensions large enough to nucleate, grow, and coalesce voids or microcracks, causing spallation. Thus, the result of an encounter between a rod and plate is determined by microscopic failure processes such as homogeneous plastic flow, shear banding, and tensile fracture in the plate and in the rod. The impact conditions and properties of

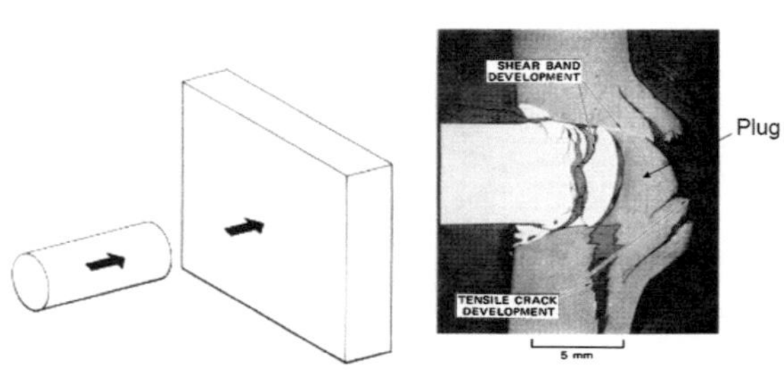

FIGURE 3-1 Impact on steel plate. Rod impacting a plate at 90 degrees (left), and cross section showing simple plugging of rolled homogeneous armor by shear instabilities and back surface spalling by nucleation, growth, and coalescence of voids and cracks (right). SOURCE: Erlich, D.C., L. Seaman, D.A. Shockey, and D.R. Curran. 1980. Development and Application of a Computational Shear Band Mode. Menlo Park, Calif.: SRI International.

both plate and rod determine which failure processes operate. Typically, however, it is a combination of simultaneously active failure modes that governs the outcome of the encounter.

Other failure modes may be invoked at higher velocities. Figure 3-2 shows a polished and etched cross section through the crater in a 1-in.-thick steel plate that has been impacted at 6 km/s by a 12.7-mm-diameter polycarbonate sphere.[2,3] Adiabatic[4] shear bands can be seen as white-etching bands of hard, untempered martensite extending into the plate (1), surfaces of strain localization that look like bands when seen edge-on. The path of the bands is followed by brittle cracks (2), which intersect with other cracks and liberate fragments. Just below the crater are spherical voids (3), a manifestation of ductile tensile failure; these are linked by shear bands. Homogeneous plastic flow (4) is made clear by the deviation of the process rolling lines from the horizontal. Ultimately, the hemispherical volume of dark-etching material just below the point of experienced $\alpha \leftrightarrow \varepsilon$ polymorphic phase change brought about by pressure (5). The grain size is refined and the transformed material is significantly hardened. The boundary of the dark-etching material is a 130 kbar isobar. Thus, five failure modes operated at once, with the stress relaxation effect of each mode affecting the behavior of the others.

The damage beneath the crater in Figure 3-2 is complex and seems at first nearly impossible to interpret, yet it reveals how the material is failing. Such damage "hieroglyphics" must nevertheless be read and understood in order to predict penetration behavior and design microstructures with enhanced protective capabilities. Key to developing a deeper understanding are laboratory experiments that isolate each specific damage mechanism. This would allow each failure mode to operate under a range of well-controlled rate, temperature, and stress state conditions, providing the opportunity to study and quantitatively describe its evolution by means of real-time observation or post-test analysis of tests interrupted at various stages of damage development.

Finding 3-1. Ballistic penetration of metals can occur by five failure modes—adiabatic shear bands, cracks, voids, plastic deformation, and phase changes—more than one or all of which can occur simultaneously.

[2]Shockey, D.A., D.R. Curran, and P.S. DeCarli. 1975. Damage in steel plates from hypervelocity impact, I: Physical changes and effects of projectile material. Journal of Applied Physics 46(9): 3766-3775.

[3]Bertholf, L.D., L.D. Buxton, B.J. Thorne, R.K. Byers, A.L. Stevens, and S.L. Thompson. 1975. Damage in steel plates from hypervelocity impact II: Numerical results and spall measurement. Journal of Applied Physics 46(9): 3776-3783.

[4]"Adiabatic" refers to any process which occurs without heat transfer.

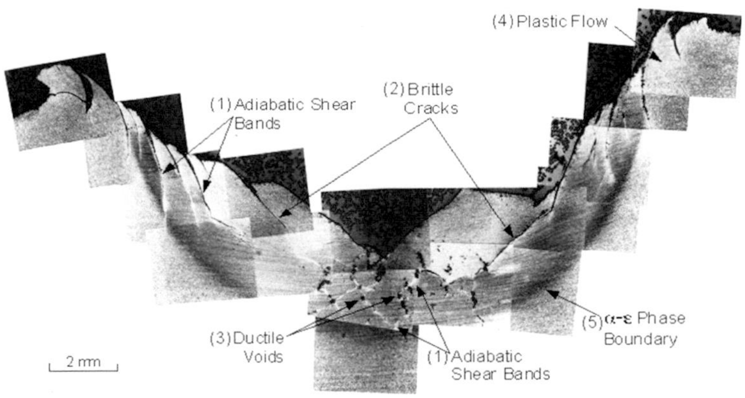

FIGURE 3-2 Polished and etched cross section through the crater in a steel plate that was impacted at 6 km/s by a 12.7-mm-diameter polycarbonate sphere. Five damage modes operated. SOURCE: Reprinted with permission from Shockey, D.A., D.R. Curran, and P.S. De Carli, Journal of Applied Physics, 46, 3766, (1975). Copyright 1975, American Institute of Physics.

PENETRATION MECHANISMS IN CERAMICS AND GLASSES

Penetration of thick sections of ceramics and glasses occurs by damaging the target material at the leading surface of the projectile and then pushing the damaged material out of the projectile path.[5] Here, too, an understanding of the damage mechanisms is key to developing ceramics and glasses with improved ballistic performance.

The damage mechanisms are readily revealed in experiments in which the projectile does not penetrate—that is, at velocities and test conditions sufficient to initiate the damage process but insufficient for ingress. Such tests produce ring cracks and radial cracks on the impacted surfaces, as well as the well-known Hertzian cone cracks, which extend into the target at divergent angles from the shot line. More important when considering penetration mechanisms, however, is the microdamage produced in the target directly ahead of the projectile, since it is the material in this location that must be extruded from the projectile path to permit penetration.

The polished cross sections taken through the shot lines after tests on SiC and TiB_2 (Figure 3-3) show the Hertzian cone cracks and, often, an obvious zone of damaged material immediately beneath the impacting projectile.[6]

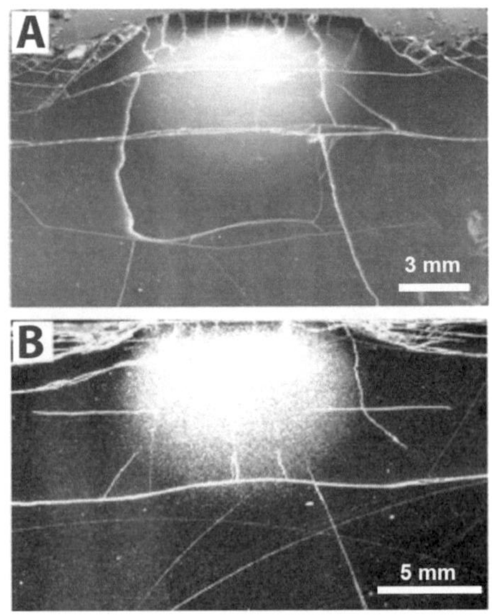

FIGURE 3-3 Polished cross sections through the shot line of a SiC (A) and a TiB_2 (B) target, showing typical microdamage immediately below the impact site after a no-penetration experiment with a long rod tungsten projectile. SOURCE: LaSalvia, J.C., and J.W. McCauley. 2010. Inelastic deformation mechanisms and damage in structural ceramics subjected to high-velocity impact. International Journal of Applied Ceramic Technology 7(5): 595-605. See also LaSalvia, J.C., R.B. Leavy, J.R. Houskamp, H.T. Miller, D.E. MacKenzie, and J. Campbell. 2010. Ballistic impact damage observations in a hot-pressed boron carbide. Pp. 45-55 in Advances in Ceramic Armor V. J.J. Swab, D. Singh, and J. Salem, eds. New York, N.Y.: John Wiley & Sons.

---

[5]Shockey, D.A., A.H. Marchand, S.R. Skaggs, G.E. Cort, M.W. Burkett, and R. Parker. 1990. Failure phenomenology of confined ceramic targets and impacting rods. International Journal of Impact Engineering 9(3): 263-275. Also in Shockey, D.A., A.H. Marchand, S.R. Skaggs, G.E. Cort, M.W. Burkett, and R. Parker. 2002. Failure phenomenology of confined ceramic targets and impacting rods. Pp. 385-402 in Ceramic Armor Materials by Design, Ceramics Transactions Column 134. J.W. McCauley, A. Rajendran, W. Gooch, S. Bless, S. Wax, and A. Crowson, eds. Westerville, Ohio: The American Ceramic Society.

[6]LaSalvia, J.C., and J.W. McCauley. 2010. Inelastic deformation mechanisms and damage in structural ceramics subjected to high-velocity impact. International Journal of Applied Ceramic Technology 7(5): 595-605. See also LaSalvia, J.C., R.B. Leavy, J.R. Houskamp, H.T. Miller, D.E. MacKenzie, and J. Campbell. 2010. Ballistic impact damage observations in a

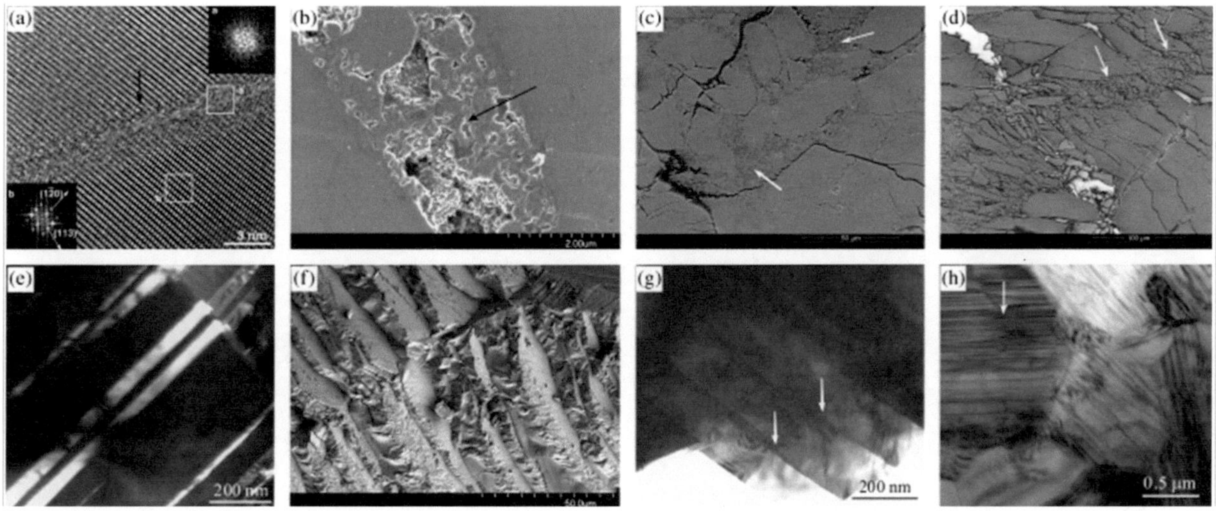

FIGURE 3-4 Damage mechanisms (see arrows) observed in several ceramics: Nanoscale amorphization bands in ballistic-generated B_4C fragments (a); shear "band" in a sphere-impacted B_4C (b); shear-induced amorphization in a sphere-impacted b-SiC (c); multiscale fragmentation and shear localization in the same material (d); twins in sintered Al_2O_3 shocked above its Hugoniot elastic limit (e); aluminum oxynitride (AlON) fracture surface showing multiscale cleavage (f); AlON fragment showing microcleavage (g); and stacking faults in a SiC (h). SOURCE: LaSalvia, J.C., and J.W. McCauley. 2010. Inelastic deformation mechanisms and damage in structural ceramics subjected to high-velocity impact. International Journal of Applied Ceramic Technology 7(5): 595-605. See also LaSalvia, J.C., R.B. Leavy, J.R. Houskamp, H.T. Miller, D.E. MacKenzie, and J. Campbell. 2010. Ballistic impact damage observations in a hot-pressed boron carbide. Pp. 45-55 in Advances in Ceramic Armor V. J.J. Swab, D. Singh, and J. Salem, eds. New York, N.Y.: John Wiley & Sons.

Closer examination of the damage zone, known as the Mescall zone (MZ),[7] provides valuable details of the material failure process. Figure 3-4 shows the variety of damage mechanisms observed beneath the projectile impact sites in several ceramics. These mechanisms include intergranular and transgranular macro- and microcracking; shear localization; solid-state amorphization; dislocation activity; twinning; stacking faults; and phase transformations.[8]

The damage process requires some time, typically 1-3 µs, and can be observed with high-speed photography. Penetration can proceed only after material in the MZ has failed and has been pushed from the projectile path. Thus the projectile dwells on the target surface before beginning to penetrate. If a projectile did not reach a certain velocity—the transition velocity—and if the ceramic held together sufficiently—that is, if it exhibited "target confinement"—the projectile would not penetrate and "interface defeat" would be said to have occurred.

Projectiles with sufficient kinetic energy to fully develop the MZ penetrate by continuously damaging the target material at the projectile tip and extruding the fragments to the side of the shot line. Insight into the extrusion process is obtained by examining cross sections of partially penetrated target blocks. Monolithic targets of a soda lime glass impacted by a hemi-nosed steel rod at velocities sufficient to penetrate partway through the target[9] retain the cracking pattern and fragments produced during penetration (Figure 3-5a). The cracks and fragments are revealed by infiltrating a damaged target with a low-viscosity epoxy, then sectioning the target with a diamond saw, usually on a plane through the shot line. Next, the surfaces of section are polished and examined by optical and scanning electron mi-

hot-pressed boron carbide. Pp. 45-55 in Advances in Ceramic Armor V. J.J. Swab, D. Singh, and J. Salem, eds. New York, N.Y.: John Wiley & Sons.

[7]The Mescall zone—first defined in Shockey, D.A., A.H. Marchand, S.R. Skaggs, G.E. Cort, M.W. Burkett, and R. Parker. 1990. Failure phenomenology of confined ceramic targets and impacting rods. International Journal of Impact Engineering 9(3): 263-275—is named after John Mescall, a scientist at the U.S. Army Materials and Mechanics Research Center, who deduced the existence of the finely comminuted volume of target directly beneath the nose of an advancing projectile from his computational simulations (Mescall, J., and C. Tracy. 1986. Improved modeling of fracture in ceramic armors. Pp. 41-54 in Army Science Conference Proceedings, 17-19 June 1986, Volume III. Washington, D.C.: Department of the Army, Deputy Chief of Staff for Research, Development & Acquisition; Mescall, J., and V. Weiss. 1983. Materials Behavior under High Stress and Ultrahigh Loading Rates. New York, N.Y.: Plenum Press).

[8]LaSalvia, J.C., and J.W. McCauley. 2010. Inelastic deformation mechanisms and damage in structural ceramics subjected to high-velocity impact. International Journal of Applied Ceramic Technology 7(5): 595-605. See also LaSalvia, J.C., R.B. Leavy, J.R. Houskamp, H.T. Miller, D.E. MacKenzie, and J. Campbell. 2010. Ballistic impact damage observations in a hot-pressed boron carbide. Pp. 45-55 in Advances in Ceramic Armor V. J.J. Swab, D. Singh, and J. Salem, eds. New York, N.Y.: John Wiley & Sons.

[9]Shockey, D.A., D. Bergmannshoff, D.R. Curran, and J.W. Simons. 2008. Physics of glass failure during rod penetration. Pp. 23-32 in Advances in Ceramic Armor IV: Ceramic Engineering and Science Proceedings, Volume 29, Issue 6. L.P. Franks, ed. Hoboken, N.J.: John Wiley & Sons.

a 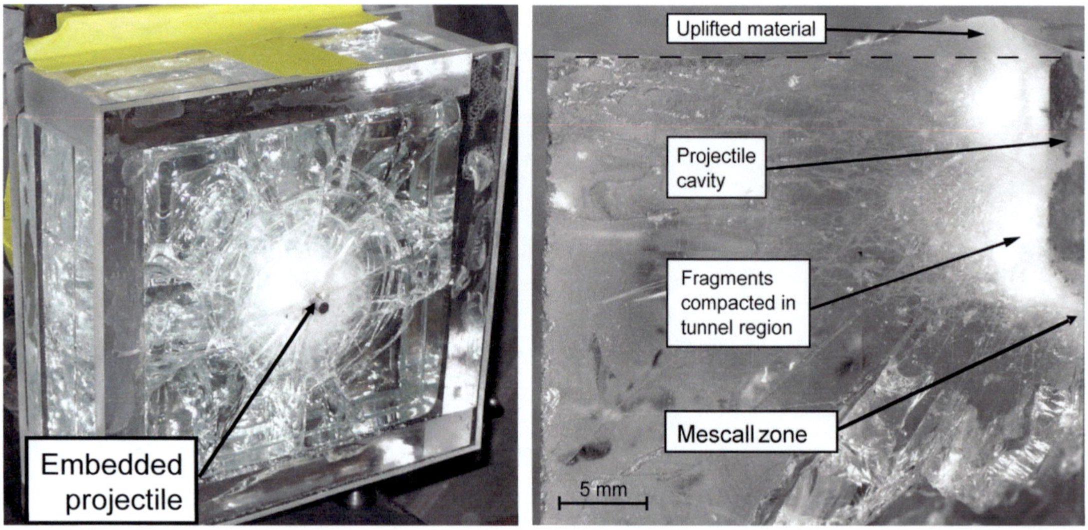 b

FIGURE 3-5 A 200 × 200 × 75 mm^3 monolithic soda lime glass target (confined on all sides with polymethyl methacrylate plates) partially penetrated by a 31.75 × 6.35-mm-diameter heminosed steel rod impacting at 300 m/s (a); a surface of section through the shot line showing damage around the projectile cavity (b). SOURCE: Shockey, D., J. Simons, and D. Curran. 2010. The damage mechanism route to better armor materials. International Journal of Applied Ceramic Technology 7(5): 566-573.

croscopy to observe cracking details and the size and shapes of fragments (Figure 3-5b).

Glass and ceramic targets show the radial cracks, ring cracks, cone cracks, and lateral cracks typical of a rod or particle impact. Target fragments about one to three projectile radii in diameter form a cylindrical zone (tunnel) surrounding the embedded projectile. An uplifted "lip" of material is often produced at the impact surface. The size and shape of MZ fragments can be determined and quantified by examining petrographic sections through the tunnel debris. The MZ of highly comminuted material at the leading edge of the penetrator shown in Figure 3-5b is smaller than what would be expected during penetration because the stresses at the tip of an arresting penetrator are smaller than those in advance of a moving penetrator.

The fragmentation and cracking patterns suggest that material ahead of the projectile is loaded, damaged, and displaced in three successive steps under consecutive tensile-, shear-, and compression-dominated stress states (Figure 3-6). A material element in the path of an advancing penetrator initially experiences tension and develops closely spaced cone cracks running at acute angles to the penetration direction. Subsequent lateral cracks break up the material between adjacent cone cracks. As the projectile moves closer, a local volume (about the size of the projectile nose) of the cracked material is overrun by a low-confinement field of high shear and is comminuted into fine fragments. Third, the projectile imposes high pressure and extrudes the comminuted material into the cracked and coarsely fragmented tunnel and to the sides of the projectile nose.

Finding 3-2. An examination of the mechanics of penetration in brittle materials reveals four important characteristics of microstructures that are key for improved body armor materials. The structures must (1) resist deformation and macro (cone and lateral) cracking; (2) be more difficult to comminute; (3) break into fragment geometries that are more resistant to flow; and (4) form more dilatant fragment beds.[10]

PENETRATION MECHANISMS IN POLYMERIC MATERIALS

Polymers such as polycarbonate are often used in armor systems as backing plates (spall shields), as intermediate layers in a laminated glass or ceramic system, as a scratch-tolerant front plate, or as a matrix to embed strong fibers.

Because the material failure mechanisms are sensitive to boundary conditions, they are somewhat determined by the application. Real-time observation with high-speed cameras shows that the penetration of polycarbonate plates by cylindrical projectiles occurs by elastic dishing, petalling, cone cracking, and plugging.[11] The projectile initially indents the surface of the plate, causing the distal plate surface to bulge and shear yielding around the impact site. As the penetrator advances, cracks form ahead of it. Depending on the projec-

[10]Shockey, D.A., J.W. Simons, and D.R. Curran. 2010. The damage mechanism route to better armor materials. International Journal of Applied Ceramic Technology 7(5): 566-573.

[11]Wright, S.C., N.A. Fleck, and W.J. Stronge. 1993. Ballistic impact of polycarbonate: An experimental investigation. International Journal of Impact Engineering 13(1): 1-20.

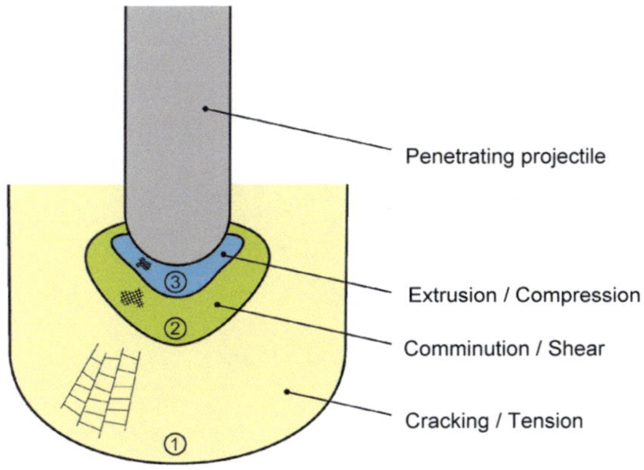

FIGURE 3-6 Three material processing zones and three stress states experienced by a material element in the path of an advancing penetrator. SOURCE: Shockey, D., J. Simons, and D. Curran. 2010. The damage mechanism route to better armor materials. International Journal of Applied Ceramic Technology 7(5): 566-573.

tile nose shape, plate perforation occurs by petalling or by plugging—that is, by pushing a cylinder of material ahead of the projectile through the distal plate surface. Evidence of melting has been observed. Material failure mechanisms may include tensile failure by nucleation, growth, and coalescence of planar cracks, spherelike voids, and shear instabilities. In glassy polymers, crazing, or the formation of oriented fibrils and intervening voids, is a common precursor to crack formation and tensile failure.

Polymer fibers are used in ballistic materials and as reinforcing elements in composite materials. A careful and detailed study of nanoscale failure phenomenology would be most useful in developing fibers with better ballistic performance. Figure 3-7 shows a fabric after it has been impacted by a platelike projectile.[12] The failure mechanisms of polymer fibers can be determined by examining the severed fiber ends with a scanning electron microscope (SEM).[13] For example, the internal structure of a 20-μ-diameter poly-p-phenylene benzobisthiazole (PBZT) fiber consists of large length-to-width, ribbonlike fibrils typically 1 μ wide, which in turn are made up of microfibrils of similar geometry but only a few nanometers wide (Figure 3-8). Figure 3-9 indicates that tensile fracture first occurred at defects such as voids and kinks and was assisted by the residual stresses that arose during processing.[14]

While the details of the tensile failure mechanism are not well known, high magnification shows that the fibrils in the fibers are stretched, suggesting tensile failure analogous to that seen in tensile tests of metals. Fiber material very likely undergoes homogeneous plastic deformation and localized plastic deformation in much the same way as metals; failure may also occur by the nucleation of voids, cracks, and shear bands.

It is not understood how the material microstructure at this level (the nano level) influences the deformation, localization, and failure behavior of the material. Failure initiators are thought to originate in material defects such as tiny voids, foreign particles, and chain entanglements (shown in Figure 3-10) resulting from chemical inhomogeneities or processing procedures.

Fiber failure modes other than tensile failure are also observed. For example, a projectile's impact on fabric backed with a stiff plate of ceramic compresses the fabric against the backing and causes transverse loads on the yarns and fibers that can result in deformation and failure. When compressed fibers are examined by SEM, they and the fibrils show flattening, kinking, and buckling.

Finding 3-3. The influence of the nano- and microstructure of polymeric materials on the deformation, localization, and failure behavior of the materials is not well understood, especially at high strain rates and high pressures.

Finding 3-4. Closing the large gap between the currently attainable and the theoretical strengths of fibers would benefit greatly from studies of ballistically (and quasi-statically) failed fibers at the nano- and micro levels to determine the mechanism(s) of material failure and identify the nanostructural features initiating the failure process or otherwise assisting it.

FAILURE MECHANISMS IN CELLULAR-SANDWICH MATERIALS DUE TO BLASTS

A cellular material sandwiched between two faceplates provides mass-efficient protection against blast loads. The structure absorbs energy and reduces the transmitted force as the walls of each cell deform and fail. Thus, to improve or tailor the response of cellular structures to blast loads, cell failure mechanisms must be understood.

Cellular materials include polymers, ceramics, and metals and metal foams; cell geometries include honeycomb and other lattices as well as stochastically random geometries. Several material properties contribute to the effective absorp-

[12]Shockey, D.A., D.C. Erlich, and J.W. Simons. 2004. Lightweight Ballistic Protection of Flight-Critical Components on Commercial Aircraft, Part 2: Large-Scale Ballistic Impact Tests and Computational Simulations, DOT/FAA/AR-04/45,P2. Available online at http://www.tc.faa.gov/its/worldpac/techrpt/ar04-45p2.pdf. Last accessed April 15, 2011.

[13]Hearle, J.W.S., B. Lomas, and W.D. Cooke. 1998. Atlas of Fibre Fracture and Damage to Textiles. Boca Raton, Fla.: CRC Press.

[14]Allen, S.R., A.G. Filippov, R.J. Farris, and E.L. Thomas. 1981. Macrostructure and mechanical behavior of fibers of poly-p-phenylene benzobisthiazole. Journal of Applied Polymer Science 26(1): 291-301.

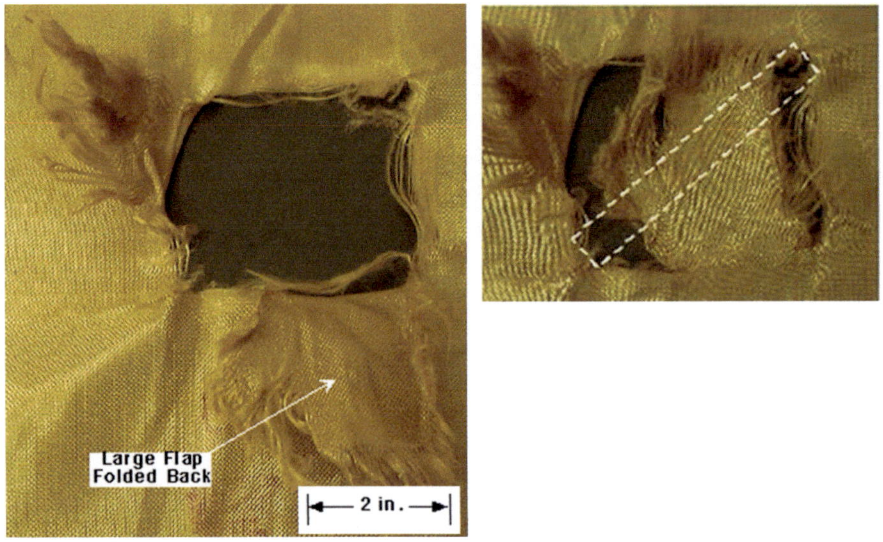

FIGURE 3-7 Post-test observation of fabric damage from a platelike projectile showing yarn breakage characteristics (left); the projectile size is shown with the fabric flap in its original position (right). SOURCE: Shockey, D.A., D.C. Erlich, and J. W. Simons. 2004. Lightweight Ballistic Protection of Flight-Critical Components on Commercial Aircraft, Part 2: Large-Scale Ballistic Impact Tests and Computational Simulations, DOT/FAA/AR-04/45,P2. Available online at http://www.tc.faa.gov/its/worldpac/techrpt/ar04-45p2.pdf. Last accessed April 15, 2011.

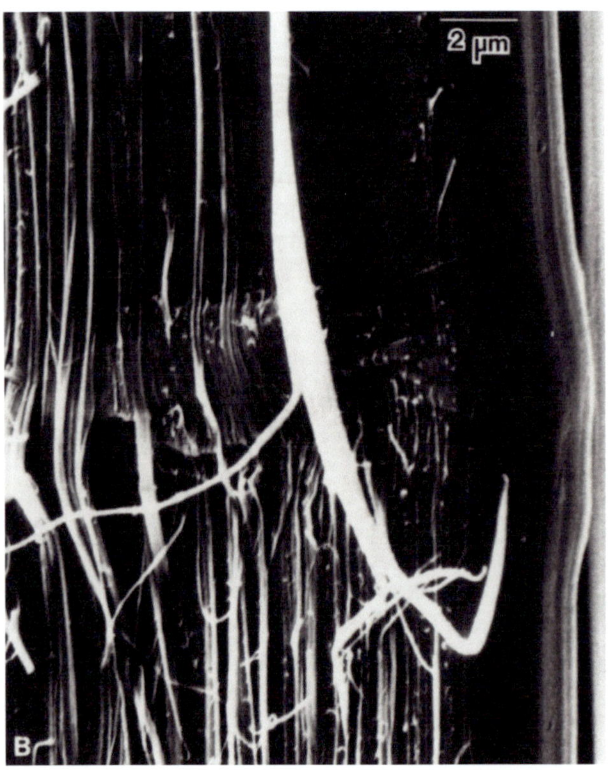

FIGURE 3-8 SEM micrograph revealing fibrillar microstructure in an as-spun PBZT fiber. SOURCE: Allen, S.R. 1983. Mechanical and morphological correlations in poly-(p-phenylene benzobisthiazole) fibers. Ph.D. Dissertation. Amherst, Mass.: University of Massachusetts.

tion of the energies of blast loads. These properties include the elastic stiffness, the yield strength, strain hardening, and the level of plateau stress at which a material compresses plastically.

In foams of ductile aluminum, discrete bands of collapsed cells establish the onset of yielding, the hardening, and the level of the plateau stress (Figure 3-10).[15] When the load is applied slowly, the bands form progressively and independently; at higher loading rates the cellular structure deforms by the advancement of a crush front from the impact surface.[16]

Figure 3-11 shows three specific failure mechanisms operating sequentially in a ductile aluminum cell under a quasi-static load.[17] First, localized plastic straining occurs at a cell node.

Next, the cell membrane plastically buckles. Finally, the cell collapses. Blast-loaded aluminum foam may fail in

[15]Bastawros, A.-F., H. Bart-Smith, and A.G. Evans. 2000. Experimental analysis of deformation mechanisms in a closed-cell aluminum alloy foam. Journal of the Mechanics and Physics of Solids 48(2): 301-322.

[16]Tan, P.J., S.R. Reid, J.J. Harrigan, Z. Zou, and S. Li. 2005. Dynamic compressive strength properties of aluminum foams Part I: Experimental data and observations. Journal of the Mechanics and Physics of Solids 53(10): 2174-2205.

[17]Bastawros, A.-F., H. Bart-Smith, and A.G. Evans. 2000. Experimental analysis of deformation mechanisms in a closed-cell aluminum alloy foam. Journal of the Mechanics and Physics of Solids 48(2): 301-322.

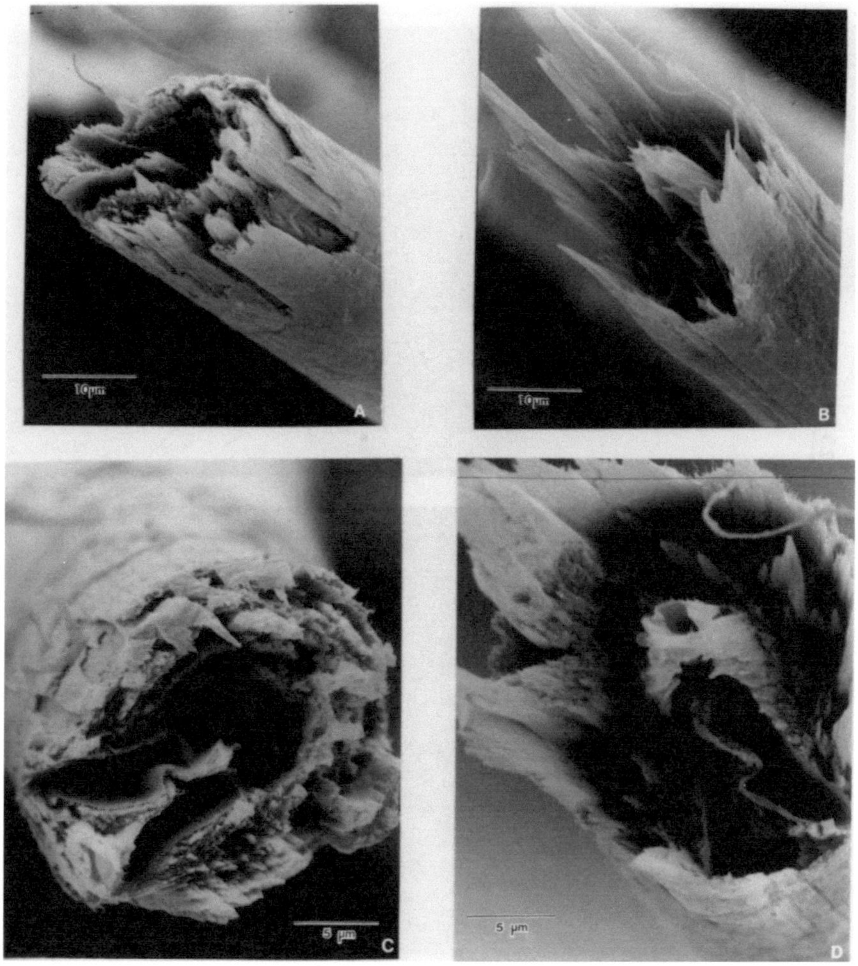

FIGURE 3-9 SEM side views (A,B) and end-on views (C,D) of matching fracture ends of a tensile-fractured PBZT fiber. SOURCE: Allen, S.R. 1983. Mechanical and morphological correlations in poly-(p-phenylene benzobisthiazole) fibers. Ph.D. Dissertation. Amherst, Mass.: University of Massachusetts.

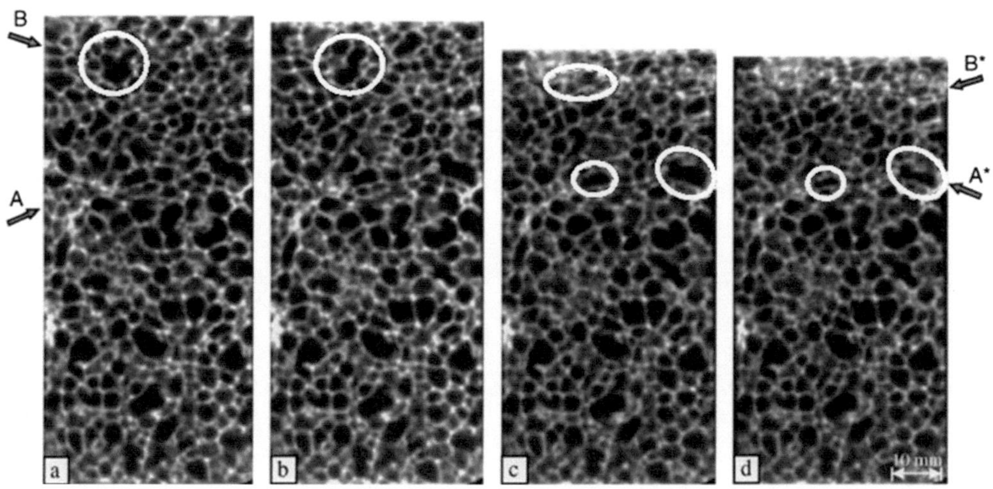

FIGURE 3-10 Sequence of computerized axial tomography scan images showing macro deformation bands in quasi-static compression-loaded ductile aluminum foam. (Dominant bands are identified with arrows, and the cells most visibly subjected to distortion are circled.) Note the buckling deformations exhibited by one of the membranes in each cell. SOURCE: Bastawros, A.-F., H. Bart-Smith, and A.G. Evans. 2000. Experimental analysis of deformation mechanisms in a closed-cell aluminum alloy foam. Journal of the Mechanics and Physics of Solids 48(2): 301-322.

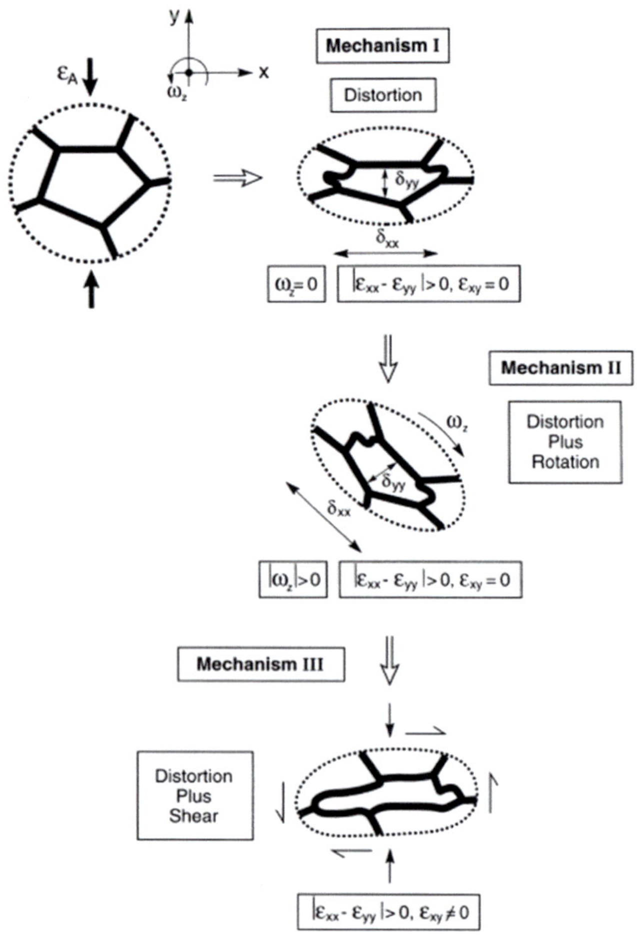

FIGURE 3-11 Sequential mechanisms responsible for cell collapse in ductile aluminum foam under quasi-static load. SOURCE: Bastawros, A.-F., H. Bart-Smith, and A.G. Evans. 2000. Experimental analysis of deformation mechanisms in a closed-cell aluminum alloy foam. Journal of the Mechanics and Physics of Solids 48(2): 301-322.

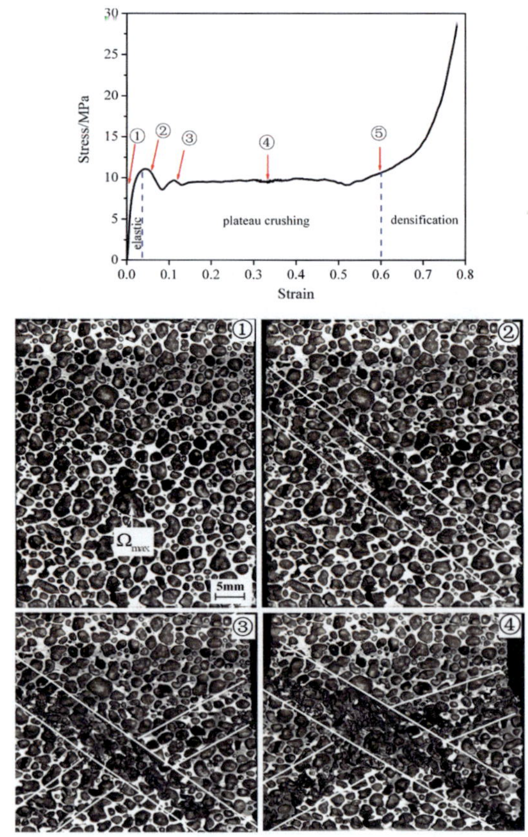

FIGURE 3-12 Stress-strain curve for a brittle aluminum foam subjected to quasi-static compression (top). Bands of fractured cells after imposed quasi-static engineering compressive strains of 0, 5.6 percent, 11.7 percent, 33.3 percent, and 60 percent, respectively (lower four images). SOURCE: Song, H-W., Q-J. He, J-J. Xie, and A. Tobota. 2008. Fracture mechanisms and size effects of brittle metallic foams: In situ compression tests inside SEM. Composites Science and Technology 68(12): 2441-2450.

tension[18] owing to a reflected tensile stress wave or recoil after compaction.

Brittle metallic foams fail at the macroscopic level by forming fracture bands similar to the deformation bands in ductile aluminum (Figure 3-12). Compression, tension, and shear cause cell walls to crack; friction and shear operate between fractured cells. At the cell membrane level, failure is by the brittle cracking of cell walls under compression, tension, and shear and by friction and shear between fractured cells (Figure 3-13).[19]

Finding 3-5. Cellular materials absorb blast energy by deformation and failure of cell walls.

CONCLUSIONS

This brief survey of how materials undergo penetration shows that

- Penetration occurs by material failure;

[18]Langdon, G.S., D. Karagiozova, M.D. Theobald, G.N. Nurick, G. Lu, and R.P. Merrett. Fracture of aluminum foam core sacrificial cladding subjected to air-blast loading. International Journal of Impact Engineering 37(6): 638-651.

[19]Song, H-W., Q-J. He, J-J. Xie, and A. Tobota. 2008. Fracture mechanisms and size effects of brittle metallic foams: In situ compression tests inside SEM. Composites Science and Technology 68(12): 2441-2450.

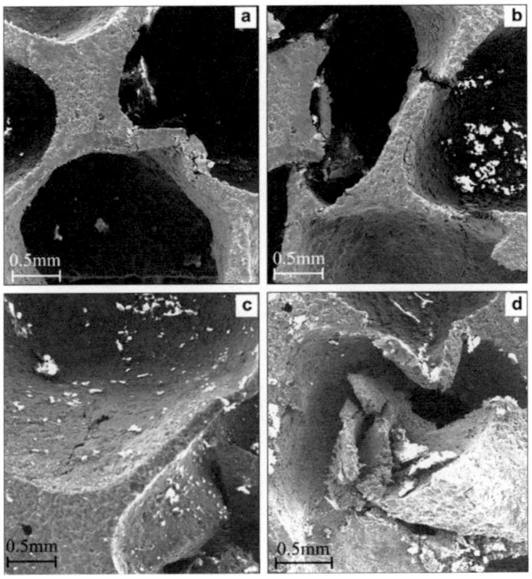

FIGURE 3-13 SEM images of failed cells in brittle aluminum foam showing failure modes (a) under compression, (b) tension and shear, (c) face cracking, and (d) friction and shear between fractured cells. SOURCE: Song, H-W., Q-J. He, J-J. Xie, and A. Tobota. 2008. Fracture mechanisms and size effects of brittle metallic foams: In situ compression tests inside SEM. Composites Science and Technology 68(12): 2441-2450.

- Material failure occurs at the microstructural level, including the chemical phase and phase composition, and the nanostructural level;
- Failure can occur in five modes—adiabatic shear bands, cracks, voids, plastic deformation, and phase changes—more than one of which can occur simultaneously;
- Material failure is a kinetic process, involving nucleation, growth, and coalescence of these failure modes; and
- Cellular materials absorb blast energy by deformation and failure of cell walls.

Recommendation 3-1. Organizations and individuals engaged in developing protection materials should seek to maximize penetration resistance. A comprehensive methodology for this should entail the following:

- Using the understanding of how armor materials fail to suggest possible microstructures—that is, those microstructures discovered through experimental/computational investigations as well as those identified during manufacturing trials that may oppose damage development;
- Designing microstructures that inhibit, disrupt, or avoid altogether the failure mechanisms operating in armor during projectile attack;
- Developing innovative laboratory tests that invoke the pertinent damage mechanisms;
- Designing tests so as to favor only one mechanism and so avoid the complication of several mechanisms operating simultaneously and influencing each other.
- Designing other tests that invoke two or more mechanisms, to investigate synergistic effects. Such tests must be conducted under well-controlled and monitored conditions of load, rate, and temperature and need to measure the governing (nonconventional) material failure properties;
- Performing other tests in which the load application is stopped at various percentages of the maximum load and the specimens sectioned and examined microscopically to observe the damage at increasing stages of development and the interaction of the damage with microstructural features;
- Prescribing microstructures that repress or interfere with failure mechanisms and interacting with processing engineers to innovate ways to achieve these microstructures;
- Choosing and implementing chemistries and processing routes;
- Performing laboratory tests and ballistic tests, noting results, adjusting initial thoughts, and exploring a second generation of chemistries and processing; and
- Continuing iterations with the goal of achieving ever better protective materials.

The time and expense of developing a superior armor material can be greatly reduced if the response of a chosen microstructure to ballistic attack can be at least approximately predicted by means of a computational simulation. A material deformation and failure model based on observations of penetrated targets and quantified by parameters derived from laboratory tests can be expected to provide more reliable results than current models. To assist in constructing a computational model of damage evolution, damage features observed in the interrupted tests should be counted and measured as a function of location in the specimen (by determining a nucleation rate). These data can be correlated with a stress and strain history obtained by computational simulations of the tests in order to develop equations that describe damage development. These equations would constitute the computational materials damage model and could be used in finite element codes to compute damage evolution during target penetration.

Thus, achievement of improved protective materials necessitates that damage development be hindered, most likely by specifying microstructures that resist the nucleation, growth, and coalescence of cracks, voids, and shear bands and by specifying as well chemistries and processing routes to achieve those microstructures.

Recommendation 3-2. Organizations and individuals engaged in developing protection materials should

- Choose materials based on their ability to inhibit or avoid material failure mechanisms, as opposed to choosing materials based on their bulk properties.
- Formulate mathematical models of damage evolution and use these in computational simulations of ballistic penetration scenarios to assist and expedite the design of improved armor materials and systems.

4

Integrated Computational and Experimental Methods for the Design of Protection Materials and Protection Systems: Current Status and Future Opportunities

There are two important challenges to be considered in improving protection systems. The first is to develop materials that are more efficient than existing materials, and the second is to design protection systems that optimally exploit existing or improved materials and in which the materials are physically arranged to optimize their protective properties. Advanced simulations and experimental methods are important for meeting both challenges.

Protection materials must be modeled on the atomic and microstructural levels such that their crystalline structure and microstructure can be computationally modeled to determine how changes at those levels affect their macrostructural (continuum) properties. Although there is no particular prescribed way to design materials with improved performance, computational methods enhance our understanding and give us insights into the synthesis and fabrication processes.

In addition to improving nano- and microstructural modeling techniques, researchers must ensure that the models can feed into new continuum models such that the net effect of the new materials can be assessed at the macroscopic level, which is the level of interest for an application. These multiscale, multiphysics computations could take the form of separate computations on the micro and macro levels or they could be integrated and performed in a single computation. Finally, the computational capabilities for complex material systems must be improved as well, such that system designs can be optimized quickly, accurately, and confidently with uncertainties quantified.

Protection materials and material systems made up of combinations of materials have attracted attention for many years. A substantial community of experimentalists, analysts, and armor designers is dedicated to improving existing protection capabilities and to discovering new materials and material combinations. This chapter takes a broad view of the underlying science base and reviews current activities with an eye to identifying opportunities in materials science and mechanics (theoretical, experimental, and computational) that could significantly advance protection performance.

The range of materials in use today for protection applications is quite remarkable, spanning metals, ceramics, and polymers. Materials for protection are combined in various ways, including ceramics constrained by metals or polymers and layered metal/ceramic/polymer systems. Some of the materials can be used as composite systems while others are protective structures in their own right. This chapter opens with a brief survey of the status of simulation capabilities for several of the most important systems, including simulations for the penetration of ceramic and metallic targets by projectiles and for the blast resistance of metallic plate structures. It should be noted at the outset that in spite of decades of concerted research efforts to develop simulation methods, the design of protection systems today still relies heavily on the make-it-and-shoot-it empirical approach. Meanwhile, simulations have reached the point where they can provide insight into system behavior and be used to point to promising possibilities. One objective of this report is to identify scientific opportunities that will elevate simulation methods to an equal partnership with empirical methods for advancing protection systems.

The following tools are needed for accurate simulation for most applications of structural materials:

- Knowledge of material response described by sound constitutive models characterizing both the deformation and failure over a wide range of strain rates, temperatures, and multiaxial stresses.
- Computational methods capable of capturing deformation and fracture under intense dynamic loads.
- Experimentation to supply basic material inputs to the constitutive models implemented in the computational codes and to provide performance data against which the simulations can be checked.

These three tools—constitutive models, computational methods, and experimentation—underlie simulation fidelity and are critical for protection materials because their behavior is pushed to the extreme. This is particularly true for simulations of ballistic penetration. Much of what is covered in this chapter applies to both ballistic and blast assaults, but for the most part the discussion will be cast in the context of ballistic assaults owing to the extreme demands they place on the theoretical and experimental knowledge of material response and on numerical simulation.

After presenting three examples of current capabilities, the committee discusses present-day experimental methods. Its discussion underscores the importance of understanding and characterizing the basic mechanisms of deformation and fracture in advancing protection materials. The committee goes on to address opportunities and challenges in experimental and computational methods.

THREE EXAMPLES OF CURRENT CAPABILITIES FOR MODELING AND TESTING

Three examples illustrate current capabilities for simulating the actual test performance of protection materials and highlight opportunities for further advances. They are (1) projectile penetration of an aluminum plate; (2) projectile penetration of ceramic plates; and (3) blast loading of steel sandwich plates. These exemplary cases demonstrate that a rational approach to armor design based on computational and experimental methods is feasible. It is not the committee's intention to cover all possible armor systems or to bound armor performance characteristics.

Projectile Penetration of High-Strength Aluminum Plates

Accurate simulation of projectile penetration of metal plates is being worked on using all three tools, and several groups have achieved predictive success. A recent study by Børvik et al.[1] addresses the penetration of plates of 7075 aluminum by two types of projectiles. The authors are from a research group in Norway noted for its emphasis on each of these three tools.

Figure 4-1 shows a blunt projectile and an ogive-nosed projectile, both of hardened steel (projectiles such as these are often used in unclassified studies) exiting a 20-mm-thick plate of AA7075-T651 aluminum. Figure 4-2 presents a plot of the exit velocity of the projectile as a function of its initial velocity before impact. As mentioned in Chapter 2, the initial velocity at which the projectile just manages to penetrate the plate with zero residual velocity is known as the ballistic limit V_0; Figure 4-3 presents the results of numerical simulations of these tests.

The constitutive relation used to characterize plastic deformation of AA7075 in the simulations of Børvik et al.[2] is the Johnson-Cook[3] relation, which has been used in many recent simulations of this type. There are six constants in this constitutive law that must be chosen to give the best possible fit to the data on the material. Supplementing the Johnson-Cook relation is an equation relating the temperature increase to plastic deformation. In addition to accounting for the effect of stress state, the constitutive model accounts for the effects of the strain rate and thermal softening on plastic deformation and can capture some aspects of adiabatic shear localization. To calibrate the constitutive laws for a given material, an extensive suite of tests must be performed, from tensile and compressive stress-strain tests up to tests at large strains in differing material orientations and temperatures, with strain rates as high as 10^4 s^{-1}. The Johnson-Cook deformation relation is supplemented by a material fracture criterion that usually employs a critical value of the equivalent plastic strain, dependent on the stress triaxiality. Stress triaxiality is the ratio of hydrostatic tension to the von Mises effective stress. A series of notched-bar tensile ductility tests was used by Børvik et al.[4] to calibrate the critical effective plastic strain at fracture as a function of stress triaxiality. As this outline makes clear, the characterization of a material for input into constitutive models is a considerable task in its own right.

To simulate the penetration of a hard, ductile metal target, the numerical method must account for large plastic strains, for dynamic effects, including inertia and material rate dependence, and for material failure in the form of shear-off or separation. The simulations reported here use the finite-element code LS-DYNA[5] for the computations. For several decades, finite-element codes have been able to model large strains, but the intense deformations encountered in penetration are challenging because they involve the difficult problem of remeshing to avoid overly distorted elements. It is also important to model the material failure response after the critical plastic strain has been attained. Current procedures usually erode an element during the final failure process, stepping down its stress to zero and finally deleting the element. In addition, it is essential to account for the pressure and friction exerted by the projectile on the plate.

The simulation challenge presented by projectile penetration owing to distortion of the meshes is evident in Figure 4-4. The blunt-nosed projectile produces shear localization through the thickness of the plate, followed by shear-off, which creates a plug of material that is pushed ahead of the projectile. In contrast, the ogive-nosed projectile pushes

[1]Børvik, T., O.S. Hopperstad, and K.O. Pedersen. 2010. Quasi-brittle fracture during structural impact of AA7075-T651 aluminum plates. International Journal of Impact Engineering 37(5): 537-551.

[2]Ibid.

[3]Johnson, G.R., and W.H. Cook. 1983. A constitutive model and data for metals subjected to large strains, high strain rates, and high temperatures. Pp. 541-547 in Proceedings of the 7th International Symposium on Ballistics, The Hague, The Netherlands. Available online at http://www.lajss.org/HistoricalArticles/A%20constitutive%20model%20and%20data%20for%20metals.pdf. Last accessed April 5, 2011.

[4]Børvik, T., O.S. Hopperstad, and K.O. Pedersen. 2010. Quasi-brittle fracture during structural impact of AA7075-T651 aluminum plates. International Journal of Impact Engineering 37(5): 537-551.

[5]See http://www.lstc.com.

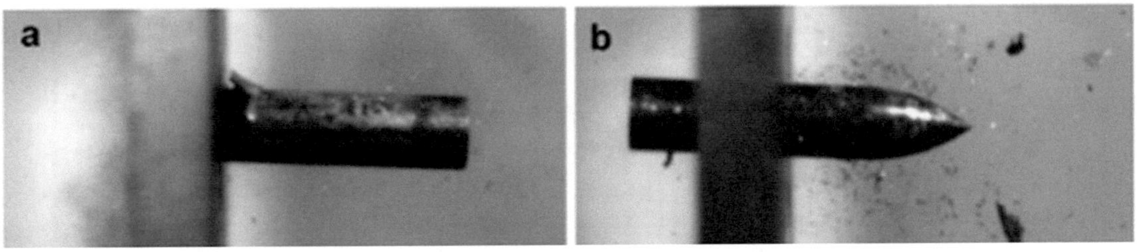

FIGURE 4-1 Blunt-nosed (a) and ogive-nosed (b) projectiles exiting a 20-mm-thick aluminum plate. SOURCE: Børvik, T., O. Hopperstad, and K. Pedersen. 2010. Quasi-brittle fracture during structural impact of AA7075-T651 aluminum plates. International Journal of Impact Engineering 37(5): 537-551.

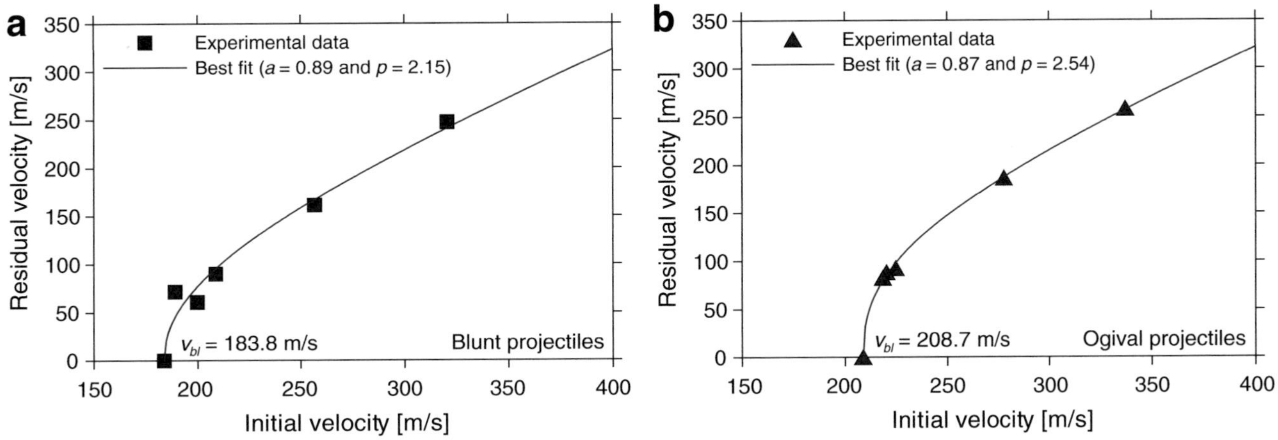

FIGURE 4-2 Experimental results for final exit (residual) velocity as a function of initial velocity for blunt-nosed (a) and ogive-nosed (b) projectiles. The smallest initial velocity producing full penetration is known as the ballistic limit, V_0. SOURCE: Børvik, T., O. Hopperstad, and K. Pedersen. 2010. Quasi-brittle fracture during structural impact of AA7075-T651 aluminum plates. International Journal of Impact Engineering 37(5): 537-551.

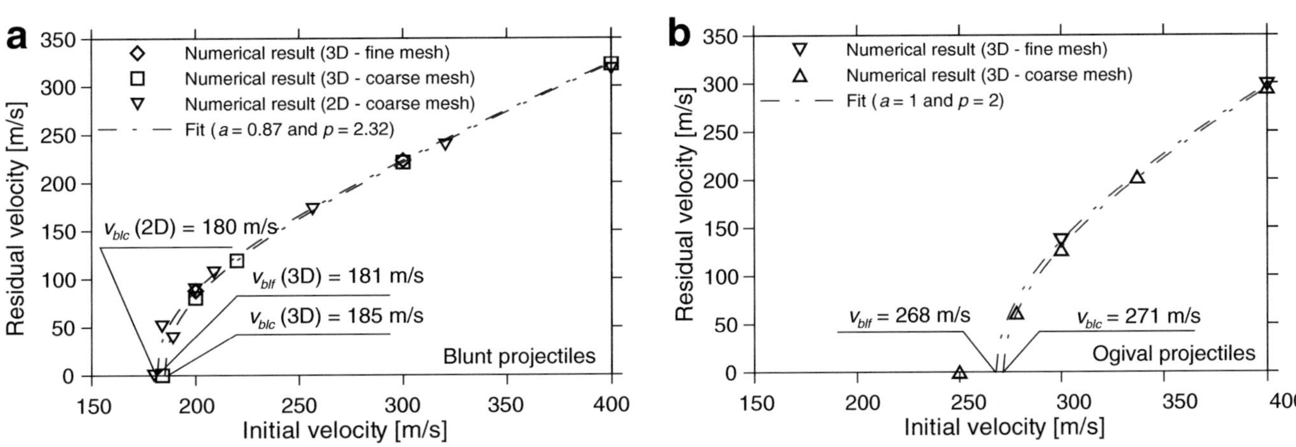

FIGURE 4-3 Numerical finite-element simulations of the ballistic behavior shown in Figure 4.2 depicting the effects of mesh refinement and the contrast between three-dimensional and two-dimensional (axisymmetric) meshing. SOURCE: Børvik, T., O. Hopperstad, and K. Pedersen. 2010. Quasi-brittle fracture during structural impact of AA7075-T651 aluminum plates. International Journal of Impact Engineering 37(5): 537-551.

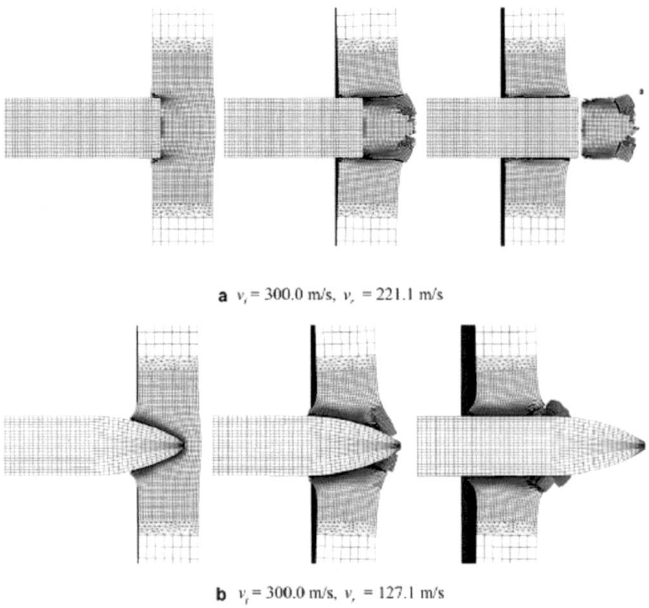

FIGURE 4-4 Simulations of penetration of a plate of AA7075-T651 showing finite-element mesh for a blunt-nosed (a) and an ogive-nosed (b) hard steel projectile. In both cases the projectile velocity prior to impact is 300 m/s; the exit speed of the blunt-nosed projectile is 221 m/s while that of the ogival projectile is 127 m/s. SOURCE: Børvik, T., O. Hopperstad, and K. Pedersen. 2010. Quasi-brittle fracture during structural impact of AA7075-T651 aluminum plates. International Journal of Impact Engineering 37(5): 537-551.

material radially outward, dissipating more energy. The numerical results in Figure 4-3 reproduce both sets of data in Figure 4-2 quite accurately, including the ballistic limits.

While AA7075 aluminum is not the most important material for projectile defeat, these 2008 simulations represent the state of the art. All the material parameters required as inputs to the constitutive and failure models have been independently measured, including those of the steel projectiles. Only the finite-element mesh layout and the element size are selected by the analyst. The predictions in Figure 4-3 depend on element size, because the constitutive model used in these simulations can predict the onset of shear localization and/or the fracture process but cannot predict the thickness of the associated failure zone. The thickness of a shear localization band is determined by a combination of factors, including microstructural length scales (see Chapter 3). These factors are not accounted for in commonly employed constitutive laws such as the Johnson-Cook relation, so they cannot set the size of these zones. As a result, the calculations give rise to a shear zone whose thickness is the width of one element. Thus, the energy dissipated in a zone of shear localization, or within any fracture process zone where the material is weakening, is proportional to the element size. Consequently, a systematic refinement of the mesh size to smaller and smaller elements will not converge to the correct physical result associated with shear localization and fracture zones having finite thicknesses. Although the thickness of the shear localization zone is estimated as 100 μ, the element size used in the simulations was 200 μ in the two-dimensional case and 500 μ in the three-dimensional case. Either element size calibration or a constitutive length parameter will continue to be an essential, non-straightforward requirement in penetration simulations.

The simulation of penetration represented by the results in Figure 4-3 must be pushed further to demonstrate the robustness of the predictive capability. Would the agreement between simulation and experiment continue to hold if plate thickness was doubled or if the target was two air-separated plates? Would the agreement hold up for projectiles impacting the plate at an oblique angle? More sophisticated constitutive models that incorporate the evolution of damage prior to failure and a material length based on mechanisms of deformation and failure hold promise for simulations that are more closely tied to fundamental material mechanisms and properties and freed from element size calibration. While the potential of such added sophistication has been demonstrated, the payoff in material protection simulations has yet to be realized.

Projectile Penetration of Bilayer Ceramic-Metal Plates

The simulation of projectile penetration of bilayer ceramic-metal plates further illustrates the need to combine good work on computation with sound experiments to investigate material and system properties in extreme conditions of strain, strain rate, and pressure. Holmquist and Johnson[6] published the results of such simulations for a bilayer plate of boron carbide backed by 6061-T6 aluminum alloy, where the simulations utilized the ceramic constitutive law of Johnson, Holmquist, and Beissel,[7] known as JHB. These simulations represent the state of the art in computations for the ballistic performance of ceramic armor components.

Experiments carried out many years ago by Wilkins[8] for the same system provide data on the ballistic limit that may be compared with the simulation results in Holmquist and Johnson.[9] Wilkins fired blunt and pointed projectiles at targets consisting of a 7.24-mm-thick boron carbide plate bonded to a 6.35-mm-thick piece of aluminum alloy as the backing plate, and the projectiles were made of very hard

[6]Holmquist, T.J., and G.R. Johnson. 2008. Response of boron carbide subjected to high-velocity impact. International Journal of Impact Engineering 35(8): 742-752.

[7]Johnson, G.R., T.J. Holmquist, and S.R. Beissel. 2003. Response of aluminum nitride (including phase change) to large strains, high strain rates, and high pressures. Journal of Applied Physics 94(3): 1639-1646.

[8]Wilkins, M.L. 1967. Second Progress Report of the Light Armor Program, Technical Report No. UCRL 50284. Livermore, Calif.: Lawrence Livermore National Laboratory.

[9]Holmquist, T.J., and G.R. Johnson. 2008. Response of boron carbide subjected to high-velocity impact. International Journal of Impact Engineering 35(8): 742-752.

steel. Ballistic limits of about 800 m/s and 700 m/s were obtained for the two kinds of projectiles.

It is notable that the ballistic limit for the cylinder is lower than that for the pointed projectile, indicating that for this target as well as the aluminum plate target discussed earlier, the cylinder is the better penetrator. However, it should be noted that this result is specific to the target and projectile configuration. An ogive-nosed projectile will penetrate considerably deeper into a thick aluminum target. Wilkins[10] observed that both projectiles cause the bilayer to bend at impact, an effect that tends to generate tensile stress at the far side of the ceramic plate. As ceramics are very poor at coping with tensile stress, the bending causes the ceramic plate to break. In the case of the cylindrical projectile, the full impact of the hit is felt by the target immediately. The bilayer begins to bend almost at once, and the ceramic plate fractures due to tensile stresses at a relatively early stage of the impact. On the other hand, the sharp nose of the pointed projectile does not immediately fully load the impact onto the target. Instead, the forces applied by the projectile to the bilayer build up gradually as the point of the projectile flattens, enabling the ceramic to remain intact for longer and to serve as better armor against the threat of the pointed projectile.

The JHB constitutive law is summarized in Figure 4-5, which shows ceramic strength versus applied pressure. In this context, "strength" is the ability of the material to support shear stress without extensive deformation. Such deformation may occur as the material yields and flows like a very viscous liquid owing to rearrangements within its internal lattice structure as it fractures and comminutes into small particles, which then flow collectively like sand. The relationships in Figure 4-5 are shown for intact material and failed material, each at two different strain rates, denoted by $\dot{\varepsilon}^*$. The connections between ceramic strength and applied pressure depicted in Figure 4-5 are used in the JHB model to represent the fact that a ceramic is strong in compression (i.e., at high pressure) and weak in tension (i.e., at negative pressure). The plot for intact material indicates that at high pressure, strength is almost insensitive to pressure. Under these conditions, a ceramic cannot fracture. Instead, at a critical level of shear stress (equal to the ceramic strength) it flows by the motion of dislocations that rearrange the internal structure of the ceramic lattice. In negative pressure (i.e., weak tension) the strength of the intact ceramic is very low and vanishes at a critical pressure, negative T. This situation reflects the fact that in tension, ceramic cracks and fractures at a low tensile stress. As the pressure applied to the ceramic is increased, it is less likely to crack and its strength increases. The plots for intact ceramic (Figure 4-5) interpolate this behavior between the extremes of tensile stress and very

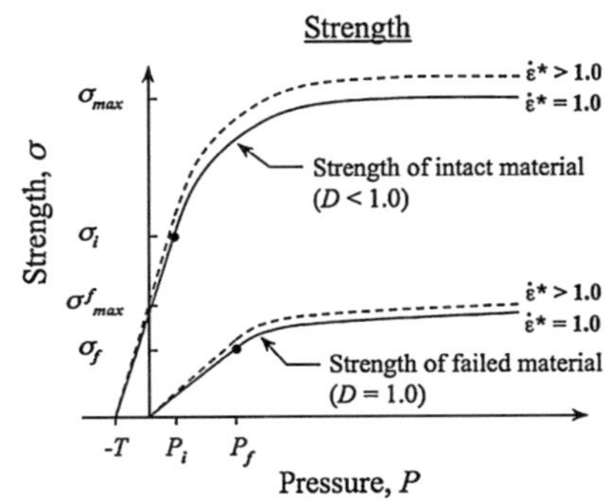

FIGURE 4-5 Ceramic strength versus applied pressure for the JHB constitutive model. The relationship is shown for intact material and failed material, each at two different strain rates, denoted by $\dot{\varepsilon}^*$. NOTE: D stands for damage. D = 1, fully damaged; D < 1 not fully damaged; D = 0 would mean no damage. As is illustrated, the damage weakens the material. SOURCE: Reprinted with permission from Johnson, G.R., T.J. Beissel, and S.R. Beissel, Journal of Applied Physics, 94, 1639, (2003). Copyright 2003, American Institute of Physics.

high pressure. The plots also indicate that the strengths will be slightly different at high and low strain rates.

It is obvious that when the ceramic cracks and fractures, it will be irreversibly damaged as it comminutes into granular material. This situation is captured in Figure 4-5, which shows that failed material has lower strengths at the same pressure than an intact material. Furthermore, the comminuted material cannot support tensile stresses, and so the plot of strength versus pressure for failed material terminates at the origin in Figure 4-5. The JHB constitutive law encompasses detailed rules for transitioning the state of the ceramic from intact to failed, and, broadly speaking, these rules implement the concept that as the material experiences deformation by flow of the fracturing material, the strength is steadily degraded. Therefore, as extensive deformation of the ceramic takes place, its strength steadily changes from the initial level appropriate for intact ceramic to that for failed ceramic.

Another feature of the JHB model as implemented in simulations of projectiles hitting the bilayer of boron carbide and aluminum alloy[11] is that once the material has failed and is subsequently, or simultaneously, placed under tension, the original continuum material is converted into a collection of individual free-flying particles. Such a condition represents the situation observed in experiments[12] where much of the

[10]Wilkins, M.L. 1967. Second Progress Report of the Light Armor Program, Technical Report No. UCRL 50284. Livermore, Calif.: Lawrence Livermore National Laboratory.

[11]Holmquist, T.J., and G.R. Johnson. 2008. Response of boron carbide subjected to high-velocity impact. International Journal of Impact Engineering 35(8): 742-752.

[12]Wilkins, M.L. 1967. Second Progress Report of the Light Armor Program, Technical Report No. UCRL 50284. Livermore, Calif.: Lawrence Livermore National Laboratory.

ceramic material is removed from the crater created by the penetration of projectile when clouds of comminuted ceramic particles form.

Constitutive laws such as the JHB model have many material parameters in them that characterize the elastic, acoustic, yielding, and fracturing behavior of the material being simulated. Holmquist and Johnson[13] calibrated most of the material parameters in their model using the results of plate impact experiments[14] involving a solid disc launched at high speed against a flat surface of the material being investigated. They augmented the information from these tests with spall data from experiments.[15] These observations reinforce the notion that successful simulation depends on the availability of experimental data to (1) characterize the parameters in the models being used, (2) validate and test the quality of the computational results, and (3) provide insights into how computational simulation should be conducted for a given material in a given situation.

Results from Holmquist and Johnson's simulations[16] show that for the initial velocities—V = 790 m/s for the pointed projectile and 700 m/s for the cylinder. At 100 μs after first hitting the bilayer at velocity V, the projectile has a residual velocity V_r. This velocity is 200 m/s for the pointed projectile and 257 m/s for the cylinder, and the projectiles will not slow down much more because the target is then offering no resistance. These results match those of Wilkins[17] in multiple ways, including appearance of the crater and the value of the ballistic limit.

A further feature of the results from the simulation is the distinct bending of the bottom aluminum layer, with prior concave upward bending of the plate being apparent in the now destroyed segment of the aluminum immediately below the penetrator. Although the ceramic layer in its residual shape is largely unbent due to the lack of a bond between the materials in the bilayer in the simulations, the ceramic will have bent like the aluminum in the early stage of projectile penetration, though to a lesser extent than the bending of the aluminum layer. Nevertheless, the results of the simulation clearly point out the importance of bending in the projectile penetration of relatively thin ceramic targets. A feature of the concave upward bending of the ceramic immediately below the projectile as it penetrates the target is a significant tensile stress at the bottom surface of the ceramic, leading to its fragmentation and comminution. In this state, the ceramic will retain the ability to resist the penetrator as long as the comminuted granular material is well contained by the aluminum backing and the penetrator itself. However, if such constraint is lost, the ceramic becomes ineffective in resisting the penetrator as it simply turns into a freely flying granular cloud. This consideration is an important element in the proper design of ceramic armor.

Note that Holmquist and Johnson[18] simulated two other types of experiment in addition to the penetration of a thin bilayer target: the impact on boron carbide plates[19] and the deep penetration of steel-jacketed boron carbide blocks.[20,21] Each of these experiments was successfully simulated with use of the JHB constitutive law, showing that the state of the art of simulation for the ballistic response of ceramic targets is quite far advanced. However, Holmquist and Johnson found it necessary to use a different set of material parameters for each distinct type of experiment. Therefore no single material model is yet able to capture the penetration and material response phenomena occurring in the cases of, for example, a ceramic under plate impact, deep penetration by a long heavy rod, and perforation of a thin bilayer target. As the authors note, this limitation of the results they obtained suggests that some important mechanisms of ceramic response are not being modeled accurately in the JHB constitutive law and that further work will be necessary to improve the constitutive laws for the response of ceramic under ballistic conditions of high strain, high strain rate, and high pressure.

All-Steel Sandwich Plates for Enhanced Blast Protection: Design, Simulation, and Testing

Traditionally, plate structures designed to withstand blast loads have employed monolithic plates. Within the past decade, the Office of Naval Research has supported efforts to explore whether all-metal sandwich plates comprised of the same material and having the same mass per area can be more effective against blasts than monolithic metal plates. Studies completed to date have considered various core types, such as honeycombs, corrugated plates, and lattice

[13]Holmquist, T.J., and G.R. Johnson. 2008. Response of boron carbide subjected to high-velocity impact. International Journal of Impact Engineering 35(8): 742-752.

[14]Vogler, T.J., W.D. Reinhart, and L.C. Chhabildas. 2004. Dynamic behavior of boron carbide. Journal of Applied Physics 95(8): 4173-4183.

[15]Wilkins, M.L. 1967. Second Progress Report of the Light Armor Program, Technical Report No. UCRL 50284. Livermore, Calif.: Lawrence Livermore National Laboratory.

[16]Holmquist, T.J., and G.R. Johnson. 2008. Response of boron carbide subjected to high-velocity impact. International Journal of Impact Engineering 35(8): 742-752.

[17]Wilkins, M.L. 1967. Second Progress Report of the Light Armor Program, Technical Report No. UCRL 50284. Livermore, Calif.: Lawrence Livermore National Laboratory.

[18]Holmquist, T.J., and G.R. Johnson. 2008. Response of boron carbide subjected to high-velocity impact. International Journal of Impact Engineering 35(8): 742-752.

[19]Vogler, T.J., W.D. Reinhart, and L.C. Chhabildas. 2004. Dynamic behavior of boron carbide. Journal of Applied Physics 95(8): 4173-4183.

[20]Orphal, D.L., R.R. Franzen, A.C. Charters, T.L. Menna, and A.J. Piekutowski. 1997. Penetration of confined boron carbide targets by tungsten long rods at impact velocities from 1.5 to 5.0 km/s. International Journal of Impact Engineering 19(1): 15-29.

[21]Lundberg, P., L. Holmberg, and B. Janzon. 1998. An experimental study of long rod penetration into boron carbide at ordnance and hypervelocities. Pp. 251-258 in Proceedings of the 17th International Symposium on Ballistics. Midrand, South Africa: South African Ballistics Organization.

trusses, with attention to design and ease of manufacturing.[22] Both air and water blast environments have been investigated, and an understanding is now in place of when fluid-structure interaction effects are important. The advancement of understanding that has been achieved is due to tightly coupled numerical simulation and experimental testing. In this section a brief overview is given of recent work by Dharmasena et al.,[23] which illustrates current capabilities and limitations along with possible opportunities.

The sequence of events occurring when a sandwich plate is struck by a blast wave is depicted in simplified form in Figure 4-6, where three relatively separate stages in the sequence can be visualized. For meter-size plates subject to intense blasts, the entire process lasts about 10 ms. In the scenario sketched in the figure, fluid-structure interaction occurs in Stage I. If the mass of the face toward the blast is sufficiently large, the blast wave bounces off the plate in much the way a rubber ball would be reflected, transmitting almost twice its incident momentum to the plate before the plate has time to displace. This is a reasonable way of viewing most air blasts striking a metal face of more than a few millimeters thick. However, a 1-cm thick metal plate struck by a water blast wave interacts with the wave in such a way that the reflection is reduced and therefore a smaller fraction of the incident wave momentum is transferred to the plate. This basic fluid-structure interaction effect for water blasts was discovered in World War II and has recently been extended to air blasts.[24]

Various core geometries have been investigated experimentally and by simulations, including hexagonal and square honeycomb cores; corrugated or folded-plate cores; and cores made of truss elements. These have generally been plates fashioned from relatively ductile steels. A folded-plate core has the advantage that it is readily manufactured. This is also true of several truss-core geometries, which have the added advantage that the core is an open structure useful for multifunctional applications.[25] Which core yields the best performance depends on the type and level of blast, whether the blast is in air or water, and whether the standoff is close or remote. A sandwich plate can be designed to capitalize on the fluid-structure interaction effect because the mass per area of the face sheet toward the blast will be less than half that of its monolithic competitor. For this effect to come into play, the core must not be overly strong, so that only the face sheet acquires momentum during the period it is impacted by the water blast.[26] For air blasts, decoupled calculations can supply a good approximation. In such cases, the pressure history is computed in the first step by treating the plate structure as rigid; this history is then applied to the structure in the second step to obtain its response. While fluid-structure effects are not usually significant in air blasts, interaction effects can be important for air blasts that entrain sand or gravel, such as those experienced by vehicles exposed to buried improvised explosive devices.

Figure 4-7 displays the deformation of clamped square-honeycomb-core sandwich plates made from stainless steel and subject to three explosive levels at close standoff in air. At the lowest level shown, the back face undergoes relatively little deformation; the high bending stiffness of the sandwich plate is very effective. At the two higher levels of blast, significant stretching of the face sheets occurs in addition to core crushing. Both core crushing and face sheet stretching absorb substantial energy. While severely deformed, these plates have not fractured.

Accurate simulations of blast-loaded structures require input material properties for the constitutive relation, knowledge of the temporal and spatial pressure pulse on the structure, and a finite element code that can cope with highly nonlinear material and geometric behavior, including internal contacting surfaces. As no fracture occurred in the test specimens, no attempt was made to model damage or fracture in the simulations.[27] Finite-strain plasticity was employed along with input of tensile data for the stainless steel as a function of strain and strain rate.

Comparisons of the simulations with the experimental results are displayed in Figure 4-8. Nearly all the experimental details are replicated. Even the buckling of the webs can be captured accurately. The back face of these sandwich plates deflects less than the equivalent mass of solid plate even in an air blast. The performance of the sandwich plate relative to a monolithic plate would be better in water blasts due to fluid-structure interaction that favors sandwich plates.

There are no "adjustable" parameters in the simulations presented above. Thus, one can conclude it is possible to carry out calculations to improve the design of plate structures against blast loads of various types. This optimistic assessment must be tempered by the following considerations:

- The ultimate blast resistance of these structures has not been determined. To do so would require subject-

[22]Wadley, H.N.G. 2006. Multifunctional periodic cellular metals. Philosophical Transactions of the Royal Society A: Mathematical, Physical and Engineering Sciences 364(1838): 31-68.

[23]Dharmasena, K.P., H.N.G. Wadley, Z. Xue, and J.W. Hutchinson. 2008. Mechanical response of metallic honeycomb sandwich panel structures to high-intensity dynamic loading. International Journal of Impact Engineering 35(9): 1063-1074.

[24]Kambouchev, N., L. Noels, and R. Radovitzky. 2006. Nonlinear compressibility effects in fluid-structure interaction and their implications on the air-blast loading of structures. Journal of Applied Physics 100(6): Article number 063519.

[25]Wadley, H.N.G. 2006. Multifunctional periodic cellular metals. Philosophical Transactions of the Royal Society A: Mathematical, Physical and Engineering Sciences 364(1838): 31-68.

[26]Liang, Y., A.V. Spuskanyuk, S.E. Flores, D.R. Hayhurst, J.W. Hutchinson, R.M. McMeeking, and A.G. Evans. 2007. The response of metallic sandwich plates to water blast. Journal of Applied Mechanics 74(1): 81-99.

[27]Dharmasena, K.P., H.N.G. Wadley, Z. Xue, and J.W. Hutchinson. 2008. Mechanical response of metallic honeycomb sandwich panel structures to high-intensity dynamic loading. International Journal of Impact Engineering 35(9): 1063-1074.

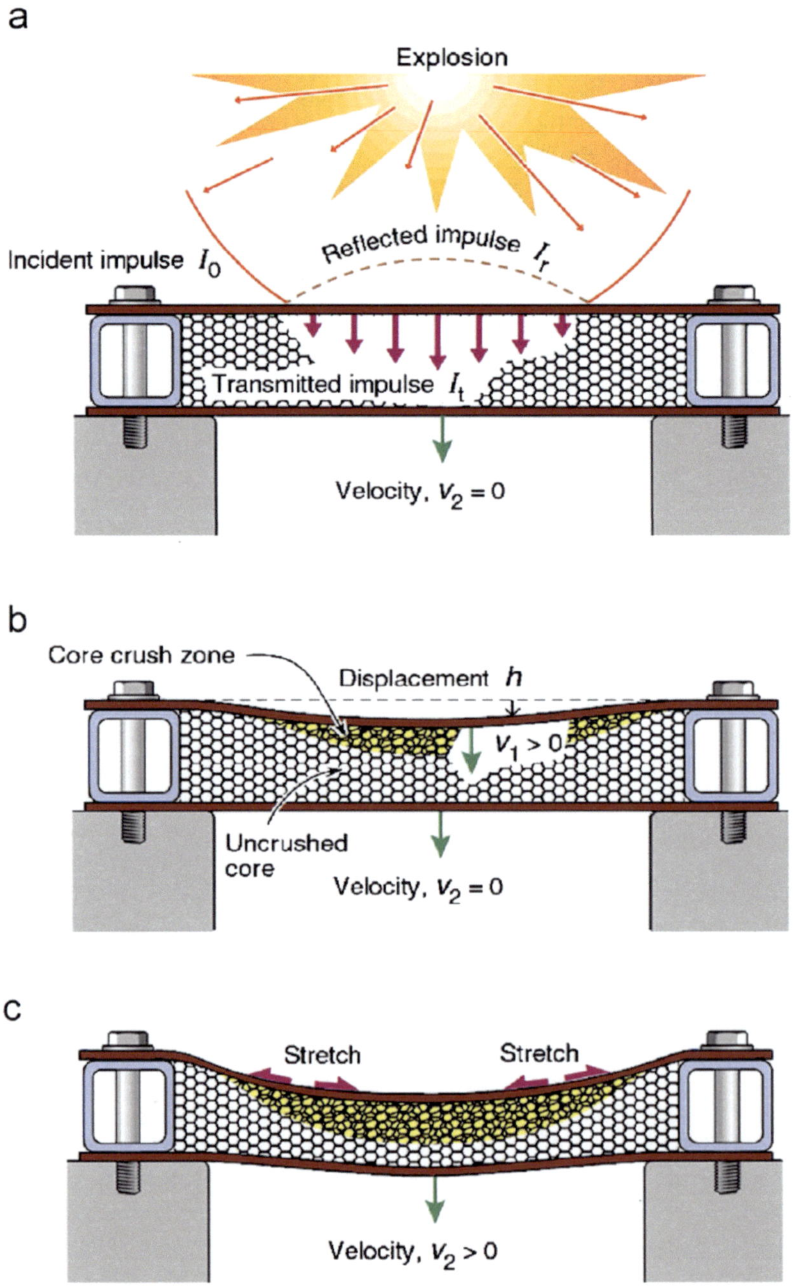

FIGURE 4-6 Schematic depicting the response of a clamped sandwich plate to blast loading: (a) Impulsive loading (Stage I); (b) core crushing (Stage II); and (c) overall bending and stretching (Stage III). SOURCE: Dharmasena, K.P., H.N.G. Wadley, Z. Xue, and J.W. Hutchinson. 2008. Mechanical response of metallic honeycomb sandwich panel structures to high-intensity dynamic loading. International Journal of Impact Engineering 35(9): 1063-1074.

ing the structures to larger blasts and taking account of fracture in the simulations. Reliable models for such simulations are not yet established for either monolithic plates or sandwich plates.

- Plate structures are susceptible to failures along welds and joints, and simulations of these events are not yet reliable. Existing continuum models can be calibrated to reproduce fractures in tension or in shear, but they cannot reliably predict both types of fractures under a wide range of stress states. Mechanistic-based fracture models are needed to expand predictive capabilities.
- Highly refined meshes were employed in the sandwich plate simulations reported above—far more refined than would be feasible for-large scale structures. New constitutive models and computational

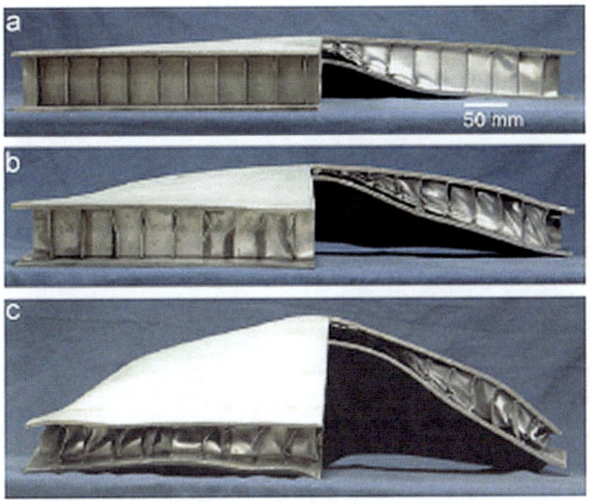

FIGURE 4-7 Half-sectional square honeycomb core test panels. The impulse load is (a) 21.5 kPa s, (b) 28.4 kPa s, and (c) 33.7 kPa s. Stainless steel sandwich plates with square honeycomb cores clamped around their edges subjected to three levels of air blast. The plates were sectioned after deformation to display the core and the relative position of the faces. The core webs are 0.76 mm thick with spacing 30.5 mm. The core thickness is 51 mm. Each face sheet is 5 mm thick. The core comprises 24 percent of the mass of the plate. The equivalent thickness of a solid plate with the same mass per area is 13.1 mm. SOURCE: Dharmasena, K.P., H.N.G. Wadley, Z. Xue, and J.W. Hutchinson. 2008. Mechanical response of metallic honeycomb sandwich panel structures to high-intensity dynamic loading. International Journal of Impact Engineering 35(9): 1063-1074.

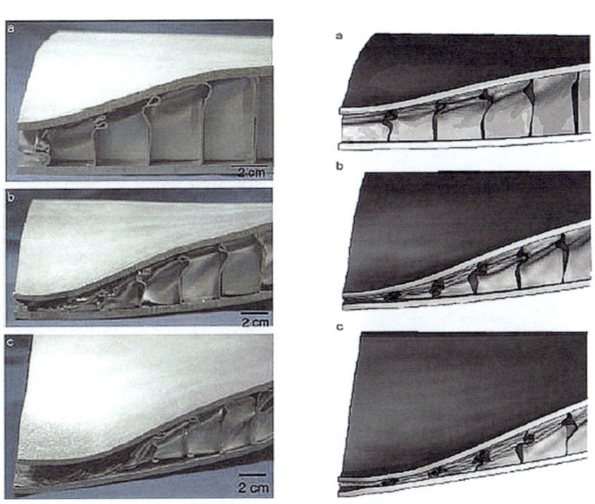

FIGURE 4-8 Comparison of experimental test specimens (on the left) deformed at the three levels of air blast shown, with simulations carried out for the same plates and level of blasts (on the right). SOURCE: Dharmasena, K.P., H.N.G. Wadley, Z. Xue, and J.W. Hutchinson. 2008. Mechanical response of metallic honeycomb sandwich panel structures to high-intensity dynamic loading. International Journal of Impact Engineering 35(9): 1063-1074.

methods are required to capture the main deformational features of the core with relatively coarse meshing.

THE STATE OF THE ART IN EXPERIMENTAL METHODS

As the discussion of the three examples above illustrates, experimental methods are at the heart of any effort to observe and characterize material behavior. This is especially true for protection materials, which experience extreme rates of loading and for which both deformation and failure must be understood and characterized. This section outlines current experimental methods relevant to protection materials and points to opportunities for advancing their capabilities.

Definition of the Length Scales and Timescales of Interest

The committee has focused on developing lightweight protective materials for future Army applications and interpreted its mandate broadly to include providing protection from threats that involve the *rapid* deposition of energy directly into a material or structure. Examples of threats of this type include direct impact by (1) an incoming projectile and (2) explosive, or blast, loading. The timescales associated with these events are of paramount importance, and the characteristic velocities associated with propagating waves, projectiles, or failure processes generate associated length scales. These scales can be envisioned in the two-dimensional space shown in Figure 4-9, where the inclined straight lines represent the domains in space and time that are affected by phenomena at each of the defined speeds. Typical components and structures in Army applications will be of the sizes represented in the blue shaded region, usually a millimeter to several centimeters. Given these sizes, the longest timescales associated with threat events are of the order of a millisecond (in the case of blast loading). Most of the controlling phenomena operate at much smaller timescales (microseconds down to nanoseconds). The characteristic length scales that control material response to threat are of the order of nanometers to hundreds of micrometers. The experimental challenges associated with this field largely arise from the need to resolve phenomena at these timescales and length scales. As shall be seen in this section, the vast majority of available experimental methods provide either high time resolution or high spatial resolution, but few provide both.

Relation to Experimental Methods

Experimental methods must be able to access the appropriate regimes in the length scale and timescale space in order to investigate any particular behavior or phenomenology. A critical issue here is that these scales should be investigated simultaneously. Because the events are transient and involve complex loading paths, it is difficult to pin down real-time

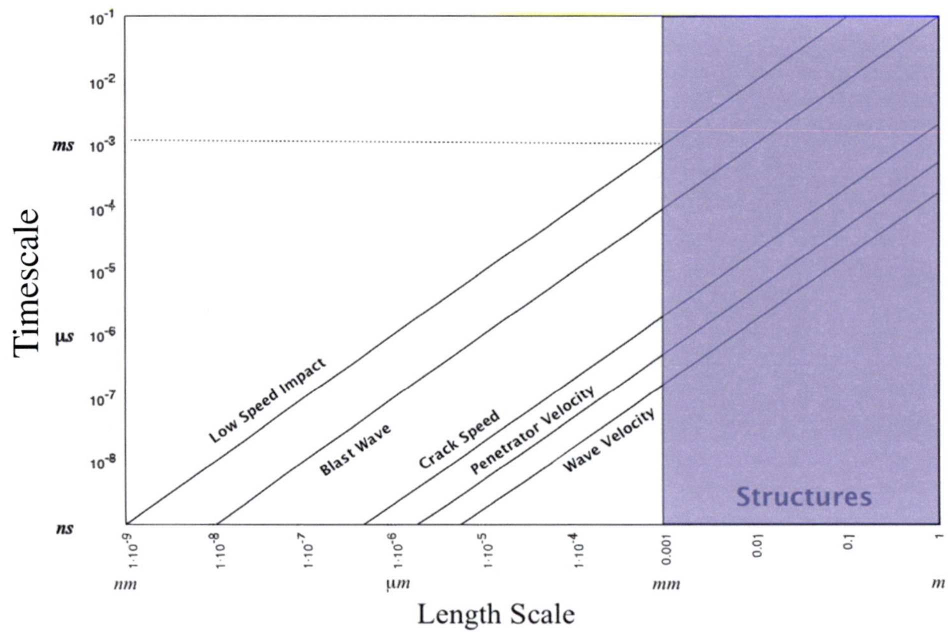

FIGURE 4-9 Length scales and timescales associated with typical threats to Army fielded materials and structures. The lines represent the velocities associated with specific phenomena observed in impact events, such as the blast wave, cracks, and stress waves. The choice of structural length scale and the particular phenomenon of interest then determines a characteristic timescale for the problem.

behavior through postmortem analysis of the material or structure. This does not imply that there are no relevant phenomena at the larger timescales (multiple milliseconds and larger). Such timescales are relevant to a number of threats, particularly those related to blast and explosive loading, and they are tightly coupled to structural dynamics, which can involve both material and geometric nonlinearity. Broadly speaking, however, the challenges in understanding and observation at these longer timescales and length scales consist largely of correctly exploring the coupling of the dynamics to the design space.

In this section, state-of-the-art experimental methods capable of exploring various regimes in the length scale–timescale space are described. To begin, experimental methods will be classified in terms of their intended applications. First, however, a broad comment is in order. One approach to understanding the interaction between threat and material is to perform a highly instrumented version of the actual threat event. Although this approach is very useful, and is indeed the most definitive metric for the effectiveness of a protective material within a specific protected system, it does not necessarily provide significant guidance for the development of radically improved protective systems. The focus of this section is, accordingly, on the more fundamental experiments associated with developing a basic understanding of the mechanisms, behaviors, and processes associated with the threat-material interaction that can lead to improved constitutive characterizations, including those for failure processes.

Classification of Experimental Methods[28]

Experimental techniques commonly called "impact experiments" often have very different objectives. Since the design of an experimental technique depends on the goal of the experiment, it must first be decided what information one wants to extract from the experiment. Typically, what are called "impact" (or "dynamic") experiments fall into one of four categories, listed here according to the complexity of the dynamics:

1. *High-strain-rate experiments.* These measure the high-strain-rate characteristics of a material;
2. *Shock physics experiments.* These aim at understanding shock wave propagation in a material or structure; they may also develop high strain rates, but the high-rate deformations vary as a function of space and time;
3. *Impact phenomenology experiments.* These experiments endeavor to understand or discover impact phenomena such as cratering efficiency or fragmentation; and
4. *Dynamic failure experiments.* These would help us to understand the processes of dynamic failure within a material or structure.

[28]Except as noted in the text, this section drawn from Ramesh, K.T. 2008. High strain rate and impact experiments. Chapter 33 in Handbook of Experimental Solid Mechanics. W.N. Sharpe, Jr., ed. New York, N.Y.: Springer.

A detailed consideration of the state of the art for each of these types of experiments, based on examples from the literature, follows.

Evaluating Material Behavior at High Strain Rates

Most of the inelastic (and particularly the plastic) deformations due to impacts at rapid velocities occur at high strain rates. The deformations may lead to large strains and high temperatures. The high-strain-rate behavior of many materials (often defined as the dependence $\sigma_f(\varepsilon,\dot{\varepsilon},T)$ of the flow stress on the strain, strain rate, and temperature) is not, however, well understood, particularly at high strains and high temperatures. Some experimental techniques have been developed to measure material properties at high strain rates. Here, the committee considers experimental techniques that develop controlled high rates of deformation in the bulk of the specimen rather than techniques that develop high strain rates just behind a propagating wave front.

The main experimental techniques for measuring the rate-dependent properties of various materials are described in Figure 4-10 (the stress states developed by the various techniques may not necessarily be identical). One outstanding recent review of these methods is that of Field et al.[29] Here, strain rates above 10^2 s^{-1} are classified as high strain rates, those above 10^4 s^{-1} are called very high strain rates, and those above 10^6 s^{-1} are ultrahigh strain rates. Strain rates at or below 10^{-3} s^{-1} are usually considered to represent quasi-static deformations, and strain rates below 10^{-6} s^{-1} are considered to represent "creep." The emphasis here is on experimental techniques for strain rates greater than 10^2 s^{-1}—that is, high (10^2-10^4 s^{-1}), very high (10^4-10^6 s^{-1}) and ultrahigh ($>10^6$ s^{-1}).

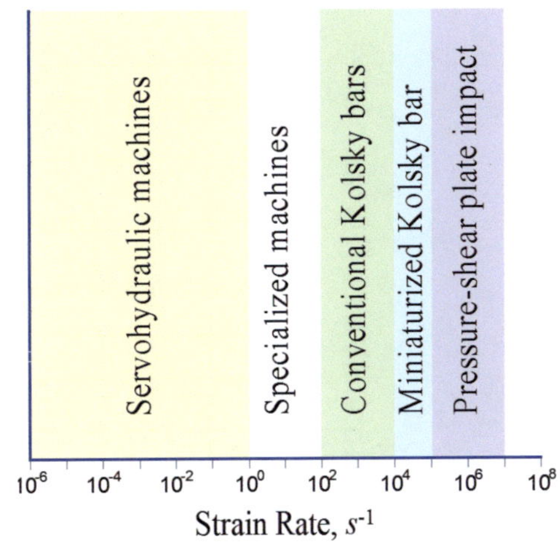

FIGURE 4-10 Experimental techniques used for the development of controlled high-strain-rate deformations in materials.

Kolsky Bars

The now-classical experimental technique in the high-strain-rate domain is the Kolsky bar, or split-Hopkinson pressure bar, experiment[30] for determining the mechanical properties of various materials (e.g., metals,[31] ceramics,[32] and polymers[33]) in the strain rate range 10^2 through 8×10^3 s^{-1} (see Figure 4-11). This technique is now in use throughout the world. Since the fundamental concept involved in this technique was developed by Kolsky,[34] the term Kolsky bar will be used here. Kolsky bar experiments may include compression, tension, torsion, or combinations of all of these.[35]

The Kolsky bar consists of two long bars, called the input and output bars, that are designed to remain elastic throughout the test. These bars sandwich a small specimen, usually cylindrical, which is expected to develop inelastic deformations. The bars are typically made of high-strength steels, such as maraging steel, with a very high yield strength and substantial toughness. Other bar materials that have been used include 7075-T6 aluminum, magnesium alloys and poly(methyl methacrylate) (for testing very soft materials), and tungsten carbide (for testing ceramics). One end of the input bar is impacted by a projectile made of a material identical to that of the bars; the resulting compressive pulse propagates down the input bar to the specimen. Several reverberations of the loading wave occur within the specimen; a transmitted pulse is sent into the output bar and a reflected pulse is sent back into the input bar. Typically, resistance strain gages are placed on the input and output bars for measuring (1) the incident pulse generated by the impacting projectile; (2) the reflected pulse from the input bar/specimen interface; and (3) the transmitted pulse through the specimen to the output bar. The strain gage signals are typically measured using high-speed digital oscilloscopes with at least 10-bit accuracy and preferably with differential inputs to reduce noise.

Many extensions and modifications to the traditional Kolsky bar system have been developed over the last five

[29]Field, J.E., S.M. Walley, W.G. Proud, H.T. Goldrein, and C.R. Siviour. 2004. Review of experimental techniques for high rate deformation and shock studies. International Journal of Impact Engineering 30(7): 725-775.

[30]Nicholas, T., and A.M. Rajendran. 1990. Material characterization at high strain-rates. Pp. 127-296 in High Velocity Impact Dynamics. J.A. Zukas, ed. New York, N.Y.: John Wiley & Sons.

[31]Nemat-Nasser, S., and J.B. Isaacs. 1997. Direct measurement of isothermal flow stress of metals at elevated temperatures and high strain rates with application to Ta and Ta-W alloys. Acta Materialia 45(3): 907-919.

[32]Chen, W., G. Subhash, and G. Ravichandran. 1994. Evaluation of ceramic specimen geometries used in a split Hopkinson pressure bar. DYMAT Journal 1: 193-210.

[33]Walley, S., and J. Field. 1994. Strain rate sensitivity of polymers in compression from low to high strain rates. DYMAT Journal 1: 211-228.

[34]Kolsky, H. 1949. An investigation of the mechanical properties of materials at very high rates of loading. Proceedings of the Physical Society: Section B 62(11): 676-700.

[35]Gray III, G.T. 2000. Classic split-Hopkinson pressure bar testing. Pp. 462-476 in ASM Handbook Volume 8: Mechanical Testing and Evaluation. H. Kuhn and D. Medlin, eds. Materials Park, Ohio: ASM International.

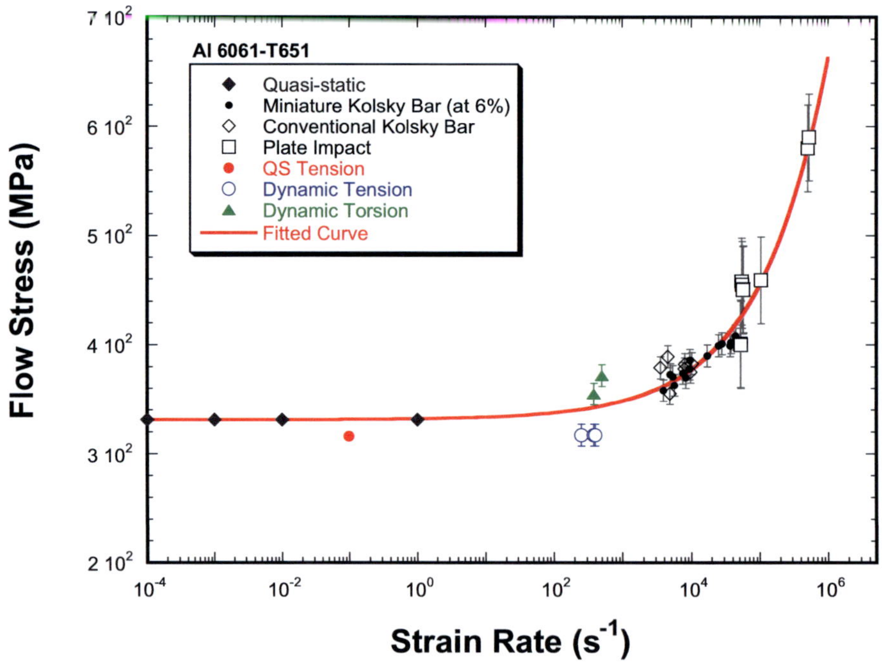

FIGURE 4-11 High-strain-rate behavior of 6061-T6 aluminum determined through servohydraulic testing, compression and torsional Kolsky bars, and high-strain-rate pressure-shear plate impact.

decades. Most of these are listed in Table I of a review[36] by Field et al., which includes an exhaustive literature set. Very high strain rates (up to 5×10^4 s^{-1}) can be attained in miniaturized systems while retaining the ability to study materials at strain rates as low as 1.0×10^3 s^{-1} (the maximum achievable strain rate is limited by an inertial correction and varies with the material being tested). Both computational and experimental results have shown that this extended capability can be attained not only without violating the requirements for valid high-rate testing but also while improving the precision and accuracy of the experimental results.[37]

High-Strain-Rate, Pressure-Shear Plate Impact

Researchers have developed techniques called high-strain-rate, pressure-shear plate impact techniques to study shearing behavior in materials experiencing homogeneous shearing deformations at exceedingly high shear rates (10^4 to 10^6 s^{-1}) and superimposed hydrostatic pressures of several gigapascals.[38] Although this approach[39] assesses a wider range of responses, it is not used as much as the Kolsky bar experiment since it necessitates much higher investment in lab equipment and personnel whose training is time-consuming. The experiment involves the impact of plates that are flat and parallel but inclined relative to their direction of approach. The specimen is a very thin (say, 100 μ), very flat plate of the material being investigated. This specimen is adhered to a hard plate (the "flyer"), which is itself mounted on a projectile that is launched through the barrel of a gas gun at an immobile target ("anvil" plate). The target is positioned in a special fixture, known as the target holder, within an evacuated chamber. The flyer and the anvil plates are aligned before impact using an optical technique. Rotation of the projectile is prevented by a key in the projectile, which glides within a matching keyway machined in the barrel.

At impact, plane longitudinal (compressive) and transverse (shear) waves are generated in the specimen and the target plate propagating at the longitudinal wave speed c_l and the shear wave speed c_s. These waves reverberate within the specimen, causing the normal stress and the shear stress to

[36]Field, J.E., S.M. Walley, W.G. Proud, H.T. Goldrein, and C.R. Siviour. 2004. Review of experimental techniques for high rate deformation and shock studies. International Journal of Impact Engineering 30(7): 725-775.

[37]Jia, D., and K.T. Ramesh. 2004. A rigorous assessment of the benefits of miniaturization in the Kolsky bar system. Experimental Mechanics 44(5): 445-454.

[38]Clifton, R.J., and R.W. Klopp. 1985. Pressure-shear plate impact testing. Pp. 230-239 in ASM Handbook Volume 8: Mechanical Testing. Materials Park, Ohio: ASM International.

[39]Other, similar approaches include the shear-compression test (SCS test) developed by Ravichandran and co-workers, which is simpler to perform but more difficult to analyze.

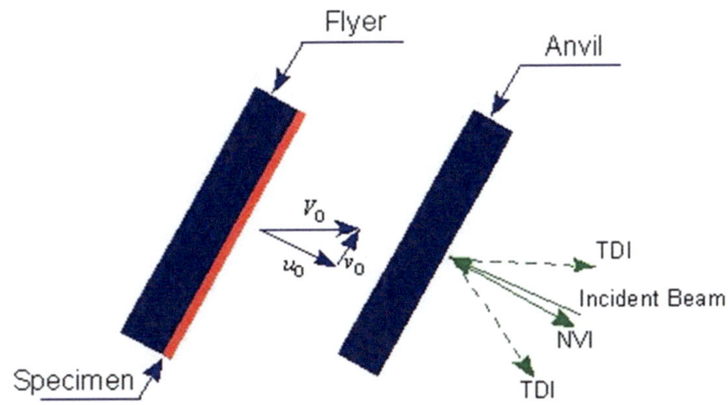

FIGURE 4-12 Schematic of the high-strain-rate, pressure-shear plate impact experiment. The specimen thickness is greatly exaggerated for clarity. TDI, transverse-displacement interferometer; NVI, normal velocity interferometer.

build up in the specimen material. As Yadev notes, information on the stress levels sustained by the specimen material is carried by the normal and transverse waves propagating into the target plate.[40] The target remains elastic, so there is a linear correspondence between the stresses and the particle velocities in the target plate. It is therefore good enough to measure the normal and transverse particle velocities in the target plate in order to obtain the specimen's stress state and deformation state. The whole experiment is over before any unloading waves from the periphery of the plates reach the point of observation. Thus only plane waves are involved, and a one-dimensional analysis is not only sufficient but also rigorously correct. Like most plate impact experiments, this is a uniaxial strain experiment in that no transverse normal strains can occur during the time of interest.

Measurements of particle velocities at the free surface of the target plate are made using laser interferometry off a diffraction grating that is photodeposited onto the rear surface.[41] The normal velocity and the transverse displacement at the center of the rear surface of the target are measured. The high-strain-rate, pressure-shear plate impact technique is capable of achieving shear rates of 8×10^4 to 10^6 s^{-1}, depending on the specimen thickness.[42] A version of this experiment that is designed to allow recovery of the specimen (for microstructural examination) after a single high-strain-rate shear loading has also been developed.[43]

The superimposed hydrostatic pressures that can be exerted during the high-strain-rate, pressure-shear plate impact experiment may be as high as 10 GPa, depending on the impedances of the flyer and target plates and the projectile velocity. The superimposed hydrostatic pressures must always be remembered when comparing high-strain-rate, pressure-shear plate impact data with data obtained using the other techniques shown in Figure 4-12, since all of the other techniques can generate essentially uniaxial stress states, typically corresponding to low hydrostatic pressures. In particular, while the effect of pressure on the flow stress of most metals is negligible in comparison with the effect of strain rate, the effect of pressure on the strength of polymers, ceramics, glasses, and amorphous materials may be substantial, even in comparison with the effect of strain rate.

Investigating Shock Physics

Experiments designed to study the propagation of large-amplitude stress waves within materials constitute a very broad class of impact experiments. The interest here is in experiments that examine the interactions of waves with materials, particularly exciting inelastic modes such as plasticity, cracking, or other kinds of damage. In contrast to the experiments in the preceding section, the experiments in this section all generate strain rates and stress states that vary in both space and time, and the wave propagation is fundamentally dispersive (i.e., the waveform changes as the wave propagates) because of material behavior.

In broad terms, wave-propagation experiments of this

[40]Yadev, S. 1995. The mechanical response of a 6061-T6 A1/A1 "2O" 3 metal matrix composite at high rates of deformation, Acta Metallurgica Et Materialia.

[41]Except as noted in the text, this section is drawn from Ramesh, K.T. 2008. High strain rate and impact experiments. Chapter 33 in Handbook of Experimental Solid Mechanics. W.N. Sharpe, Jr., ed. New York, N.Y.: Springer.

[42]See, for example, Frutschy, K.J., and R.J. Clifton. 1998. High-temperature pressure-shear plate impact experiments on ofhc copper. Journal of the Mechanics and Physics of Solids 46(10): 1723-1744.

[43]Jia, D., A.M. Lennon, and K.T. Ramesh. 2000. High-strain-rate pressure-shear recovery: A new experimental technique. International Journal of Solids and Structures 37(12): 1679-1699.

type fall into two categories: bar wave experiments[44] and plate impact experiments, or more specifically, uniaxial-stress-wave propagation experiments and uniaxial-strain-wave propagation experiments. The plate impact experiments are far more common since they can explore a wider range of the phenomena that arise in impact events. In the timescales associated with ultrahigh-strain-rate experiments, uniaxial strain conditions are sampled. Such results are difficult to compare with results obtained at high and very high strain rates (typically obtained with uniaxial stress experiments), particularly if the material has pressure-dependent properties.

The strain rates developed in large-amplitude wave propagation experiments, where shocks are developed, can be on the order of 10^6 to 10^8 s^{-1}, but they only exist for a short time behind a propagating wave front, and because of inelastic dissipation, as well as reflections from surfaces, the strain rates will vary with position in the impact plate. The temperatures behind the wave front may be substantial and must be accounted for as well. Comparisons of material properties estimated using wave propagation experiments and high-strain-rate experiments (the distinction made in this chapter) can therefore require careful parsing of experimental conditions.

A shock wave generated during a plate-impact experiment propagates at a shock speed U_S that varies with the particle velocity u_p, and it is commonly observed that these two variables are related linearly or nearly so: $U_S = U_0 + su_p$, where U_0 and s are material-specific parameters, with the first being essentially the sound wave speed in the material. Large numbers of experiments have been performed to determine these parameters in various materials. A summary of such data is presented in Meyers[45]; another useful reference is Gray.[46] The shock wave propagation literature is extensive and includes a large number of conference proceedings from the biannual meetings of the American Physical Society Topical Group on Shock Compression of Condensed Matter published by the American Institute of Physics and a series on the shock compression of solids. Experimental details are often emphasized in these conference proceedings.

The main experimental issues associated with shock wave plate impact experiments are (1) the development of gun-launching facilities at the appropriate velocities; (2) the accurate measurement of projectile velocity; (3) the measurement of the stress state within the specimen, typically through the use of stress gauges; and (4) the measurement of the particle velocities in the targets, typically through the use of interferometers such as the velocity interferometry system for any reflector (VISAR). Shock wave experiments at very high velocities, pressures, and strain rates can also be accomplished using a "laser shock" apparatus, where the interaction of a metal film with a high-power laser pulse generates the wave.[47] Such experiments are of very short duration (typically only a few nanoseconds) and are very hard to control when it is desired to generate a planar shock.

The gun-launch facilities associated with shock physics experiments are typically extremely specialized facilities run by a small number of companies and the national laboratories, and extraordinary precautions must be taken to ensure safety. Most of these facilities offer gas guns, light gas guns, or powder guns; for the higher velocities, multistage guns are typically required. Since kinetic energy increases with the square of the velocity, reaching higher velocities typically requires the use of lower-mass sabots and flyers. Velocities greater than 10 km/s have been achieved with ~1 g flyers using multistage guns.

The typical results obtained from shock experiments include the determination of the Hugoniot elastic limit (this is essentially a uniaxial strain version of the traditional elastic limit), the determination of the Hugoniot curve itself, and (depending on the instrumentation) the determination of the hydrostat and possibly of the shear strength. Note that there are independent methods (such as the diamond anvil cell) for measuring hydrostat response. Another problem with shock experiments is that it is quite difficult, although not impossible, to do recovery experiments. Additionally, it is difficult to separate strength and failure from overall thermodynamic response (lateral gages can be introduced but have their own attendant issues), because diagnostics typically provide only the longitudinal stress, which is a combination of the spherical and deviatoric stresses.

A sophisticated capability exists within the national laboratories for analyzing the results of shock wave experiments in fine detail, and much of the current understanding of the high-pressure behavior of materials comes from such experiments. These methods are generally incapable of determining the rate-dependent shear strength of materials, except under very special conditions. This leads to one of the primary difficulties in understanding material behavior under the extreme conditions developed in the armor problem: The experimental techniques available are generally either most sensitive to material behavior under high pressure or under high strain rate, but they rarely provide accurate information under *combined* high pressure and ultrahigh strain rate. As the shock is passing a material point, the stress increases very

[44]Cazamias, J.U., W.D. Reinhart, C.H. Konrad, and L.C. Chhabildas. 2001. Bar impact tests on alumina (AD995). Pp. 787-790 in Shock Compression of Condensed Matter—2001: Proceedings of the Conference of the American Physical Society, Topical Group on Shock Compression of Condensed Matter, Atlanta, Ga., June 25-29. AIP Conference Processing volume 620. M.D. Furnish, N.N. Thadhani, and Y. Horie, eds. New York, N.Y.: Springer.

[45]Meyers, M.A. 1994. Dynamic Behavior of Materials. New York, N.Y.: John Wiley & Sons.

[46]Gray III, G.T. 2000. Shock wave testing of ductile materials, Pp. 530-538 in ASM Handbook Volume 8: Mechanical Testing and Evaluation. H. Kuhn and D. Medlin, eds. Materials Park, Ohio: ASM International.

[47]Kimberley, J., J. Lambros, I. Chasiotis, J. Pulskamp, R. Polcawich, and M. Dubey. 2010. Mechanics of energy transfer and failure of ductile microscale beams subjected to dynamic loading. Journal of Mechanics and Physics of Solids 58(8): 1125-1138.

rapidly (ultrahigh strain rates). Once the shock has passed, the material is under a state of high pressure but at relatively low strain rates since there is no relative particle motion. Upon release, however, the material again experiences high strain rates, but now starting at high pressures. Further, under uniaxial strain conditions, self-confinement conditions prevail. Efforts to bridge this gap are recommended.

A great deal can be learned about material behavior (the constitutive law rather than the failure process) by analyzing the results of a suite of experiments that include all of those discussed above, combined with a microscopic analysis of the deformation mechanisms active within the material before loading and postmortem. This capability would be significantly enhanced by in situ determination of the active mechanisms, an essential requirement for ceramics and polymers.

The Department of Defense (DoD) has traditionally been a user of armor analysis codes developed by the Department of Energy (DoE) (LS-DYNA, CTH, ALE3D, LAMMPS, and others) rather than a developer of new algorithms and production codes. There are individual exceptions where multiscale, multiphysics methods are being developed, but this is not a concerted effort that will furnish the tools needed to address the future of protection material systems. There have been significant advances in the models and algorithms addressing the limitations of existing armor codes, which must be integrated into existing or completely new codes to achieve the next level of understanding of armor material response.

Investigating Dynamic Failure Processes

Turning now to the experimental study of dynamic failure processes, many are active mechanisms within the impact events, as discussed earlier. In a broad sense, failure processes consist of brittle fracture at various length scales, void growth associated with ductile fracture, void collapse, the development of adiabatic shear bands, and a variety of structural instabilities such as necking.

For the case of the dynamic brittle fracture process in ceramics, an example of the state of the art in the understanding of the dynamic failure process is that provided in Paliwal et al.[48] High-speed photography with a modified compression Kolsky bar technique was used to observe the dynamic failure of uncoated and Cr-coated, transparent polycrystalline aluminum oxynitride (AlON) undergoing uniaxial high-strain-rate compression. High-speed photographs were correlated in time with stress measurements in the specimen (Figure 4-13).

In the fully transparent samples, dynamic activation, growth, and coalescence of cracks and resulting damage zones from spatially separated internal defects were directly observed and correlated with the macroscopic loss of load-carrying capacity and the ultimate catastrophic failure of this material. Identical experiments on the coated material showed only the dynamic progressive failure on the specimen surface, not the origin of the failure images at the internal defects. Therefore, the actual failure mode differs from what is suggested by the photographs of the opaque ceramic undergoing dynamic compression. By means of high-speed photographs on transparent AlON, these authors obtained real-time data on the damage kinetics, which suggest that the cause of the final failure for AlON under dynamic loading was the formation of a damage zone that propagates unstably, *not* splitting parallel to the loading axis.

This is an example of the value of using a model material—the transparency of the AlON allows determination of internal failure processes that would be otherwise inaccessible—and demonstrates that modeling approaches that only generate axial splitting modes do not properly describe the dynamic failure processes, even though they may capture surface features. Nor would the postmortem examination of fragments in such experiments provide this critical information on dynamic failure processes in the material.

An excellent example of an experimental technique that provides critical information on the dynamic failure process of void growth is provided by the work of Chhabildas and co-workers, who use a line VISAR to examine the process of spallation.[49] Spallation is the process of dynamic void growth within a "spall plane" generated by converging rarefaction fans in a shock experiment designed to generate local tension. Experiments designed to generate such a spall plane normally use a VISAR—which measures the particle velocity at a single point on the rear surface of the target—to determine the "spall strength" of the target material. This spall strength is used in a number of armor design approaches as a measure of when the material will fail under hydrostatic tension. Furnish et al.[50] used a line VISAR—which measures the particle velocity of a number of points (rather than a single point) along a line on the rear surface of the target—instead of a single-point VISAR to make the measurement. Their results show (Figure 4-14) that there is a stochastic character to the spall process, and that the spall strength of the material is not a single number but a result of the specific microstructural defect distribution within the target. This latter idea was articulated in Wright and Molinari[51] and Wright and Ramesh[52] in terms of a model of

[48]Paliwal, B., K.T. Ramesh, and J.W. McCauley. 2006. Direct observation of the dynamic compressive failure of a transparent polycrystalline ceramic (AlON). Journal of the American Ceramic Society 89(7): 2128-2133.

[49]Furnish, M.D., L.C. Chhabildas, W.D. Reinhart, W.M. Trott, and T.J. Vogler. 2009. Determination and interpretation of statistics of spatially resolved waveforms in spalled tantalum from 7 to 13 GPa. International Journal of Plasticity 25(4): 587-602.

[50]Ibid.

[51]Wright, T.W., and A. Molinari. 2005. A physical model for nucleation and early growth of voids in ductile materials under dynamic loading. Journal of the Mechanics and Physics of Solids 53(7): 1476-1504.

[52]Wright, T.W., and K.T. Ramesh. 2008. Dynamic void nucleation and growth in solids: A self-consistent statistical theory. Journal of the Mechanics and Physics of Solids 56(2): 336-359.

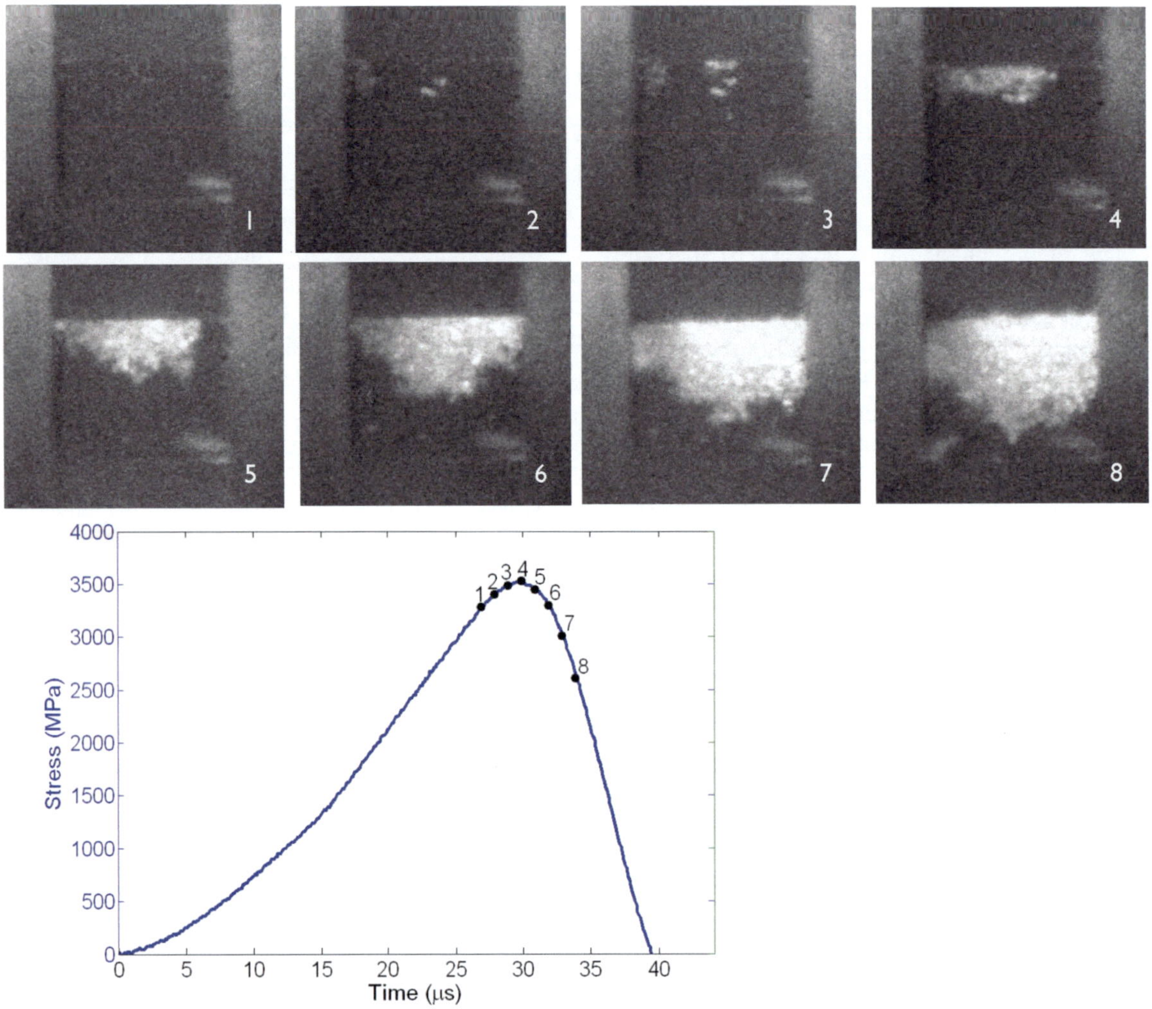

FIGURE 4-13 Photographs taken by a high-speed camera (interframe times of 1 μs and exposure times of 100 ns) of the dynamic failure process in uncoated transparent AlON. The stress-time and damage-time curve at the bottom corresponds to the photographs at the top (the times at which each photograph is taken are shown through the matched numbers on the stress-time plot). Note how the damage begins at internal flaws in the material; subsequent damage interactions lead to cooperative growth of a damage front.

dynamic void growth and interaction in a material containing a distribution of defects.

Investigating Impact Phenomenology

The experiment on the projectile impact of an aluminum plate described earlier in this chapter is an excellent example of an experiment designed to investigate impact phenomenology. Such experiments are highly instrumented and highly controlled versions of the real impact and are valuable for determining the sometimes unexpected couplings that can occur between material properties, failure processes, and system behavior. Such experiments, if they are designed to promote a detailed and specific understanding of the impact phenomenology, are particularly useful when performed on model material systems. A large fraction of the impact phenomenology experiments described in the literature has the different objective of providing a broad and generalized evaluation of the performance of the material system under a specific threat. While these performance evaluation experiments have a critical role to play in the evaluation of armor systems, it is difficult to use them to extract guidelines for the design of improved armor systems. The combination of a very experienced investigator and a large database of experimental data can be a powerful tool in armor development, but this should not be the primary approach to armor development.

Several recent developments in experimental methods hold great promise for addressing complex protection materials problems. These include improved temporal and spatial resolution, the development of high-speed cameras and associated triggering electronics, and the coupling of com-

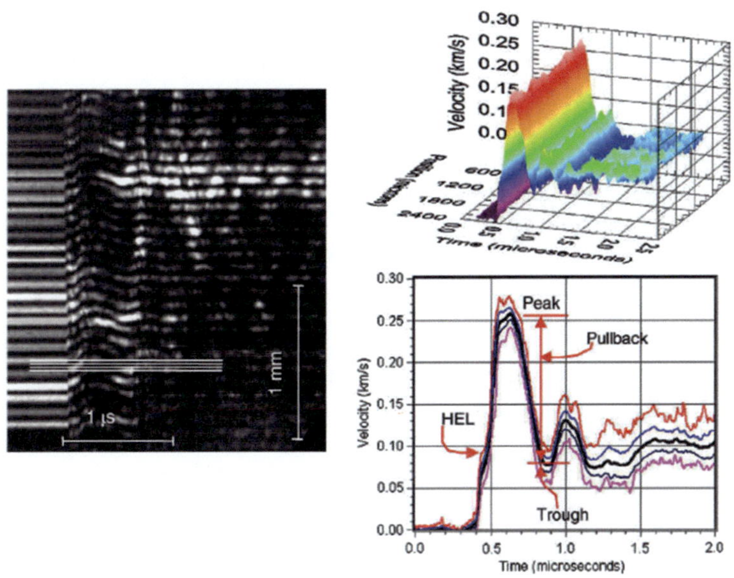

FIGURE 4-14 Line VISAR figure showing spallation in polycrystalline tantalum. The critical recent development for two failure processes is the in situ real-time observation of the active mechanism. Postmortem evaluations require the assumption of a mechanism and then the use of circumstantial evidence to verify the assumed mechanism, which can lead to erroneous conclusions in impact problems. SOURCE: Furnish, M.D., L.C. Chhabildas, W.D. Reinhart, W.M. Trott, and T.J. Vogler. 2009. Determination and interpretation of statistics of spatially resolved waveforms in spalled tantalum from 7 to 13 GPa. International Journal of Plasticity 25(4): 587-602.

putational and experimental capabilities. A variety of very sophisticated experimental techniques has been developed for addressing major parts of the problem.

However, some important technical gaps remain. An example is the characterization of the high-strain-rate response of brittle armor materials such as ceramics and glasses under combinations of high pressure and shear representative of ballistic penetration. Currently available experimental techniques are generally most sensitive to material behavior under either high pressure or high strain rate, but they rarely provide accurate information under *combined* high pressure and ultrahigh strain rate. There is also a need to develop techniques that address the variety of paths taken by material elements in the pressure/strain rate space during impact events.

Finding 4-1. Several recent developments in experimental methods hold great promise for addressing complex protection materials problems. However, some important technical gaps remain, including the following:

- The in situ and real-time determination of the active failure processes during the impact event.
- Experimental techniques that provide accurate information under combined high pressure and ultrahigh strain rate.
- Techniques that address the variety of paths taken by material elements in the pressure/strain rate space during the impact event.

MODELING AND SIMULATION TOOLS

Modeling and simulation (M&S) has long been considered an invaluable tool for analyzing engineering systems in a wide range of technology areas. The expected role of M&S is to provide a quantitative description of the physical system response that can be used to assess system performance and inform potential improvements. Such has been the case in protection materials technology, where, as discussed in this chapter, significant effort has been devoted in the last 50 years to developing the basic science, algorithms, simulation software, and hardware infrastructure to meet this goal. However, owing to the intricacies and unique physical complexities associated with the response of materials subject to extreme loading conditions, and to the dependence of protection materials performance on such details, the full potential of M&S has so far not been realized. In addition, the role of M&S is currently undergoing a significant revision as the result of efforts to develop rigorous M&S-based uncertainty quantification (UQ) methodology for the assessment and certification of complex systems. In this new view, the role of M&S is no longer to provide best-practices predictions of the response of the system, often in the form of isolated "hero calculations," but to provide predictions with rigorously quantified uncertainties. This paradigm shift could lead to profound change in the way M&S is conducted, in its interaction with experimental science, and in the manner in which protection systems are assessed and qualified.

One may review the state of the art in M&S technologies for the analysis of protection materials by identifying the main technology gaps and challenges. The committee suggests directions in which these technologies could be further developed so as to have a bigger impact on protection materials. A path forward is recommended to advance the simulation-aided design of the next generation of protection materials and systems.

Background and State of the Art

Why M&S Is Needed

The fundamental macroscopic properties of materials influencing armor performance—high strength, elastic stiffness, and ductility—are well known even if the manner in which they combine to create the most effective performance is still far from certain. A clear avenue for improving armor performance is thus to continue the quest for ever lighter, stronger, and stiffer yet more malleable materials (Chapter 5). The correlation of material properties with armor performance is usually difficult to establish exactly, as performance depends on specific details of the threat and the armor system and on their complex interactions. For example, for a given threat and a certain mass of protective material with predetermined macroscopic properties, armor performance is found to be strongly dependent on the layout of the material in the armor system—in other words, the material system. In addition, as discussed in the earlier section on mechanisms, the performance is influenced by more subtle properties of the material response such as strain hardening and rate dependency and by certain aspects of the material's mesostructure—for example, topological, composite, or cellular arrangement—and its microstructure—for example, grain size, anisotropy, residual stresses, defect characteristics, and adherence between components.

The foregoing suggests that details of the mechanical response of the material system such as wave propagation, localized plastic deformation and fracture, crack propagation, and others play an important role in determining armor performance because they affect the ability to erode projectiles, diffuse or divert the load, distribute damage away from the impact location, dissipate energy at sufficiently fast rates, and delay failure. A key unanswered question in armor applications is this: Given constraints on the weight (and/or, perhaps, cost) of candidate high-performance protection materials, what is the optimal, possibly hierarchical, structural layout that maximizes performance for a given threat? It is expected that even an approximate answer to this question will greatly benefit threat defeat and/or weight reduction. There is a universe of possibilities among which the answer will be found. A science-based approach with the ability to quantitatively describe the details of the physical event constitutes a sine qua non to achieve this goal.

This is precisely what M&S does well: It can assess the role of operative, possibly competing, mechanisms that influence macroscopic behavior. M&S is also an invaluable tool for helping to interpret material testing experiments. Experiments play two fundamental roles in M&S. On the one hand, they provide the input data on material properties and behavior for the simulations. On the other hand, lab-scale and field tests with adequate instrumentation provide quantitative data that can be compared with simulation results to validate the models. A validated computational framework can then be used to explore the design parameter space via simulation of material properties/behavior, structural topology, and geometry and dimensions to obtain solutions with improved performance and, conceivably, solutions of the inverse problem of finding the optimal solution for the given problem constraints.

As seen in this chapter, the promise of M&S has been only partially achieved thus far. In brief, current M&S technology is able to describe basic aspects of the interaction of a threat—for example, a kinetic energy penetrator—and a target armor material, including momentum transfer, dissipation of the kinetic energy of the insult by plastic deformation, friction, and, to some extent, material damage. The predictive capability of M&S has been used for improving the basic understanding of penetration mechanics beyond what experiments and simple analytical models can provide, and in some cases for guiding armor designs. However, a number of key phenomena as well as mathematical and numerical aspects of M&S are beyond the ability of existing technology and need to be addressed for M&S to achieve its potential.

Application of M&S to Protection Materials

The standard approach in M&S has five elements:

- *Computational framework.* The computer software that encodes the fundamental mathematical formulations and algorithms for solving the initial boundary value problem governing the dynamics of the physical event.
- *Constitutive models.* These are integrated into the computational framework and describe the material response in mathematical form at each material point in the problem domain. As such, they are responsible for capturing the relevant mechanisms of material deformation and failure.
- *Model calibration.* Owing to the fundamentally phenomenological character of the constitutive models in use, the specific response associated with different materials is encoded in material parameters, which must be calibrated to experiments. In most cases the calibration is applicable for only a limited range of the material response—for example, strain rate—and different calibrations are required, depending on the problem.
- *Model verification and validation (V&V).* This is the

process by which the fidelity of the computational framework is established. One of the most important phases of the V&V process is to conduct simulations of specially designed experiments and to assess the ability of the computational framework to quantitatively reproduce specific features of the experiments by comparing simulation and test results.

- *Production runs—large-scale simulations of problems of interest.* This corresponds to the stage when the validated M&S framework is used for simulating an application of interest in protection materials. The outcome consists of spatially and temporally resolved numerical values of the continuum fields describing the physical problem. These results are postprocessed and analyzed to draw conclusions about the role of the various threat defeat mechanisms and about the suitability of the protection design.

Despite the advances made based on this paradigm, significant modifications and enhancements will be required to fully reap the potential benefits of M&S in protection materials. Two key components missing in this picture are (1) multiscale, multiphyscis material models and algorithms incorporating information about the subscale or microstructural response, especially for improving the description of material damage and failure; and (2) the quantification of uncertainty for the overall problem analysis, including simulations and experiments.

Computational Framework

Decades of research and development have given us mathematical formulations, computational algorithms, and computer software which, to varying degrees, depending on the method and the problem, possess many of the desired attributes sought in a computational framework for numerically solving the fundamental continuum equations governing the response of protection materials. These responses include versatility, robustness, efficiency, and scalability. The history and state of the art of the so-called hydrocodes developed by the DoE and the DoD up to the early 1990s are discussed in detail in Benson's review.[53] Some of those legacy hydrocodes are still the workhorse tools used in armor applications. That history may be summarized as follows.

The vast majority of the codes in use for the analysis of protective materials employ explicit second-order accuracy for time integration and first-order spatial accuracy. The success of low-order explicit methods can be explained by their simplicity, robustness, and scalability. However, low-order methods pose a key limitation for the proper description of some features of material response where important opportunities for improvements in protection performance may be found—for example, multiscale structured materials, including composites, fabrics, phononic band-gap topological materials, and highly nonlinear granular chains. New classes of high-order accuracy implicit or semi-implicit algorithms have emerged from academic research that could be incorporated in existing or new codes to gain new levels of physical detail in protection material simulations.

The finite-element method has been the traditional Lagrangian approach. Its main advantage is that the description of material state and history, as well as the evolution of material boundaries and interfaces, is a natural outcome of the simulation. Its main disadvantage is the distortion of the mesh elements induced by large deformations, which invalidates or breaks the numerical method. A variety of remedies for this problem have been proposed, including adaptive remeshing and element "erosion," each with its own limitations. Particle-based Lagrangian discretizations avoid this problem as the neighborhood of interacting particles is allowed to evolve freely. However, this introduces discontinuous jumps in the continuum notion of the gradient fields (strains, for example) associated with the particles, which result in convergence problems. Lagrangian meshless methods[54] have emerged as a way to combine the advantages of particle methods and finite elements. The recent peridynamic formulation of the continuum problem[55,56] is also a promising Lagrangian approach that results in a nonlocal particle method with a rigorous mathematical framework and a natural introduction of a characteristic length, as required for modeling material damage.

The main advantage of Eulerian formulations, by contrast, is that the computational grid does not distort and thus allows for unconstrained deformations. These formulations originally found their main application in fluid dynamics. The ability to describe a solid material's "strength" was subsequently added to these formulations, but not without significant difficulty. Since the governing equations involve the material-time derivative of the stress tensor, the constitutive models need to be formulated in rate form in terms of frame-indifferent stress-rate measures. In addition, the material state and history, including the elastic response, need to be convected with the flow, which introduces an additional source of complexity and errors. This approach includes the tracking of free boundaries and material interfaces, all of which require special treatment, in contrast to the Lagrangian approach. Owing to the low order of the advection algorithms used, the associated dispersion errors grow over time and the

[53]Benson, D.J. 1992. Computational methods in Lagrangian and Eulerian hydrocodes. Computer Methods in Applied Mechanics and Engineering 99(2-3): 235-394.

[54]Belytschko, T., and J.S. Chen. 2007. Meshfree and Particle Methods. New York, N.Y.: John Wiley and Sons.

[55]Foster, J.T., S.A. Silling, and W.W. Chen. 2010. Viscoplasticity using peridynamics. International Journal for Numerical Methods in Engineering 81(10): 1242-1258.

[56]Silling, S.A., O. Weckner, E. Askari, and F. Bobaru. 2010. Crack nucleation in a peridynamic solid. International Journal of Fracture 162(1-2): 219-227.

convergence is poor. Another disadvantage of Eulerian codes is that the fixed grid must cover the entire region of interest.

The arbitrary Lagrangian-Eulerian formulation,[57] a combination of both approaches, attempts to exploit the advantages of each approach by allowing high distortions to be represented in a Lagrangian framework. Combinations of finite-element and meshless-particle methods have been developed, as have combinations of finite-element and Eulerian methods. The rest of this section summarizes the history and evolution of the codes developed and utilized for armor applications.

The HEMP[58] code was developed in the early 1960s by Wilkins, at the Lawrence Livermore Laboratories.[59] This was a two-dimensional Lagrangian code based on an explicit finite-difference formulation that could handle large strains, elastic-plastic flow, wave propagation, and sliding interfaces. It was a significant new computational capability at that time. The TOODY code was a similar finite-difference code developed at Sandia National Laboratories.[60] In the late 1960s, Wilkins and others used the HEMP code for the design and analysis of light armor, which included ceramic, metallic, and composite components. The first of five progress reports was entitled *An Approach to the Study of Light Armor*.[61] This highly influential work underpinned much of the subsequent numerical work.

During the same time frame implicit finite-element methods were introduced. However, finite-element methods for analyzing fast dynamic response only became practical when explicit methods for integrating time were introduced. Three different codes emerged in the 1970s implementing this approach: EPIC,[62] HONDO,[63] and WHAMS.[64] A determinant advantage of finite-element over finite-difference methods is their natural ability to represent complex geometries.

Numerous Eulerian codes that incorporated the effect of material strength (CTH, HULL, JOY, MESA) soon emerged as contenders to finite-element codes.[65] Of the several Eulerian codes available today, the CTH code[66] has been widely distributed by Sandia Laboratories and is the most commonly used code for impact and penetration computations for protection structures and materials.

Because of some of the limitations associated with Eulerian codes, Lagrangian approaches for severe distortions continued to be developed. In 1987, two three-dimensional erosion algorithms were published[67,68] that allowed highly distorted elements to be discarded (eroded) and the interfaces to be automatically updated as the solution progressed. Most of the current Lagrangian finite-element codes used for impact and penetration (EPIC, DYNA, LSDYNA, PRONTO, and PRESTO) now have some form of an erosion option. Although this approach introduces some inaccuracies, it allows problems with very severe distortions to be simulated in a Lagrangian framework.

One well-known limitation of element erosion is that it gives the wrong energy-release rate when a crack propagates at an angle to the mesh.[69] The reason for this failure of consistency and convergence is that, in conventional erosion implementations, the crack is forced to zig-zag through the mesh, with the result that the fracture energy is overestimated by a geometrical factor. However, it has been recently shown that a local averaging of the energy in the computation of the energy-release rate eliminates the mesh bias and results in convergent approximations.[70]

Of the particle methods, the smoothed particle hydrodynamics approach, which included material strength, was introduced in Libersky and Petschek.[71] After much initial enthusiasm it was soon discovered that the smoothed-particle hydrodynamics algorithm (and other similar particle algorithms that carried all of the variables at the nodes) had some

[57]Hirt, C.W., A.A. Amsden, and J.L. Cook. 1974. An arbitrary Lagrangian-Eulerian computing method for all flow speeds. Journal of Computational Physics 14(3): 227-253.

[58]In this report, this and other such codes are referred to by the abbreviated forms familiar to the community associated with the topics discussed.

[59]Wilkins, M.L. 1964. Calculation of elastic-plastic flow. Pp. 211-263 in Methods in Computational Physics, Volume 3: Fundamental Methods in Hydrodynamics. New York, N.Y.: Academic Press.

[60]Bertholf, L.D., and S.E. Benzley. 1968. TOODY II: A Computer Program for Two-Dimensional Wave Propagation, Technical Report SC-RR-68-41. Albuquerque, N.M.: Sandia Laboratories.

[61]Wilkins, M.L., C.A. Honodel, and D. Sawle. 1967. Approach to the Study of Light Armor, Technical Report UCRL-50284. Livermore, Calif.: Lawrence Radiation Laboratory.

[62]Johnson G.R. 1976. Analysis of elastic-plastic impact involving severe distortions. Journal of Applied Mechanics 43 Ser E(3): 439-444.

[63]Key, S.W., Z.E. Beisinger, and R.D. Krieg. 1978. HONDO II—A Finite Element Computer Program for the Large Deformation Dynamic Response of Axisymmetric Solids, Technical Report SAND78-0422. Albuquerque, N.M.: Sandia Laboratories.

[64]Beltyschko, T., and R. Mullen. 1978. WHAMS: A program for transient analysis of structures and continua. Pp. 151-212 in Structural Mechanics Software Series, Volume 2. N. Perrone and W. Pilkey, eds. Charlottesville, Va.: University Press of Virginia.

[65]Immele, J.D., C.E. Anderson, R.J. Asaro, S.G. Cochran, L.W. Davison, J.C. Foster, G. Johnson, G. Randers-Perhson, and J. Short. 1989. Report of the Review Committee on Code Development and Material Modeling, LA-UR-89-3416. Arlington, Va.: Defense Advanced Research Projects Agency.

[66]McGlaun, J.M., S.L. Thompson, and M.G. Elrick. 1990. CTH: A three-dimensional shockwave physics code. International Journal of Impact Engineering 10(1-4): 351-360.

[67]Johnson, G.R., and R.A. Stryk. 1987. Eroding interface and improved tetrahedral element algorithms for high-velocity impact computations in three dimensions. International Journal of Impact Engineering 5(1-4): 411-421.

[68]Belytschko, T., and J.I. Lin. 1987. A three-dimensional impact-penetration algorithm with erosion. International Journal of Impact Engineering 5(1-4): 111-127.

[69]Negri, M. 2007. Convergence analysis for a smeared crack approach in brittle fracture. Interfaces and Free Boundaries 9(3): 307-330.

[70]Schmidt, B., F. Fraternali, and M. Ortiz. 2009. Eigenfracture: An eigendeformation approach to variational fracture. Multiscale Modeling and Simulation 7(3): 1237-1266.

[71]Libersky, L.D., and A.G. Petschek. 1991. Smooth particle hydrodynamics with strength of materials. Pp. 248-257 in Advances in the Free-Lagrange Method, Lecture Notes in Physics Volume 395. H.E. Trease, M.J. Fritts, and W.P. Crowley, eds. Berlin, Germany: Springer-Verlag.

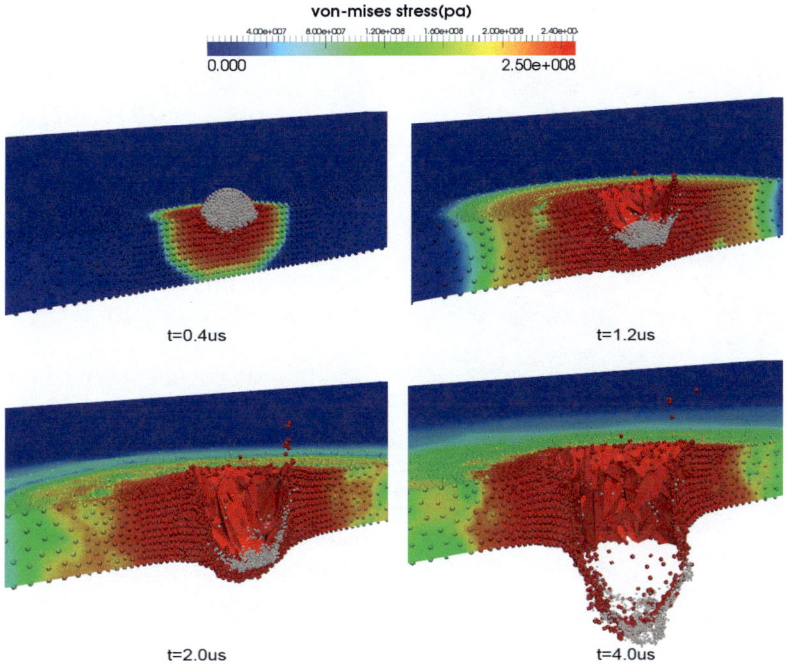

FIGURE 4-15 Optimal transportation mesh-free simulation of a steel plate perforated by a steel projectile striking at various angles. Top: evolution of the perforation process. Bottom: perforated configurations for several incidence angles.

limitations in terms of instabilities, accuracy, and efficiency. For the past 20 years there has been a strong emphasis on the development of a wide range of meshless-particle algorithms for a wide range of applications. A promising new direction combines elements of optimal transportation theory with meshfree (max-ent) interpolation of the fields, material-point sampling of material states, and provably convergent erosion schemes to account for fracture (see Figure 4-15). The optimal transportation mesh-free method is an example of this approach. Compared to other particle methods, optimal transportation exhibits strong and provable convergence properties and eliminates tension instabilities that afflict traditional particle methods.

An approach based on the conversion of elements into particles has shown to effectively deal with the problem of element distortion.[72] As Johnson notes, with this approach the entire initial geometry is represented by elements, and then, as the elements on surfaces or interfaces become highly distorted, they are converted into meshless particles. All of the element variables are transferred to the particle, and the particle is attached to the face of an adjacent element (if one exists). With this approach, most of the problem is represented by accurate and efficient elements, with only the highly distorted regions represented by particles. Very severe distortions can be represented in a Lagrangian framework.

An alternative approach to alleviate deformation-induced mesh distortion in Lagrangian finite-element algorithms is to adaptively and continuously regenerate the mesh during the simulation. An additional advantage of adaptive remeshing methods is the ability to optimally refine the mesh for maximum accuracy. This idea was applied successfully to penetration mechanics problems in axisymmetric conditions. Recent advances in computational geometry and mesh optimization have enabled the extension of this idea to three dimensions. Figure 4-16 shows its application to simulating the oblique impact of a spherical-nosed steel penetrator on an aluminum target.

One of the issues with adaptive remeshing approaches is the error introduced in the transfer of field variables from the old to the new mesh, which tends to produce artificial diffusion. This problem can be somewhat alleviated by adapting the mesh locally instead of completely regenerating it. Another important issue is scalability in parallel calculations. It is well established in computational geometry that algorithms involving general topological changes in the data structures are very hard to implement in parallel and are usually inherently nonscalable because they require the propagation (communication) of unstructured and evolving data among processors. As a result, parallel calculations are efficient only up to a few tens of processors at best.

Another significant concern is that different codes produce different answers for the same problem, a telling indication of the lack of convergence in the solution. This is illustrated in Figure 4-17, where five different computa-

[72]Johnson, G.R., and R.A. Stryk. 2003. Conversion of 3D distorted elements into meshless particles during dynamic deformation. International Journal of Impact Engineering 28(9): 947-966.

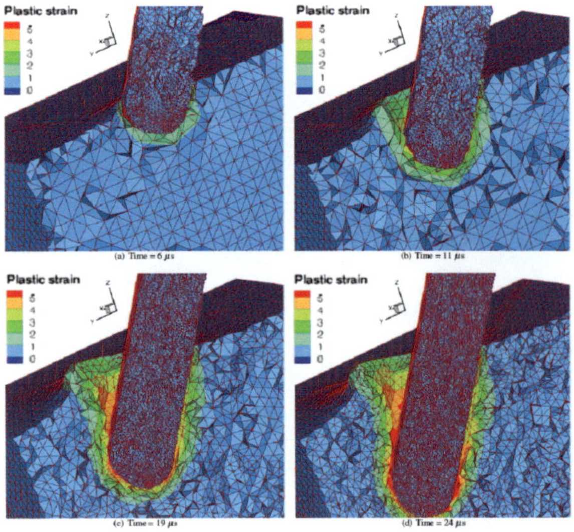

FIGURE 4-16 Example of a Lagrangian finite-element simulation that uses adaptive re-meshing and refinement to eliminate element distortion and to optimize the mesh.

tional approaches are used for the same problem (a tungsten projectile impacting a steel target at an impact velocity of 1,615 m/s), and all five approaches used exactly the same material models. The five approaches use (1) a finite-volume (Eulerian) algorithm, (2) an erosion algorithm that discards highly distorted elements, (3) a generalized particle algorithm where the entire problem is represented by particles, (4) a conversion algorithm that converts distorted elements into particles, and (5) a hybrid algorithm where the pressures are computed with particles and strength is computed with elements. On the one hand it is encouraging that several different approaches can provide good general agreement with the experimental results. On the other hand, there are noticeable differences in the size of the fragments and the response of the tungsten projectile.

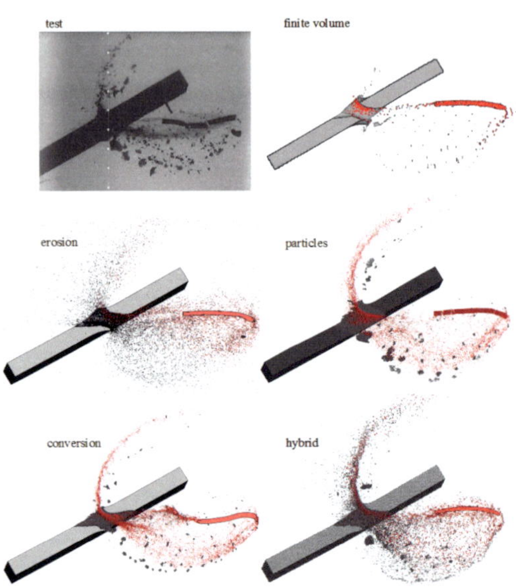

FIGURE 4-17 A comparison of results from five computational approaches for a tungsten projectile impacting a steel target at 1,615 m/s. SOURCE: Beissel, S.R., C.A. Gerlach, T.J. Holmquist, and J.D. Walker. In press. Comparison of numerical methods in the simulation of hypervelocity impact. Proceedings of 11th Hypervelocity Impact Symposium, Freiburg, Germany 2010.

Finding 4-2. Although much progress has been made in developing computational frameworks for the analysis of protection materials, all of the algorithms have strengths and weaknesses.

Additional Challenges in Computational Framework Another important issue for future threats that may subject protective materials to conditions well in the nonlinear shock physics regime has to do with the numerical treatment of shock-type discontinuities. It has been widely established that special computational methods are needed to address the jump discontinuities associated with shocks that arise in materials under extreme compressive loadings.[73] A standard approach in hydrodynamic calculations using the existing low-order methods is introducing a viscous term into the equations to smooth out the shock. Although this method has proven to be a robust and simple approach for capturing shock, it introduces errors and problems. "Shockless heating" and "wall heating" occur in strong shocks, leading to errors in the energy, both behind and in front of the shock. Concomitant with errors in the energy are errors in the density and the shock speed. Emerging high-order implicit or semi-implicit methods developed by the fluid mechanics community for compressible turbulence, in which both the compressibility shock and viscous effects are important, have significant potential for solid materials as well.

A critical missing component in the protective material simulation tools in use today is the ability to represent material damage and failure explicitly. The conventional approach is to make use of so-called continuum damage models (see the section "Constitutive Models" below). Such models have been effective for describing damage in an average, or "smeared," sense but are unsuitable for capturing (1) the discrete nature of material fracture and (2) crack nucleation and propagation, according to the laws of fracture mechanics. A promising class of approaches for doing so is based on the "discrete crack" model of fracture.[74] In this approach,

[73]Benson, D.J. 1992. Computational methods in Lagrangian and Eulerian hydrocodes. Computer Methods in Applied Mechanics and Engineering 99(2-3): 235-394.

[74]Radovitzky, R., A. Seagraves, M. Tupek, and L. Noels. 2011. A scalable 3D fracture and fragmentation algorithm based on a hybrid, discontinuous Galerkin, cohesive element method. Computer Methods in Applied Mechanics and Engineering 200 (1-4): 326-344.

crack initiation and propagation are modeled explicitly by the introduction of surfaces of discontinuity within the material. At these surfaces of discontinuity, fracture processes can be described by cohesive zone models (CZMs) of fracture[75,76] via a phenomenological traction-separation law. The key advantage of CZMs is their ability to encode in the calculation well-established laws of fracture mechanics governing the nucleation, propagation, branching, and coalescence of cracks.

The "cohesive element" method[77,78] is the most popular implementation of this concept. In this method, crack openings are represented as displacement jumps at the interelement boundaries using "interface" or "cohesive" finite elements. Camacho and Ortiz[79] presented the first formulation of this method for impact problems involving extensive fracture and fragmentation. They demonstrated that the extrinsic CZM was successful at capturing conical crack patterns in ceramic plate impact as long as finely resolved meshes were employed in the calculations. Cohesive elements provide a notable alternative to erosion and are one of the key innovations brought to ballistic calculations in the 1990s. This development finally enabled the robust and reliable tracking of sharp cracks and complex fracture and fragmentation properties. Cohesive-element calculations have proven highly predictive and have been extensively validated in a number of areas of application by, for example, Bjerke and Lambros[80] and Chalivendra et al.[81]

A full three-dimensional description of crack patterns in ceramic plate impact has recently been enabled by a new extension of CZM based on a discontinuous Galerkin reformulation of the continuum problem.[82] The main advantages of this method are its inherent scalability (demonstrated up to 4,096 cores on DoD platforms and problems involving 3 billion degrees of freedom) and its accuracy in describing wave propagation. Figure 4-18 shows the ability of the method to capture conical as well as radial and lateral cracks in ceramic plate impact.

One of the issues commonly attributed to CZM based on interface elements is that the set of available paths for crack propagation is constrained by the mesh, which is a form of mesh dependency. A variety of methods have been put forth to enable arbitrary crack paths in simulations for the purpose of reducing mesh dependency. Essentially, these approaches allow surfaces of discontinuity to propagate through the interior of volumetric elements (see, for example, the extended finite-element method,[83,84] the embedded localization line method,[85,86,87] and the cohesive segments method).[88] In this family of methods, however, highly refined meshes are still necessary to resolve the size of the fracture process zone in brittle materials. Another issue is the possibility of describing crack branching, especially in three dimensions. So far, these types of methods have only been implemented for single-processor computations. Their scalability is marred by the same problem of propagating topological changes across processors alluded to in the discussion of parallel adaptive remeshing. Efforts are currently under way to include methods of the extended finite-element type in existing codes: LS-DYNA[89] and Abaqus.[90]

Finding 4-3. A critical missing component of protective material simulation tools in use today is the ability to represent material damage and failure explicitly.

Constitutive Models

In addition to having numerical algorithms it is essential to have computational models to accurately represent the response of the materials. Numerous computational material models have been developed during the past 20 years, but

[75]Dugdale, D.S. 1960. Yielding of steel sheets containing slits. Journal of the Mechanics and Physics of Solids 8(2): 100-104.

[76]Barenblatt, G.I. 1962. The mathematical theory of equilibrium cracks in brittle fracture. Advances in Applied Mechanics 7(C): 55-129.

[77]Ortiz, M., and S. Suresh. 1993. Statistical properties of residual stresses and intergranular fracture in ceramic materials. Journal of Applied Mechanics, Transactions ASME 60(1): 77-84.

[78]Xu, X.P., and A. Needleman. 1994. Numerical simulation of fast crack growth in brittle solids. Journal of the Mechanics and Physics of Solids 42(9): 1397-1434.

[79]Camacho, G.T., and M. Ortiz. 1996. Computational modelling of impact damage in brittle materials. International Journal of Solids and Structures 33(20-22): 2899-2938.

[80]Bjerke, T.W., and J. Lambros. 2003. Theoretical development and experimental validation of a thermally dissipative cohesive zone model for dynamic fracture of amorphous polymers. Journal of the Mechanics and Physics of Solids 51(6): 1147-1170.

[81]Chalivendra, V.B., S. Hong, I. Arias, J. Knap, A. Rosakis, and M. Ortiz. 2009. Experimental validation of large-scale simulations of dynamic fracture along weak planes. International Journal of Impact Engineering 36(7): 888-898.

[82]Radovitzky, R., A. Seagraves, M. Tupek, and L. Noels. 2011. A scalable 3D fracture and fragmentation algorithm based on a hybrid, discontinuous Galerkin, cohesive element method. Computer Methods in Applied Mechanics and Engineering 200 (1-4): 326-344.

[83]Belytschko, T., and T. Black. 1999. Elastic crack growth in finite elements with minimal remeshing. International Journal for Numerical Methods in Engineering 45(5): 601-620.

[84]Moës, N., J. Dolbow, and T. Belytschko. 1999. A finite element method for crack growth without remeshing. International Journal for Numerical Methods in Engineering 46(1): 131-150.

[85]Dvorkin E.N., A.M. Cuitiño, and G. Goia. 1990. Finite elements with displacement interpolated embedded localization lines insensitive to mesh size and distortions. International Journal for Numerical Methods in Engineering 30(3): 541-564.

[86]Simo, J.C., J. Oliver, and F. Armero. 1993. An analysis of strong discontinuities induced by strain-softening in rate-independent inelastic solids. Computational Mechanics 12(5): 277-296.

[87]Armero, F., and C. Linder. 2009. Numerical simulation of dynamic fracture using finite elements with embedded discontinuities. International Journal of Fracture 160(2): 119-141.

[88]Remmers, J.J.C., R. de Borst, and A. Needleman. 2003. A cohesive segments method for the simulation of crack growth. Computational Mechanics 31(1-2 SPEC): 69-77.

[89]See http://www.lstc.com.

[90]See http://www.simulia.com.

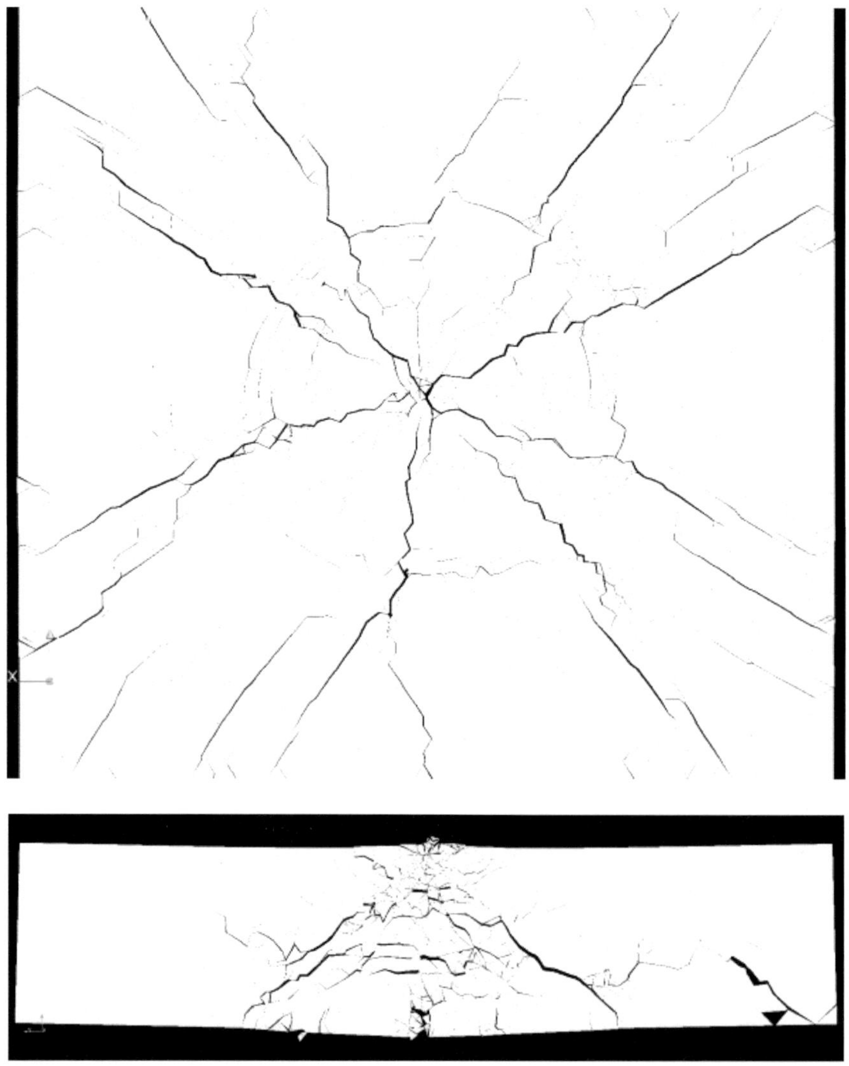

FIGURE 4-18 Prediction of conical, radial, and lateral crack patterns in ceramic plate impact by the recent cohesive zone/discontinuous Galerkin method.

only a few have taken root in the application to protection materials. These models have ranged from those for simple dynamic flow stress and dynamic failure strain to very complex models that include microstructural details. Currently, some models for materials are advanced enough to provide helpful and meaningful results, such as those illustrated in the first section of this chapter, but details of failure are not sufficiently robust to allow the predictive design of material systems to protect against specific threats.

For projectile-target interaction computations, the materials are usually modeled using phenomenological models that compute strength and failure as a function of strain, strain rate, temperature, and pressure. For metals the most commonly used strength models are the Johnson-Cook model,[91] the Zerilli-Armstrong models,[92] the Steinberg-Guinan-Lund models,[93] the Bodner-Partom models,[94] and

[91]Johnson, G.R., and W.H. Cook. 1983. A constitutive model and data for metals subjected to large strains, high strain rates and high temperatures. Available online at http://www.lajss.org/HistoricalArticles/A%20constitutive%20model%20and%20data%20for%20metals.pdf. Last accessed April 7, 2011.

[92]Zerilli, F.J., and R.W. Armstrong. 1987. Dislocation-mechanics-based constitutive relations for material dynamics calculations. Journal of Applied Physics 61(5): 1816-1825.

[93]Steinberg, D.J., S.C. Cochran, and M.W. Guinan. 1980. A constitutive model for metals applicable at high-strain rate. Journal of Applied Physics 51(3): 1498-1504.

[94]Bodner, S.R., and Y. Partom. 1975. Constitutive equations for elastic-viscoplastic strain-hardening materials. Journal of Applied Mechanics, Transactions ASME 42 Ser E(2): 385-389.

the Mechanical Threshold Stress model.[95] The Johnson-Cook model is a phenomenological model, and the others are more physically based. Generally these models require a characterization of a specific material, and it is not possible to predict the strength from a microstructural description of the material. There are fewer failure models available, with the Johnson-Cook failure model[96] being the most widely used. Although this is primarily a phenomenological model, it includes some physically based features of the ductile fracture mechanism.[97] More mechanistic constitutive models of damage and fracture based on ductile void growth,[98] such as the Gurson model,[99] have been proposed. However, they are still to be incorporated and widely adopted in production codes for ballistic analyses. Recent contributions have proposed improvements to the characterization of failure in the Johnson-Cook model[100] and the Gurson model.[101]

For ceramics there are fewer models available. A unique feature of ceramics, compared to other materials (such as metals), is that they have such high compressive strengths that they cannot be tested with typical laboratory stress-strain tests.[102] Instead, their material properties must be inferred from plate impact tests and/or penetration tests. This characteristic has made it difficult to directly obtain failure data under high (compressive) pressures and to obtain the (shear) strength of failed ceramic under high pressures. The JHB phenomenological model[103] used in the illustrative example in the beginning of the chapter has an intact strength, a failed strength, a failure component based on plastic strain and pressure, and bulking. Recently Deshpande and Evans proposed a mechanism-based model to compute damage and failure in ceramics based on microstructural parameters such as fracture toughness, crack growth rates, flaw size, and densities.[104] This new work sets an important direction for the development of constitutive models of material failure for other protection materials.

Issues with Models of Material Damage and Failure Material damage and failure in existing M&S codes is described at the constitutive model via so-called continuum damage models. In these approaches, damage is considered a state variable of the material, whose history evolves according to prescribed phenomenological laws. Either the elastic or plastic response of the material is "softened" as material damage progresses in an irreversible manner. These laws describe the effect of the operative driving forces and mechanisms such as stress intensity and triaxiality, which depend on the material type (brittle or ductile) and characteristics such as defect size and porosity. Damage models require additional parameters that must be calibrated to experiments or that sometimes have physical meaning, such as initial porosity, defect size and distribution, toughness, and so forth.

Damage is characterized by a reduction in the material's load-carrying capacity after reaching damage threshold conditions. This is always accompanied by a localization of the deformation in narrow regions, which is a precursor to failure. There is a fundamental mathematical problem with continuum damage models and any other model describing weakening material response—for example, the models of de Borst and Sluys[105] and Sluys et al.[106] In the region where softening occurs, the governing equations of the dynamic problem change their mathematical character in a fundamental way, from hyperbolic to elliptic. For elliptic equations, waves cannot propagate as their speeds become imaginary, and the softening region collapses to a vanishing width. This, in turn, implies that no energy is dissipated by the softening material, which is far from the real material response. What happens in reality is that there is always a physical process that limits the localization process and introduces a characteristic length scale in the problem, which is not considered in the classical continuum equations.

In the presence of softening, the numerical solution of the conventional continuum problem provides an erroneous resolution of the physical phenomenon. The element or grid size effectively sets the length scale necessary to regularize the problem as it imposes a lower bound for the localization zone width. However, this is just an illusion, because the solution does not converge as the mesh is refined. In the limit

[95]Follansbee, P.S., and U.F. Kocks. 1988. A constitutive description of the deformation of copper based on the use of the mechanical threshold stress as an internal state variable. Acta Metallurgica 36(1): 81-93.

[96]Johnson, G.R., and W.H. Cook. 1985. Fracture characteristics of three metals subjected to various strains, strain rates, temperatures and pressures. Engineering Fracture Mechanics 21(1): 31-48.

[97]Hancock, J.W., and A.C. Mackenzie. 1976. On the mechanism of ductile failure in high-strength steels subjected to multi-axial stress states. Journal of the Mechanics and Physics of Solids 24(2-3): 147-160.

[98]McClintock, F.A. 1968. A criterion for ductile fracture by the growth of holes. Journal of Applied Mechanics 35(2): 363-371.

[99]Gurson, A.L. 1977. Continuum theory of ductile rupture by void nucleation and growth: Part I, yield criteria and flow rules for porous ductile media. Journal of Engineering Materials and Technology, Transactions of the ASME 99 Ser H(1): 2-15.

[100]Bao, Y., and T. Wierzbicki. 2004. On fracture locus in the equivalent strain and stress triaxiality space. International Journal of Mechanical Sciences 46(1): 81-98.

[101]Nahshon, K., and J.W. Hutchinson. 2008. Modification of the Gurson Model for shear failure. European Journal of Mechanics-A/Solids 27(1): 1-17.

[102]Johnson, G.R. 2011. Numerical algorithms and material models for high-velocity impact computations. International Journal of Impact Engineering 38(6): 456-472.

[103]Johnson, G.R., T.J. Holmquist, and S.R. Beissel. 2003. Response of aluminum nitride (including phase change) to large strains, high strain rates, and high pressures. Journal of Applied Physics 94(3): 1639-1646.

[104]Deshpande, V.S., and A.G. Evans. 2008. Inelastic deformation and energy dissipation in ceramics: A mechanism-based constitutive model. Journal of the Mechanics and Physics of Solids 56(10): 3077-3100.

[105]de Borst, R., and L.J. Sluys. 1991. Localisation in a Cosserat continuum under static and dynamic loading conditions. Computer Methods in Applied Mechanics and Engineering 90(1-3): 805-827.

[106]Sluys, L.J., R. de Borst, and H.-B. Muhlhaus. 1993. Wave propagation, localization and dispersion in a gradient-dependent medium. International Journal of Solids and Structures 30(9):1153-1171.

when the mesh size goes to zero, the dissipated energy in the localization zone is zero. This numerical manifestation of the ill-posedness of the mathematical problem is what is usually referred to as "damage-induced mesh dependency." A common approach to circumvent this problem in existing codes is to calibrate the material model parameters for a given mesh size. In other words, not only the model parameters but also the mesh size are tied to a specific application. The illustrative example in the introduction to this chapter for the projectile penetrating the aluminum plate was approached in this manner. This is clearly a significant limitation.

A proper mathematical treatment of softening material response necessarily involves the modification of the classical governing equations in a way that the physically relevant length scale is introduced. A number of generalizations of the classical formulation have been proposed to this end. They involve either the introduction of higher-order derivatives in the constitutive model (gradient models, as, for example, Aifantis[107] and Fleck and Hutchinson[108]) or the spatial averaging of strains (nonlocal models such as Bazant et al.[109]). Both generalizations reflect the fact that micromechanical processes in the localization zone have an inherently nonlocal character. In the particular case of gradient-type softening or damage models, it can be shown that an internal length scale exists and that the resulting set of governing equations is well posed, having wave speeds that remain real in the softening regime. The immediate computational consequence of this reformulation is that softening-induced mesh dependence is eliminated.

These models have not permeated production computational frameworks, primarily for two reasons: (1) new (high-order) algorithms and computer codes are required because the existing algorithmic frameworks cannot accommodate the higher-order derivatives and their field continuity requirements and (2) additional constitutive parameters have emerged that in many cases do not have a clear physical meaning or a discerning experiment that can be used to calibrate them. Multiscale modeling might be one way to address this issue.

Multiscale Modeling: Issues with Phenomenological Models and the Need to Incorporate Microstructural Information

The difficulty of correlating material properties with armor performance can often be explained by the inability of macroscopic constitutive descriptions to account for details of the material's microstructure or for the associated micromechanical responses that affect global behavior. The ability to consistently incorporate the effect of micromechanical features on material response would enable rational microstructure design.

There is therefore a critical need to develop descriptions of material behavior directly rooted in the first principles of micromechanics, a long-standing aspiration of solid mechanics. This requires new mathematical frameworks; multiscale, multiphysics constitutive models; and numerical algorithms. Multiscale modeling is a rational and systematic way to construct hierarchical models for the behavior of complex material with the least amount of empiricism and uncertainty. In this approach, the pertinent unit processes at every length scale in the hierarchy of material behavior are identified. The processes at any scale are the average of the unit processes taking place at the length scale just below. The modeling effort for systems in which these relations are well defined simply involves analyzing each unit mechanism in turn and computing the averages, which eventually results in a full description of the material's macroscopic behavior. This inductive process ceases at the atomic scale, at which point the fundamental theories describing atomic bonds take over.

For instance, as part of the Caltech advanced simulation and computing program, a full multiscale model of material response was developed for tantalum.[110] The multiscale hierarchy that underlies metal plasticity is shown schematically in Figure 4-19. The foundational theory on which the hierarchy rests is quantum mechanics and, in particular, the electronic structure of metals. Quantum mechanics encapsulates the fundamental laws that govern the behavior of materials at the angstrom scale. In their density-functional-theory approximation, quantum mechanical calculations can characterize the structure and properties of crystal lattices and isolated crystal defects, especially when coarse-graining techniques are employed.[111] Fundamental properties of dislocations such as kink structure and mobility can be evaluated using molecular dynamics and empirical potentials.[112] These properties can be used to formulate theories of linear-elastic dislocation dynamics. Dislocation dynamics models—for example, the models of van der Giessen and Needleman[113]

[107]Aifantis, E.C.. 1984. On the microstructural origin of certain inelastic models. Journal of Engineering Materials and Technology, Transactions of the ASME 106(4): 326-330.

[108]Fleck, N.A., and J.W. Hutchinson. 1993. A phenomenological theory for strain gradient effects in plasticity. Journal of the Mechanics and Physics of Solids 41(12): 1825-1857.

[109]Bazant, Z.P., T.B. Belytschko, and T.-P. Chang. 1984. Continuum theory for strain-softening. Journal of Engineering Mechanics 110(12): 1666-1692.

[110]Cuitiño, A.M., L. Stainier, G. Wang, A. Strachan, T. Cain, W.A. Goddard III, and M. Ortiz. 2001. A multiscale approach for modeling crystalline solids. Journal of Computer-Aided Materials Design 8(2-3): 127-149.

[111]Gavini, V., K. Bhattacharya, and M. Ortiz. 2007. Quasi-continuum orbital-free density-functional theory: A route to multi-million atom nonperiodic DFT calculation. Journal of the Mechanics and Physics of Solids 55(4): 697-718.

[112]Cuitiño, A.M., L. Stainier, G. Wang, A. Strachan, T. Cain, W.A. Goddard III, and M. Ortiz. 2001. A multiscale approach for modeling crystalline solids. Journal of Computer-Aided Materials Design 8(2-3): 127-149.

[113]Van Der Giessen, E., and A. Needleman. 1995. Discrete dislocation plasticity: A simple planar model. Modelling and Simulation in Materials Science and Engineering 3(5): 689-735.

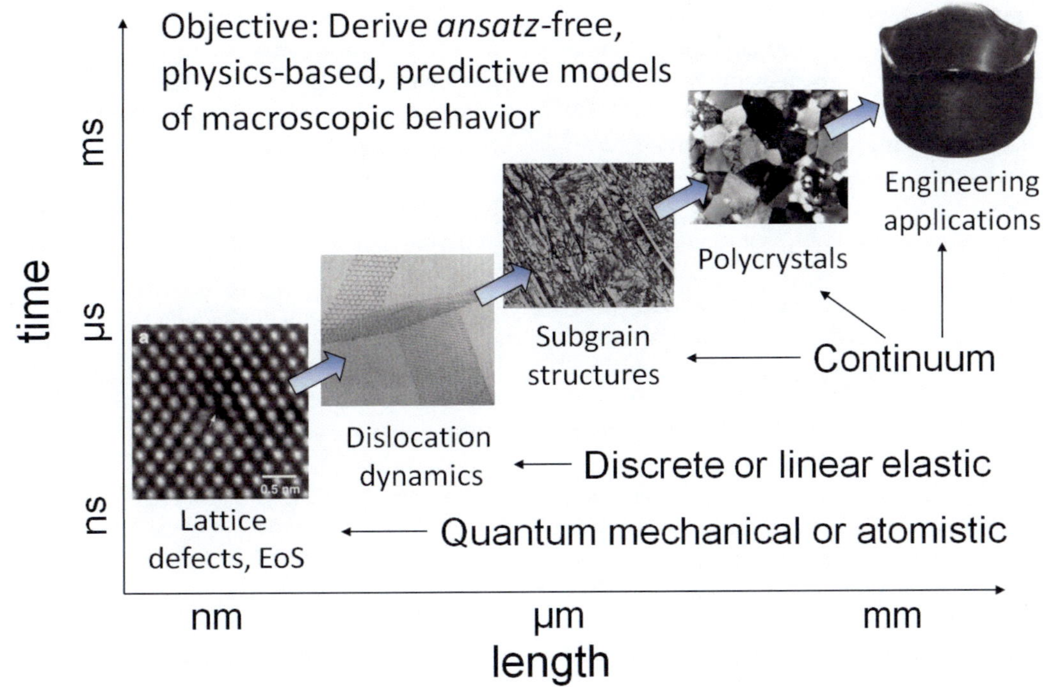

FIGURE 4-19 Multiscale hierarchy for metal plasticity. The arrows indicate upscaling directions across length scales.

and Arsenlis et al.[114]—have the potential for characterizing the work-hardening characteristics of metals. However, to date, such models are restricted to small deformations and impossibly high dislocation densities and deformation rates. The deformation of individual grains is often strongly heterogeneous and entails the formation of lamellar dislocation structures. Variational formulations of plasticity based on incremental energy minimization have proven effective at predicting such structures and characterizing the effective behavior of the material, including well-established scaling relations such as those of Hall-Petch and Taylor.[115] Finally, the direct simulation of polycrystalline behavior, in which the polycrystalline structure is resolved by the mesh, is within the reach of present petascale computing power.[116,117] Large-scale simulations and a detailed experimental validation process showed that this multiscale approach not only reproduced the observed effective response of polycrystalline metals but also captured local details of the deformation and grain interactions.

It is, however, easier to expound the multiscale paradigm than to carry it out. Today, the unit mechanisms can be analyzed and the effective behavior characterized based either on numerical schemes or on a motley assortment of analytical tools, such as mean-field theories, statistical mechanics, transition-state theory, or homogenization. The great breadth of the field and its current state of development mean that multiscale modeling generally cannot be easily formalized as a self-contained, unified theory and therefore remains as much art as a science. As a result, there is a tendency to base multiscale modeling on purely numerical schemes such as molecular dynamics, kinetic Monte Carlo, quasicontinuum, and direct numerical simulation of polycrystals. One common multiscale paradigm is "information-passing"—that is, computing material constants that are then used to inform upscale models. An important limitation of this type of multiscale analysis is that it does not provide insight into, nor does it supply, the functional form of the models governing material behavior at the various scales of interest. A competing paradigm consists of running several schemes, each operating on a different length scale and feeding average information to the upper scales, as part of the same calculation, which is referred to as "concurrent multiscale computing." However, this paradigm is self-limiting owing to the inordinate volume of computing that it generates, and

[114]Arsenlis, A., W. Cai, M. Tang, M. Rhee, T. Oppelstrup, G. Hommes, T.G. Pierce, and V.V. Bulatov. 2007. Enabling strain hardening simulations with dislocation dynamics. Modelling and Simulation in Materials Science and Engineering 15(6): 553-595.

[115]Ortiz, M., and E.A. Repetto. 1999. Nonconvex energy minimization and dislocation structures in ductile single crystals. Journal of the Mechanics and Physics of Solids 47(2): 397-462.

[116]Zhao, Z., R. Radovitzky, and A. Cuitino. 2004. A study of surface roughening in fcc metals using direct numerical simulation. Acta Materialia 52(20): 5791-5804.

[117]Zhao, Z., M. Ramesh, D. Raabe, A.M. Cuitiño, and R. Radovitzky. 2008. Investigation of three-dimensional aspects of grain-scale plastic surface deformation of an aluminum oligocrystal. International Journal of Plasticity 24 (12): 2278-2297.

to the difficulty in interpreting and learning from the vast amounts of numerical data that it generates. Thus, whereas much of multiscale computing is driven by the rapid pace of development of computational platforms, the goal of "full physics"—that is, of employing solely fundamental theories in calculations and brute computational force—remains elusive at present.

An appealing alternative to computational multiscale schemes is to derive models of effective behavior across length scales analytically. In recent years, powerful techniques for characterizing such effective, or macroscopic, behavior, including relaxation and gamma convergence, have been developed in the context of the "modern calculus of variations" by, among others, Müller.[118] What these methods do is to exhaustively evaluate all the possible subscale behaviors, or microstructures, that may develop in the material in response to macroscopic deformation and to determine the optimal, or "softest," material response enabled by those microstructures. Examples of a material with microstructures that can be treated in this manner include martensite, subgrain dislocation structures, dislocation walls and networks, ferroelectric domains, shear bands, spall planes, and others. By using the relaxed, or macrocoscopic, material model in calculations, the microstructural length scale is effectively pushed down to the subgrid level and need not be accounted for in the calculations explicitly, at enormous computational savings.[119] Remarkably, the calculations still capture the exact macroscopic behavior exactly, since the effect of all possible microstructures has in effect been precomputed in the course of determining the relaxed model. Finally, the upscaling of the material behavior happens without a loss of information, since the optimal microstructures can always be reconstructed from the macroscopic solution. This ability to reconstruct microstructures from the macroscopic response may be critical in applications where the extreme values of the microscopic deformation and temperature fields, and not just their average values, are of consequence.

Unfortunately, explicit relaxations are known for only a handful of material models, although the list of such models continues to grow. Despite this paucity of explicit results, relaxation and related methods illustrate the important role that analytical methods can play in the field of multiscale analysis. Indeed, when used in simulations, each material model that is added to the list of explicitly known relaxations, or homogenizations, saves vast volumes of computation and, perhaps more importantly, makes feasible calculations that would otherwise be intractable using sheer brute force. Clearly, for this important effort to be effective, stronger coordination and collaboration between experimentalists and modelers must be encouraged. In summary, mathematical analysis could give simulations a great competitive advantage and should be an important part of a balanced approach to multiscale modeling. Efforts are under way at the Army Research Laboratory to apply this paradigm to protection materials.

Model Verification and Validation

Verification and validation are defined in the DoE plan for the Strategic Computing & Simulation Validation & Verification Program as follows:

- *Verification.* The process of determining that a computer simulation correctly represents the conceptual model and its solution.
- *Validation.* The process of determining the degree to which a computer simulation is an accurate representation of the real world.

The DoE Accelerated Strategic Computing Initiative has defined V&V requirements for the computer codes used as part of the national nuclear Stockpile Stewardship Program. One of the requirements is to develop a well-defined plan for V&V for each code. The idea is that a successfully executed V&V plan will certify the suitability of a computer code for a particular application. This paradigm is now commonplace for large-scale simulation efforts at DOE Defense Programs laboratories.

Although the idea has taken hold that some form of V&V is required in protection material simulation codes and some efforts have been made, a rigorous formalism and framework such as those established at DOE would greatly benefit the DoD research community (see Figure 4-20).

The next step beyond V&V is UQ to determine the uncertainties that affect not only simulations but experiments as well. It is widely accepted that experimental results are accompanied by systematic and random errors. UQ attempts to quantify these errors in a meaningful way. Each computation involves both numerical and physical parameters that have ranges, and distributions, of values. UQ techniques quantify the effect on the simulation outcomes of these parameter variations. Such sensitivity information is directly relevant to design.

Production Runs—Large-Scale Simulations of Problems of Interest

Simulations of protective material performance are commonly conducted on multiprocessor parallel computers available in DoD as part of the High Performance Computing Modernization Program. The DoD platforms belong to the so-called teraflop generation (10^{12} flops, where a flop is the number of floating point operations per second) (www.

[118]Müller, S. 1999. Variational models for microstructure and phase transitions. Pp. 85-210 in Calculus of Variations and Geometric Evolution Problems, Springer Lecture Notes in Math 1713. F. Bethuel, G. Huisken, S. Mueller, K. Steffen, S. Hildebrandt, and M. Struwe, eds. Berlin, Germany: Springer-Verlag.

[119]Conti, S., P. Hauret, and M. Ortiz. 2007. Concurrent multiscale computing of deformation microstructure by relaxation and local enrichment with application to single-crystal plasticity. Multiscale Modeling and Simulation 6(7): 135-157.

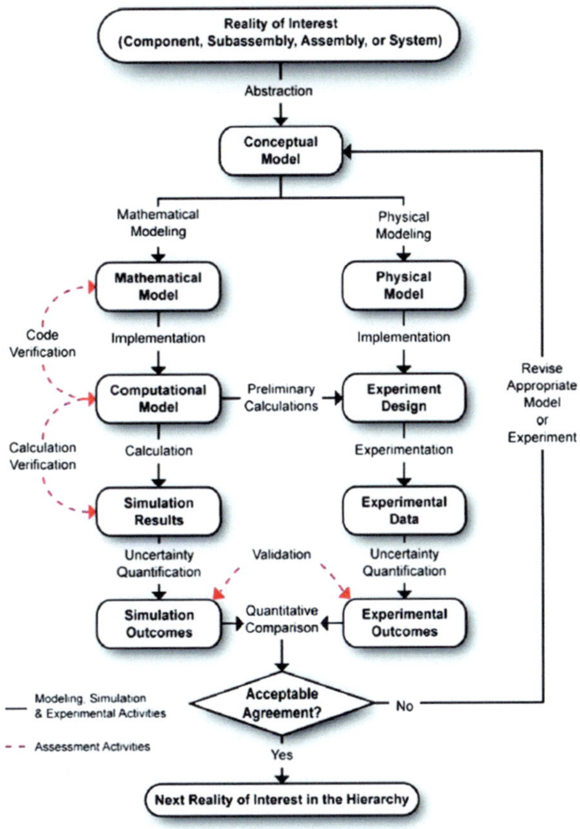

FIGURE 4-20 V&V process. SOURCE: Reprinted from ASME V&V 10-2006, by permission of the American Society of Mechanical Engineers. All rights reserved.

top500.org) and have on the order of 10^4 cores (processing units). Existing hydrocodes are reasonably scalable in the range of hundreds to a few thousand processors. But production armor simulations seldom need more than a few hundred processors. A typical large simulation involves a few million degrees of freedom and requires tens of gigaflops. Although this resolution makes it possible to conduct the simulations in three dimensions, in most cases much higher resolution is necessary to obtain results that converge.

Record-breaking platforms have recently achieved the petaflop (10^{15} flops) scale and involve between 10^5 and 3×10^5 cores. After conducting a study of the key technology challenges for exascale computing, the Defense Advanced Research Projects Agency (DARPA), announced the Omnipresent High Performance Computing Program (OHPC), aimed at building computers that exceed current petascale computers to achieve the mind-boggling speed of one quintillion (1,000,000,000,000,000,000) calculations per second (1 exaflop).[120] Such computers are needed, according to DARPA, to "meet the relentlessly increasing demands for greater performance, higher energy efficiency, ease of programmability, system dependability and security."[121] They will, among other things, offer unique opportunities for the simulation-based design of protective materials.

Quantification of Margins and Uncertainties

The extreme-scale computing world is quickly moving into the exascale, largely driven by the DoE (Figure 4-21). The unprecedented computing power is bringing about not just incremental improvements in capacity, fidelity, and resolution but also a paradigm shift in predictive science. The new main goal of this science is to make predictions with rigorously quantified uncertainties, so that the system can be certified or qualified. Specifically, in physics-based quantification of margins and uncertainties (QMU) the goal is to rigorously quantify means and uncertainties in the response of complex systems by maximizing the use of physical and computational models and minimizing use of experiments.[122,123,124,125,126,127] The development of such approaches is driven by applications in which experimental data are prohibitively expensive or cannot be obtained in the laboratory under the operating conditions of the device.

---

[120]Kogge, P., K. Bergman, S. Borkar, D. Campbell, W. Carlson, W. Dally, M. Denneau, P. Franzon, W. Harrod, K. Hill, J. Hiller, S. Karp, S. Keckler, D. Klein, R. Lucas, M. Richards, A. Scarpelli, S. Scott, A. Snavely, T. Sterling, R.S. Williams, and K. Yelick. 2008. ExaScale Computing Study: Technology Challenges in Achieving Exascale Systems, September 28. Available online at http://www.er.doe.gov/ascr/Research/CS/DARPA%20exascale%20-%20hardware%20%282008%29.pdf. Last accessed April 7, 2011. On June 20, 2010, DARPA announced the Omnipresent High Performance Computing Program (OHPC). See https://www.fbo.gov/index?s=opportunity&mode=form&id=3ba522c52b23884843a6639c8cbd1154&tab=core&_cview=0.

[121]Dillow, C. 2010. DARPA Wants to Usher in the Age of Exaflop Computing. Available at http://www.popsci.com/technology/article/2010-06/darpa-wants-usher-age-exaflop-computing. Accessed May 2, 2011.

[122]National Research Council. 2008. Evaluation of Quantification of Margins and Uncertainties Methodology for Assessing and Certifying the Reliability of the Nuclear Stockpile. Washington, D.C.: The National Academies Press.

[123]Eardley, D., H. Abarbanel, J. Katz, J. Cornwall, S. Koonin, P. Dimotakis, D. Long, S. Drell, D. Meiron, F. Dyson, R. Schwitters, R. Garwin, J. Sullivan, R. Grober, C. Stubbs, D. Hammer, P. Weinberger, R. Jeanloz, and J. Kammerdiener. 2005. Quantification of Margins and Uncertainties (QMU), JSR-04-330, March. Available online at http://www.stanford.edu/group/uq/docs/jason_qmu_margins.pdf. Last accessed April 7, 2011.

[124]Helton, J.C. 2009. Conceptual and Computational Basis for the Quantification of Margins and Uncertainty, SAND2009-3055, June. Available online at http://www.scribd.com/doc/27238941/Conceptual-and-Computational-Basis-for-the-Quantification-of-Margins-and-Uncertainty. Last accessed April 7, 2011.

[125]Pilch, M., T.G. Trucano, and J.C. Helton. 2006. Ideas Underlying Quantification of Margins and Uncertainties (QMU): A White Paper, SAND2006-5001, September. Available online at http://www.stanford.edu/group/uq/docs/qmu_ideas.pdf. Last accessed April 7, 2011.

[126]Sharp, D.H., and M.M. Wood-Schultz. 2003. QMU and nuclear weapons certification: What's under the hood. Los Alamos Science 28: 47-53.

[127]Lucas, L., H. Owhadi, and M. Ortiz. 2008. Rigorous verification, validation, uncertainty quantification and certification through concentration-of-measure inequalities. Computer Methods in Applied Mechanics and Engineering 197(51-52): 4591-4609.

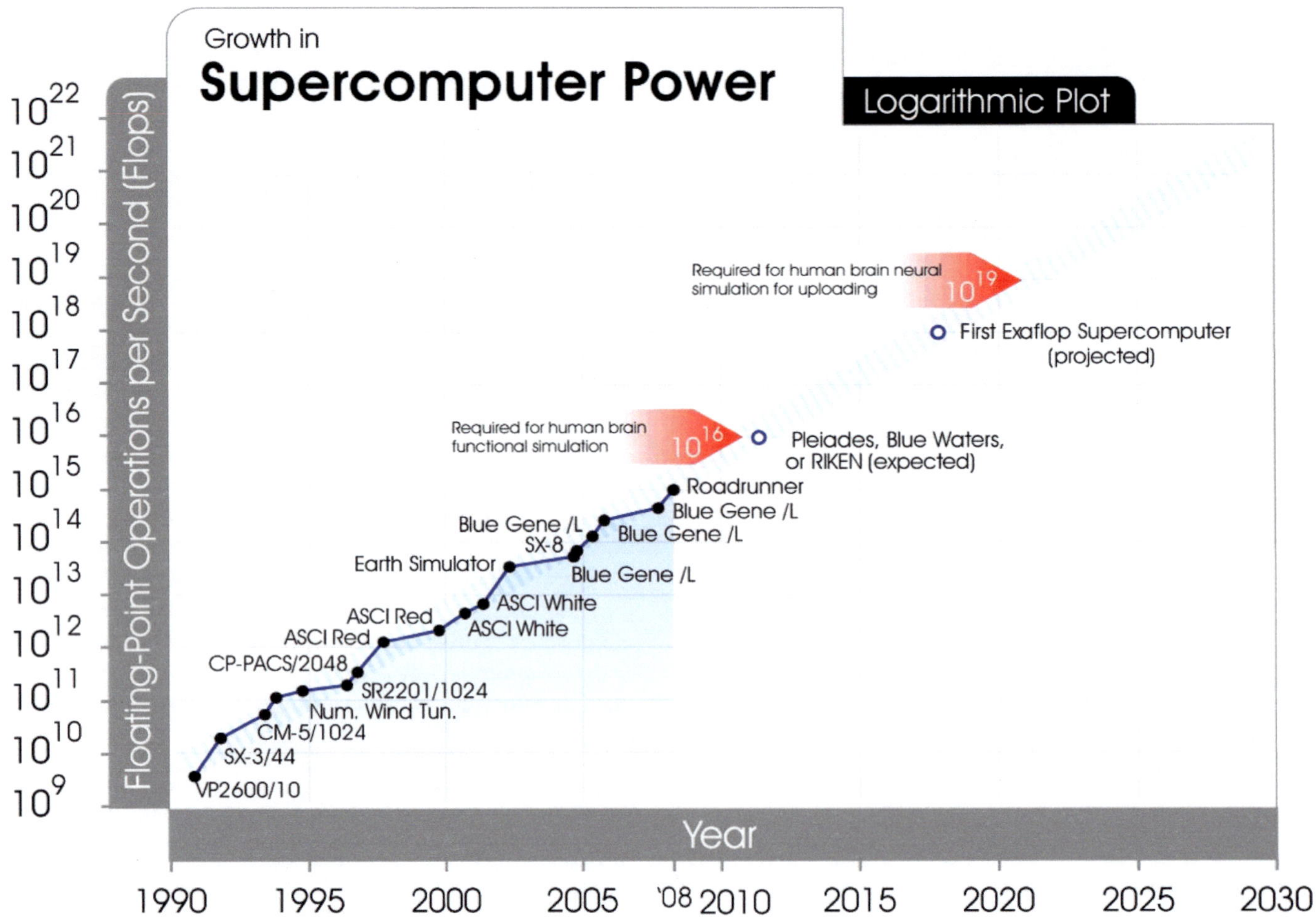

FIGURE 4-21 Growth in supercomputer powers as a function of year. SOURCE: Courtesy of Ray Kurzweil and Kurzweil Technologies, Inc. Available online at http://www.kurzweilai.net/growth-in-supercomputing-power.

QMU is also thought of as a tool for making high-consequence decisions about the design, certification, and deployment of high-value assets whose failure to perform safely and reliably could cause severe economic losses or loss of life. QMU radically alters the picture of predictive science and extreme-scale computing in many ways: by insisting on rigorously quantified uncertainties in the predictions as a measure of the confidence that decision makers can place in such predictions; by injecting probability and statistics into the calculations; by insisting on a global view of the response of the system over its entire operating range, thus breaking away from the "hero calculation" mode; and by the very tight coupling between simulations and validation or integral experiments that is required in order to establish confidence in the physics models.

In the context of protection materials, QMU, as enabled by extreme-scale computing, holds the promise of physics-based qualification of armor and protection systems. In this approach, computational models would be used to compute the variability of the response of a protection system given the randomness of inputs to the model and, potentially, the stochasticity of the response. In general, providing a measure of the maximum variability of the protection system over its operating range as computed by the model would require solving global optimization problems over input parameter space. This variability in turn provides a measure for the uncertainty in the response of the protection system—that is, a measure of how well the response of the system can be pinned down under operating conditions given the randomness of the system. Such global optimization calculations are inordinately intensive, hence the need for extreme-scale computing. However, the model uncertainty is only one part of the uncertainty budget: The level of confidence that can be placed in the physics and in the computations themselves, and the level of confidence that can be placed in the experimental data, need to be rigorously evaluated. The evaluation of both these terms in the uncertainty budget requires experimental data, either data-on-demand or archival (legacy) data.

Finding 4-4. The formulation of rigorous quantification of margins and uncertainties (QMU) protocols leading to the high-confidence certification of complex systems poses a challenge. However, the benefits of the application of QMU to the design and qualification of protective systems are potentially enormous.

NEW PROTECTION MATERIALS AND MATERIAL SYSTEMS: OPPORTUNITIES AND CHALLENGES

Polymeric materials such as some polycarbonates and Kevlar have demonstrated capabilities for certain protection applications, including transparent face shields and body armor. Polyethylene-based fiber materials such as Dyneema and Spectra have some properties, such as very high specific strength, that make them very effective in ballistic and blast applications when they are employed either as a single material or in combination with ceramics and metals in the form of a composite armor system. Preliminary ballistic tests have indicated their promise, especially for some of the higher strength versions of these fibers that are not yet available commercially. Nor are the constitutive laws and property inputs available that are needed to characterize these fibers in the range of strains and strain rates relevant to ballistic or blast simulations. Even the constitutive laws and material properties needed to characterize an established fiber such as Kevlar for these purposes are not fully in place. Thus far, these materials have been assessed largely based on projectile testing alone—make a target and shoot it. Efforts to employ constitutive models of fibers and yarns in simulations of protection systems have been published in recent years, including an assessment of lightweight fragment barriers for commercial aircraft[128,129] and a multiscale model of impacts on textile fabrics.[130]

Improving the properties of specific materials used in protection systems is one route to improved ballistic performance but may not open up opportunity for significant advances for some of the most widely employed protection materials given their maturity (some polymers are clear exceptions). By contrast, there is almost certainly scope for major advancement in the design of protection material systems made up of combinations of metals, ceramics, and polymers. Given the potential of polymer/ceramic/metal composite material protection systems and given the huge number of material and architectural combinations that should be considered, there is strong motivation to select among the multitude of combinations using simulation rather than testing alone, since the latter is time consuming, expensive, and not necessarily insightful. It was noted earlier in this chapter that, to perform effectively, a block of ceramic must be packaged in such a way that it deforms and fractures under a state of high compression. Metals and polymers, and combinations thereof, have been employed as packaging materials for ceramic protection systems.

Finding 4-5. Accurate simulation of the performance of armor protection systems under various ballistic threats and multiple hits necessitates advances in numerical methods, as outlined in the section "Modeling and Simulation Tools," as well as a better understanding of mechanisms of deformation and fracture coupled with better constitutive descriptions.

Computational Materials Methods

The focus in this chapter has been on characterizing materials with respect to their performance as protection materials using observation and the experimental and computational methods of mechanics. These methods can be used to evaluate new materials, and they are essential for establishing material properties that would enhance protection performance; however, they cannot be used to design new materials nor are they able to predict fundamental material parameters such as modulus, hardness, or toughness. These more fundamental objectives are in the realm of computational materials science. Computational materials methods have been covered in a variety of reports, among them *Integrated Computational Materials Engineering: A Transformational Discipline for Improved Competitiveness and National Security*, published by the NRC in 2008,[131] and a report by DOE.[132] In view of these widely available documents, the techniques of computational materials science, other than those already outlined above, will be described only briefly in the current report. *Integrated Computational Materials Engineering* is drawn on heavily for what is provided in the next few paragraphs, in many cases verbatim. That report also recommends a computationally enabled way forward for improving the development and insertion cycle for new materials across the entire spectrum of materials science and engineering and is therefore an important forerunner of the current document. Indeed, the field of protection materials is recognized by the committee as a good chance to implement many of the concepts described in *Integrated Computational Materials Engineering*.

In addition to methods designed to simulate structural

[128]Shockey, D.A., D.C. Erlich, and J.W. Simons. 1999. Lightweight fragment barriers for commercial aircraft, paper presented to the 18th International Symposium on Ballistics, San Antonio, Tex. Available online at http://www.sri.com/psd/fracture/as_pdf/18th_int_symposium_ballistics99.pdf. Last accessed April 7, 2011.

[129]King, M.J., P. Jearanaisilawong, and S. Socrate. 2005. A continuum constitutive model for the mechanical behavior of woven fabrics. International Journal of Solids and Structures 42(13): 3867-3896.

[130]Nilakantan, G., M. Keefe, T.A. Bogetti, and J.W. Gillespie, Jr. 2010. Multiscale modeling of the impact of textile fabrics based on hybrid element analysis. International Journal of Impact Engineering 37(10): 1056-1071.

[131]National Research Council. 2008. Integrated Computational Materials Engineering: A Transformational Discipline for Improved Competitiveness and National Security. Washington, D.C.: The National Academies Press.

[132]Department of Energy. 2005. Opportunities for Discovery: Theory and Computation in Basic Energy Sciences. Available online at http://www.sc.doe.gov/bes/reports/files/OD_rpt.pdf. Last accessed April 7, 2011.

materials behavior at the macroscopic level, as described above, the computational materials scientist has a host of techniques for simulation for a variety of purposes. The wide variety of tools available reflects the fact that materials response and behavior involve a multitude of physical and chemical phenomena whose accurate treatment in models requires the spanning of many orders of magnitude in length and time. Further, computational simulation is used to tackle a wide variety of materials attributes and phenomena, including thermodynamic, kinetic, and structural properties, to bring advances in areas such as materials processing, microstructural evolution, structure-property relationships, materials stability and corrosion, and stiffness and strength. Moreover,

> . . . the length scales in materials response range from nanometers of atoms to the centimeters and meters of manufactured products. Similarly, time scales range from the picoseconds of atomic vibrations to the decades over which a component will be in service. Fundamentally, properties arise from the electronic distributions and bonding at the atomic scale of nanometers, but defects that exist on multiple length scales, from nanometers to centimeters, may in fact dominate properties. It should not be surprising that no single modeling approach can describe this multitude of phenomena or the breadth of scales involved. While many computational materials methods have been developed, each is focused on a specific set of issues and appropriate for a given range of lengths and times. Consider length scales from 1 angstrom to 100 microns. At the smallest scales scientists use electronic structure methods to predict bonding, magnetic moments, and transport properties of atoms in different configurations. As the simulation cells get larger and the times scales longer, empirical interatomic potentials are used to approximate these interactions. Optimization and temporal evolution of electronic structure and atomistic methods are achieved using conjugate gradients, molecular dynamics, and Monte Carlo techniques. At still larger scales, the information content of the simulation unit decreases until it becomes more efficient to describe the material in terms of the defect that dominates at that length scale. These units might be defects in the lattice (for example, dislocations), the internal interfaces (for example, grain boundaries), or some other internal structure, and the simulations use these defects as the fundamental simulation unit in the calculation. [133]

Table 3-1 (Table 4-1 in the current report) from the above-mentioned NRC report *Integrated Computational Materials Engineering*[134] lists a variety of computational materials methods, some of them standard and already adopted in materials development and industry activities, and others strictly research tools. As noted in the source study from which it was taken,

> . . . the table is not intended to be complete but rather to exemplify the methods available for modeling materials characteristics. This table indicates typical inputs and outputs of the software and examples of widely used or recognized codes. Electronic structure methods employ different approximate solutions to the quantum mechanics of atoms and electrons to explore the effects of bonding, chemistry, local structure, and dynamics on the mechanisms that affect material properties. Typically, tens to hundreds of atoms are included in such a calculation and the timescales are on the order of nanoseconds. In atomistic simulations, arrangements and trajectories of atoms and molecules are calculated. Generally based on models to describe the interactions among atoms, simulations are now routinely carried out with millions of atoms. Length scales and timescales are in the nanometer and nanosecond regime, and longer length scales and timescales are possible in the case of molecular system coarse graining from "all-atom" to "united atom" models (that is, interacting clusters of atoms). Dislocation dynamics methods are used to study the evolution of dislocations (curvilinear defects in the lattice) during plastic deformation. The total number of dislocations is typically less than a million, and strain rates are large compared to those measured in standard laboratory tests. Thermodynamic methods range from first-principle predictions of phase diagrams to complex database integration methods using existing tabulated data to produce phase diagrams and kinetics data.[135]

Microstructural evolution methods predict material stability and evolution at the microscopic level based on free-energy functions, elastic parameters, and kinetic databases.

> Micromechanical and mesoscale property models include solid mechanics and FEA methods that use experimentally derived models of materials behavior to explore microstructural influences on properties. The models may incorporate details of the microstructure (resolving scales at the relevant level). Results may be at full system scale. *Mesoscale structure models* include models for solidification and solid state deformation using combinations of the previous methods to predict favorable processing conditions for specific microstructural characteristics. Methods for code and systems integration offer ways to connect many types of models and simulations and to apply systems engineering strategies. Statistical tools are often used to gain new understanding through correlations in large data sets. Other important ICME tools include databases, quantifiable knowledge rules, error propagation models, and cost and performance models.[136]

Finding 4-6. Computational methods have considerable potential for aiding the architectural design of composite protection packages, but they require robust constitutive

[133]National Research Council. 2008. Integrated Computational Materials Engineering: A Transformational Discipline for Improved Competitiveness and National Security. Washington, D.C.: The National Academies Press. P. 69.

[134]National Research Council. 2008. Integrated Computational Materials Engineering: A Transformational Discipline for Improved Competitiveness and National Security. Washington, D.C.: The National Academies Press.

[135]Ibid., p. 71.

[136]Ibid.

TABLE 4-1 Mode or Method, Required Input, Expected Output, and Typical Software Used in Materials Science and Engineering

Class of Computational Materials Model/Method	Inputs	Outputs	Software Examples
Electronic structure methods (density functional theory, quantum chemistry)	Atomic number, mass, valence electrons, crystal structure and lattice spacing, Wyckoff positions, atomic arrangement	Electronic properties, elastic constants, free energy vs. structure and other parameters, activation energies, reaction pathways, defect energies and interactions	VASP, Wien2K, CASTEP, GAMES, Gaussian, a=chem., SIESTA, DACAPO
Atomistic simulations (molecular dynamics, Monte Carlo)	Interaction scheme, potentials, methodologies, benchmarks	Thermodynamics, reaction pathways, structures, point defect and dislocation mobility, grain boundary energy and mobility, precipitate dimensions	CERIU2, LAMMPS, PARADYN, DL-POLY
Dislocation dynamics	Crystal structure and lattice spacing, elastic constants, boundary conditions, mobility laws	Stress-strain behavior, hardening behavior, effect of size scale	PARANOID, ParaDis, Dis-dynamics, Micro-Megas
Thermodynamic methods (CALPHAD)	Free-energy data from electronic structure, calorimetry data, free-energy functions fit to materials databases	Phase predominance diagrams, phase fractions, multicomponent phase diagram, free energies	Pandat, ThermoCalc, Fact Sage
Microstructural evolution methods (phase-field, front-tracking methods, Potts models)	Free-energy and kinetic databases (atom mobilities), interface and grain boundary energies, (anisotropic) interface mobilities, elastic constants	Solidification and dendritic structure, microstructure during processing, deployment, and evolution in service	OpenPF, MICRESS, DICTRA, 3DGG, Rex3D
Micromechanical and mesoscale property models (solid mechanics and finite-element analysis)	Microstructural characteristics, properties of phases and constituents	Properties of materials—for example, modulus, strength, toughness, strain tolerance, thermal/electrical conductivity, permeability; possibly creep and fatigue behavior	OOF, Voronoi Cell, JMatPro, FRANC-3D, ZenCrack, DARWIN
Microstructural imaging software	Images from optical microscopy, electron microscopes, X-rays, etc.	Image quantification and digital representations	Mimics, IDL, 3D Doctor, Amira
Mesoscale structure models (processing models)	Processing thermal and strain history	Microstructural characteristics (for example, grain size, texture, precipitate dimensions)	PrecipiCalc, JMat Pro
Part-level finite-element analysis, finite difference, and other continuum models	Part geometry, manufacturing processing parameters, component loads, materials properties	Distribution of temperatures, stresses and deformation, electrical currents, magnetic and optical behavior, etc.	ProCast, MagmaSoft, CAPCAST, DEFORM, LS-Dyna, Abaqus
Code and systems integration	Format of input and output of modules and the logical structure of integration, initial input	Parameters for optimized design, sensitivity to variations in inputs or individual modules	iSIGHT/FIPER, QMD, Phoenix
Statistical tools (neural nets, principal component analysis)	Composition, process conditions, properties	Correlations between inputs and outputs; mechanistic insights	SPLUS, MiniTab, SYSTAT, FIPER, PatternMaster, MATLAB, SAS/STAT

SOURCE: National Research Council. 2008. Integrated Computational Materials Engineering: A Transformational Discipline for Improved Competitiveness and National Security. Washington, D.C.: The National Academies Press.

characterizations of component materials. Experimental data and constitutive characterizations of some materials used in composite armor systems are woefully inadequate. This is especially true for some promising polymers. Properties must be measured and characterized over the range of microstructural feature sizes and the range of strains, strain rates, and stress states relevant to blast and penetration events. Close communication between experimentalists who measure the high-stress, high-strain-rate properties of materials and the modelers who use these data is strongly encouraged.

Overall Recommendations

Recommendation 4-1. The Department of Defense should pursue an initiative for protection materials by design by exploiting the capabilities of advanced computational and experimental methods. The initiative will (1) enable improved understanding of fundamental material deformation and fracture mechanisms governing protection materials performance and (2) provide guidance for changes in material processing.

Recommendation 4-2. The protection materials by design initiative should also use advanced computational and experimental methods to simulate the ballistic and blast performance of candidate material protection systems.

Recommendation 4-3. The protection materials by design initiative should include a concerted effort to develop the next generation of Department of Defense advanced protection codes that incorporate experimentally validated, high-fidelity scientific models, as well as the necessary high-performance computing infrastructure. Progress in this direction will require the development of high spatial and temporal resolution (with 10-μ resolution in space and microsecond resolution in time) capabilities for in situ visualization of deformation and failure mechanisms during the impact event.

Recommendation 4-4. As part of the initiative, a program should be established with primary focus on code validation and verification; multiscale, multiphysics material models; integrated simulation/experimental protocols; prediction with quantified uncertainties; and simulation-based qualification to help advance predictive science for protection materials and material systems.

Recommendation 4-5. The initiative should identify a series of unclassified protection material challenge problems comprising simulation and experimental validation whose solution would convincingly demonstrate the effectiveness of protection materials by design. One such canonical problem might be the characterization of the high-strain-rate response of brittle armor materials such as ceramics and glasses under combinations of high pressure and shear representative of ballistic penetration, followed by a demonstration of the effectiveness of the new characterization in simulating the performance of a particular ceramic armor package.

5

Lightweight Protective Materials: Ceramics, Polymers, and Metals

OVERVIEW AND INTRODUCTION

The history of improving protection while reducing the weight of armor has been a remarkable materials success story. Over the last half-century, new choices of materials such as ceramics, polymers, and polymer fibers and lower density metals have significantly decreased the weight of the armor needed for the protection of personnel and vehicles. Figure 1-2 in Chapter 1 illustrates the revolutionary reductions in the areal density of vehicle armor as advanced materials have become available, starting with rolled homogeneous armor and advancing to complex composite systems. There have been similar advances in lightweight materials for personnel protection as well. As described in Chapter 2, armor systems are designed and fabricated using suitable combinations of ceramics, metals, polymers, fibers, and composites to meet specific threat requirements. The choice of materials, as well as their geometry and the means by which they are assembled, is a key factor in armor design. Each material component serves a specific purpose not only in defeating the kinetic energy of projectiles or mitigating a blast but also in maintaining the structural armor's integrity.

To provide a basic understanding of current armor materials and to anticipate areas where there could be revolutionary improvements in armor materials, this chapter examines the synthesis and processing of each of the main types of materials, with particular emphasis on the resultant material structure from the atomic to the macro scale. Potential new compositions and the tailoring of microstructures to discover material behaviors that could dramatically enhance armor performance are highlighted, as are the challenges involved in achieving such advances.

The schematic in Figure 5-1 depicts a notional armor structure,[1] consisting of both dense and porous ceramics, fibers, environmental coatings, polymer binders, and adhesive joints. The complex tile architecture presented in Figure 5-1 uses several materials and different assembly methods for those materials such that the layers perform their protective functions during the projectile impact. This chapter will examine how achieving improved material behavior but also minimizing manufacturing cost requires a deep scientific and engineering understanding of the desirable structures and compositions of advanced protective materials as well

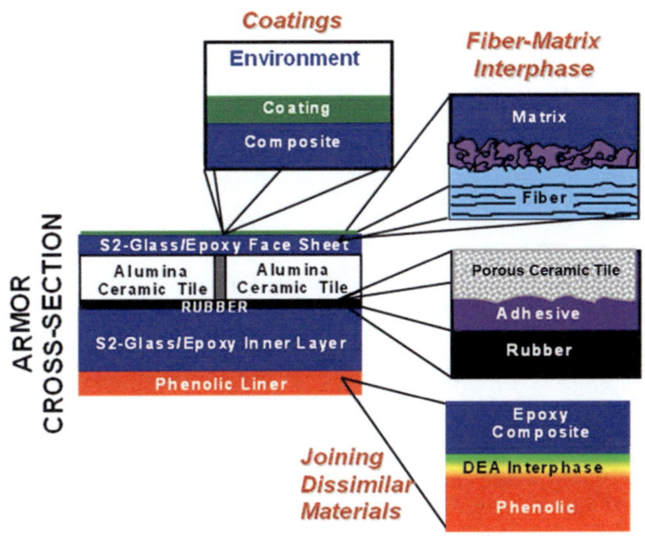

FIGURE 5-1 Schematic presentation of the cross section of an armor tile typically used for armored vehicles showing the complexity of the armor architecture. Different classes of materials, such as dense and porous ceramics, fiber composites, thermoplastic polymers, and adhesives are used for the tile assembly. DEA, diethanolamine. SOURCE: James W. McCauley, Chief Scientist, Weapons and Materials Research Directorate, Army Research Laboratory (ARL) fellow, ARL, "Armor Materials 101-501: Focus on Fundamental Issues Associated with Armor Ceramics 'Kinetic energy passive armor,'" presentation to the committee on March 9, 2010.

[1] James W. McCauley, Weapons and Materials Research Directorate, Army Research Laboratory (ARL) fellow, ARL, "Armor materials 101-501: Focus on fundamental issues associated with armor ceramics 'kinetic energy passive armor,'" presentation to the committee, March 9, 2010.

as how to make and process them. That said, as explained in Chapter 3, the requisite material properties that are to be optimized cannot be measured by the usual quasi-static measures of mechanical behavior. However, even at lower strain rates, conducting mechanical tests at small scale—that is, at the microstructural level, on the order of nanometers or microns—will likely shed light on the deformation mechanisms under known loading states and can provide information that is very useful for parallel modeling efforts, keeping in mind that the ultimate goal is real-time measurements of many properties on ballistic timescales.

As shown in Chapter 4, the behavior of an assembly in the face of a particular threat is not the simple sum of the behaviors of its component parts. Thus, an integrated experimental and modeling approach that allows clear variation of crystal and material microstructures and subsequent high-rate dynamic characterization of the material behavior by itself and as part of an armor system may enable the development of ever lighter and more effective protection materials.

A more rapid development of materials and their successful insertion into armor necessitates attention to such basic issues as the reduction of voids and impurities along with attention to the challenges of advanced designs and creating and synthesizing new material compositions, new phases, and preferred microstructures. This chapter discusses the main issues surrounding several important classes of protection materials. The accompanying set of appendixes goes into considerable detail—especially on the synthesis and processing of ceramics, cermets, and polymers—because these classes of materials have the best potential for significant improvements if the interrelationships can be elucidated between synthesis, processing methods, and the resultant structures, along with the corresponding high-rate measurement of material behavior. For the reader to appreciate the issues, the selected materials are introduced at the atomic, molecular, micro, and macro scales before describing the synthesis and processing methods. Finally, areas of potential innovation that may bring transformational changes in the design and performance of armor materials are described, along with the challenges to be overcome.

CERAMIC ARMOR MATERIALS

High-temperature refractory ceramic materials offer a unique combination of physical and mechanical properties that in turn can offer favorable protection against high-velocity armor-piercing bullets (see Chapter 2). Ceramics feature high hardness, high elastic modulus, low density, sufficient flexure, and good compressive strengths, but relatively low fracture toughness. The Hugoniot elastic limit (HEL)—the maximum uniaxial dynamic stress that a material can withstand elastically—represents the nominal potential of a ceramic as an armor-grade material.[2] However, it is almost mandatory for the candidate material to also possess a residual plastic behavior greater than the HEL, because the greatest velocity threats typically induce stresses that are higher than the HEL of materials that are commonly available. Properties such as hardness and modulus are determined by the chemical and phase compositions and microstructure of the material. Besides composition, many ceramic material properties can be influenced by the relative amounts of the various possible phases/polytypes, average grain size, grain-size distribution, and grain morphologies, as well as minor-phase content.

One of the most important aspects of ceramic materials that makes them suitable for ballistic protection is the strong covalent bonding between lightweight atoms located in the first quarter of the periodic table of elements. The elements include beryllium, boron, carbon, oxygen, magnesium, aluminum, and silicon. Indeed, the most developed and best explored armor ceramics are Al_2O_3 (aluminum oxide, or alumina), B_4C (boron carbide), and SiC (silicon carbide). However, these three materials are but a small portion of the ceramics that could be used for armor application. For example, novel boron icosahedra containing higher borides, ternary B–C–Si and B–C–N systems, and homologous Al(Mg)–B–C(N) compounds have yet to be explored.

Because ceramics are relatively brittle materials, they are sensitive to flaws, and flaws adversely affect materials performance. If flaws are prevalent, it is often difficult or almost impossible to assess the intrinsic properties and behaviors of materials. Thus, it is critical to be able to process ceramics to near-theoretical maximum density, eliminating most of the void-type defects in order to explore the fundamental behavior. Such defects are often responsible for ceramic armor failure from the shock wave of a ballistic impact, which causes cracks to nucleate at the defect sites and then grow and coalesce, causing massive failure. As noted by Lankford,[3] the ceramic would never fail (in penetration) if it could be constrained such that it would undergo plastic flow. Of course the presence of defects will keep the ceramic from reaching the stress levels necessary to activate plasticity mechanisms, and simple, practical improvement in performance can be realized by employing nondestructive evaluation analysis to reveal the larger defects in the material. Better compaction technology and sintering techniques should result in a more uniform and higher density component. Upgrades in powder quality (purity, uniformity of particles) and improvements in the formulation of sintering aids can also help eliminate voids and porosity and retain homogeneous microstructure. Highly nonuniform grain-size distributions and the presence of grain boundary phases due to poor compositional quality of the starting powders can also adversely affect performance. Agglomerated particles

[2]Fanchini, G., J.W. McCauley, and M. Chhowalla. 2006. Behavior of disordered boron carbide under stress. Physical Review Letters 97(6): Article number 035502.

[3]Lankford Jr., J. 2004. The role of dynamic material properties in the performance of ceramic armor. International Journal of Applied Ceramic Technology 1(3): 205-210.

due to poor mixing of sintering aids in the powders[4,5] and extraneous carbon additions or poor mixing of the carbon reduce the grain growth of nearby SiC grains and leave large carbon inclusions inside the fine SiC matrix.[6]

Early in the Vietnam conflict, the Department of Defense (DoD) approached the Los Alamos National Laboratory and the Lawrence Livermore National Laboratory with a request for lightweight body armor for ground troops. John Taylor at Los Alamos and Mark Wilkins at Lawrence Livermore began investigating ceramics for protection against small arms fire. Coors Ceramics was asked to fabricate an alumina molded body panel, but ground troops in the jungles of Vietnam found it too heavy and would only wear the armor on guard duty at a fixed post. Later, Wilkins et al.[7] demonstrated a relationship between hardness, compressive strength, and ballistic performance and showed that bulk properties alone were not a sufficient basis for the design of armor. They argued that some trade-off between the various properties would be necessary to derive benefits from other key properties such as fracture toughness and plasticity.[8,9] Their early work eliminated most silicate-based ceramics from consideration owing to their low hardness and the fact that mullite and other alumina ceramics containing silicate seemed to fail under lesser ballistic attack than did high-purity alumina. Wilkins et al. further focused their research on other oxides such as aluminum magnesium spinel; carbides such as silicon and boron carbides; borides such as titanium diboride; and a few nitrides, including aluminum nitride. Alumina emerged as today's most widely used ceramic armor, combining good mechanical behavior with relatively low cost. Because alumina is manufactured in quantities of millions of pounds throughout the world, it is much less expensive than either SiC or, especially, B_4C. The densities of B_4C (2.52 g/cm^3) and SiC (3.29 g/cm^3) are considerably less than that of Al_2O_3 (3.98 g/cm^3). However, because of its easy sinterability and the lower cost of the raw powders, alumina is still preferred for use in vehicle applications, where the extra weight can be tolerated, while the lighter B_4C and SiC ceramics are now used in body armor.

Alumina nanoceramics that can reach the theoretical maximum density present an opportunity to probe the effects of microstructure on material behavior in a cost-effective material. While B_4C and SiC ceramics require temperatures of 2150°C to 2200°C and, typically, applied pressure to carry out sintering to achieve to full density, alumina can be easily sintered into complex shapes to full density at 1500°C to 1600°C by pressureless sintering. Indeed, Al_2O_3 nanopowders can be sintered at 1100°C to 1200°C to full density while retaining their nanograin microstructure.[10,11] Krell's work on Al_2O_3 indicated a Hall-Petch relationship, whereby decreasing the grain size yielded an increase in hardness.[12] Chen et al.[13] suggested the importance of effective plasticity on the ballistic behavior of alumina.

Of the other ceramics named above, SiC and B_4C are the leading opaque ceramic materials for next-generation body and vehicle armor systems. Their favorable characteristics relative to alumina (Al_2O_3) are lighter weight, higher hardness, and higher stiffness.

A central tenet of materials science and engineering is that composition, crystal structure, and microstructure influence the mechanical behavior of the material. According to McCauley,[14]

> . . . the fundamental factors that affect the intrinsic material characteristics [are] related to crystal physics, i.e., elastic properties and anisotropy, phase transformation, and deformation mechanisms along with the development of new materials and transformational processing methods [that can] yield large 25-40 percent improvements in ceramic performance.

A recent case in point is the great improvement in the mechanical performance of B_4C-SiC layered particulate ceramics achieved by Orlovskaya et al. by introducing high compressive thermal residual stresses to their outer surface layer.[15]

---

[4]Bakas, M., V.A. Greenhut, D.E. Niesz, J. Adams, and J. McCauley. 2003. Relationship between defects and dynamic failure in silicon carbide. Ceramic Engineering and Science Proceedings 24(3): 351-358.

[5]Bakas, M., V.A. Greenhut, D.E. Niesz, J. Adams, and J. McCauley. 2008. Relationship between defects and dynamic failure in silicon carbide. Chapter 52 in 27th Annual Cocoa Beach Conference on Advanced Ceramics and Composites: A: Ceramic Engineering and Science Proceedings, Volume 24, Issue 3. W.M. Kriven and H.-T. Lin, eds. Hoboken, N.J.: John Wiley & Sons.

[6]Raczka, M., G. Gorny, L. Stobierski, and K. Rozniatowski. 2001. Effect of carbon content on the microstructure and properties of silicon carbide-based sinters. Materials Characterization 46(2-3): 245-249.

[7]Wilkins, M.L., C.F. Cline and C.A. Honodel. 1969. Light Armor, UCRL-71817, July 23. Livermore, Calif.: Lawrence Radiation Laboratory, University of California.

[8]Ibid.

[9]Wilkins, M.L., R.L. Landingham, and C.A. Honodel. 1971. Fifth Progress Report of Light Armor Program, UCRL-50980, January. Livermore, Calif.: Lawrence Radiation Laboratory, University of California.

[10]Krell, A. 1996. The influence of shaping method on the grain size dependence of strength in dense submicrometre alumina. Journal of the European Ceramic Society 16(11): 1189-1200.

[11]Bakas, M., V.A. Greenhut, D.E. Niesz, J. Adams, and J. McCauley. 2003. Relationship between defects and dynamic failure in silicon carbide. Ceramic Engineering and Science Proceedings 24(3): 351-358.

[12]Krell, A., P. Blank, H.W. Ma, T. Hutzler, and M. Nebelung. Processing of high-density submicrometer Al_2O_3 for new applications. Journal of the American Ceramic Society 86(4): 546-553.

[13]Chen, M.W., J.W. McCauley, D.P. Dandekar, and N.K. Bourne. 2006. Dynamic plasticity and failure of high-purity alumina under shock loading. Nature Materials 5(8): 614-618.

[14]James W. McCauley, Weapons and Materials Research Directorate, Army Research Laboratory (ARL) fellow, ARL, "Armor materials 101-501: Focus on fundamental issues associated with armor ceramics 'kinetic energy passive armor,'" presentation to the committee, March 9, 2010.

[15]Orlovskaya, N., M. Lugovy, V. Subbotin, O. Radchenko, J. Adams, M. Chheda, J. Shih, J. Sankar, and S. Yarmolenko. 2005. Robust design and manufacturing of ceramic laminates with controlled thermal residual stresses for enhanced toughness. Journal of Materials Science 40(20): 5483-5490.

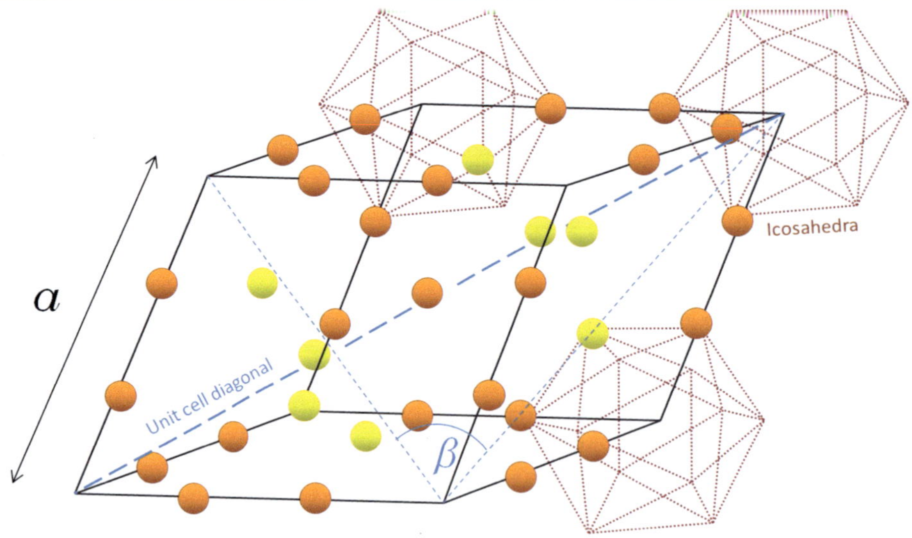

FIGURE 5-2 Rhombohedral unit cell structure of B_4C showing $B_{11}C$ icosahedra and the diagonal chain of C–B–C atoms. Boron atoms are represented as red spheres and carbon atoms as white spheres.

Crystalline Ceramics: Phase Behavior, Grain Size or Morphology, and Grain Boundary Phases

Chemical composition, crystalline structure, and stability under elevated temperatures and under stress play an important role in determining both the quasi-static properties of these materials and their dynamic deformation and failure behavior. An examination of B_4C and SiC will give readers a sense of the complexity of the atomic bonding and crystalline unit cells in these simple binary ceramics and will introduce them to intrinsic crystal defects such as stacking faults, twins, and grain boundaries, which they need to know about to understand some aspects of the ballistic performance of these two important protection materials.

Crystal Structure of Boron Carbide

Because it is not possible to precisely control the stoichiometry of boron carbide in commercially synthesized powders, it is important to understand how composition influences the atomic structure and the corresponding microstructure and properties. Boron carbide can be considered as a prototype of the interstitial compounds of rhombohedral boron, which include $B_{12}C$, $B_{12}C_2Al$, $B_{12}S$, $B_{12}O_2$, $B_{12}As_2$, $B_{12}P_2$, B_3Si, and B_4Si. Interestingly, the stoichiometric compound B_4C does not exist, and the denomination "boron carbide" refers to the whole homogeneity range extending from $B_{4.3}C$ at the carbon-rich limit to $B_{\sim 11}C$ at the boron-rich limit,[16] a range of 8.8 mol percent to approximately 20 mol percent C. None of the unit cells of the interstitial compounds can be defined precisely. Instead, the materials are made up of composition-dependent, statistically distributed, and nearly isomorphous elementary cells, whose commonality is the 12-atom slightly distorted icosahedra at each cell vertex and the mostly 3-atom linear chains on the main diagonal parallel to the crystallographic *c*-axis. The unit cells thus comprise B_{12} and $B_{11}C$ icosahedra, while the chains comprise C–B–C, C–B–B, or B–□–B (the symbol □– indicates an atom vacancy) since the similarly sized C and B atoms readily substitute for each other. The general structure formula is $(B_{12})_n(B_{11}C)_{1-n}(CBC)_p(CBB)_q(B\text{–}\square\text{–}B)_{1-p-q}$.[17] The second constituent—for example, C, Al, or O—occupies sites on the diagonal chain (see the unit cell shown in Figure 5-2).[18] For the approximately stoichiometric B_4C material, the icosahedra are $B_{11}C$ and the chains are C–B–C. Boron carbide (13.3 mol percent C) melts congruently at 2490°C and forms a eutectic mixture with carbon at 2375°C–2400°C at a composition of 29 mol percent C (see the B–C phase diagram, Figure 5-3).[19] The extremely rigid framework arises from the covalently bonded icosahedra and the chain units of co-

[16]Kuck, S., and H. Werheit. 2000. Boron Compounds. Pp. 1-491 in Non-Tetrahedrally Bonded Binary Compounds II, Landolt-Börnstein: Numerical Data and Functional Relationships in Science and Technology, New Series, subvolume 41. D. O. Madelung, ed. New York, N.Y.: Springer.

[17]Werheit, H., H.W. Rotter, S. Shalamberidze, A. Leithe-Jasper, and T. Tanaka. 2010. Gap-state related photoluminescence in boron carbide. Physica Status Solidi B 1-5. Available online at http://onlinelibrary.wiley.com/doi/10.1002/pssb.201046342/pdf. Last accessed March 31, 2011.

[18]Emin, D. 1988. Structure and single-phase regime of boron carbides. Physical Review B 38(9): 6041-6055.

[19]Thevenot, F. 1990. Boron carbide: A comprehensive review. Journal of the European Ceramic Society 6(4): 205-225.

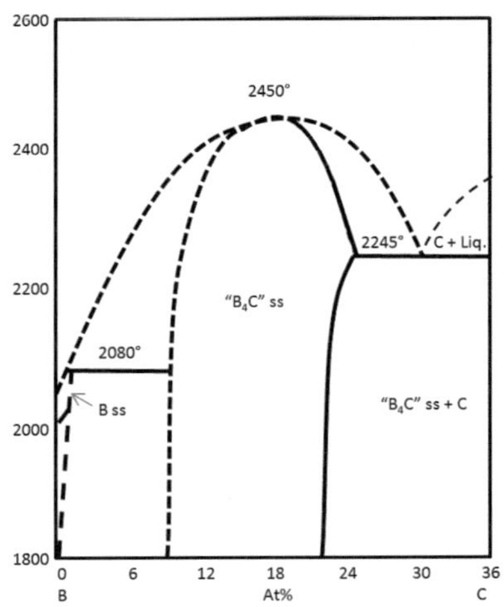

FIGURE 5-3 The boron-carbon phase diagram over the range 0-36 at % carbon. The cross-hatched region is commonly referred to as "B_4C." Different phase diagrams for the B–C system were reported in the past and there is no currently agreed upon reference phase diagram that can be reliably used to determine the correct stoichiometry and equilibrium phases.

valently bonded atoms and is responsible for the material's refractory nature and extreme hardness.[20]

The average structure, measured by x-ray diffraction pattern or by nuclear magnetic resonance, varies as the boron content is varied.[21] The theoretical density increases linearly with increasing carbon content, extending from 2.465 g/cm^3 for $B_{10.4}C$ to 2.52 g/cm^3 for B_4C. Kwei et al.[22] showed theoretically that the central boron atom in the C–B–C chain is relatively loosely held and that these locations can form vacancies along the three-atom chain, leading to a decrease in thermal conductivity.[23] Aselage et al. found a significant drop in elastic modulus when the carbon concentration fell below 13.3 percent, reflecting a change in stiffness of the most compressible structural unit, the icosahedra (when $B_{11}C \rightarrow B_{12}$). Very little is known about (1) the relative ratio of B_{12}, $B_{11}C$, and C–B–C, C–B–B, B–□–B structural units in boron carbide or (2) the rates of growth of the different crystal structures and their mutual transformations in the solid state as a function of pressure, temperature, and time.

The current working-phase diagram (Figure 5-3) for boron carbide shows that it is not a so-called line compound.[24,25] Moreover, the details of the phase boundaries and relative amounts of the polytypes have not yet been firmly established.

Boron Carbide Amorphization

The maximum contact pressure generated by a projectile incident on a ceramic depends on the velocity, bulk modulus, density, and yield strength of the projectile.[26] The impact can also lead to a rapid increase in the local temperature. When the pressure and/or the temperature exceeds a critical threshold, amorphization (the transition from the crystalline phase to the amorphous phase) or other phase transformations (crystal A to crystal B) can occur in certain materials. Boron carbide possesses the highest HEL of ceramic materials (~17-20 GPa), surpassing all of its denser competitors such as silicon carbide and alumina by a factor of 2.[27,28,29] High HEL would suggest that boron carbide could outperform other armor materials. However, when the impact pressures exceed 20 GPa, an abrupt drop in shear strength occurs, leading to a much lower dynamic performance for B_4C than that expected from its hardness and HEL.[30,31] The loss in performance of B_4C under high-velocity impact is currently believed to be related to the formation of amorphous bands inside the crystalline grains and a related weakening of the bonds. These amorphous bands were discovered using high-resolution transmission electron microscopy (TEM) to analyze fragments of B_4C ballistic tiles that had previously been subjected to supercritical impact velocities and pressures (in excess of 20-23 GPa). TEM images revealed the formation of 2-3-nm-wide intragranular amorphous bands that occur parallel to specific crystallographic planes and contiguously with apparent cleaved fracture surfaces (see Figure 5-4). At subcritical impact velocities, the amorphous bands were never observed; instead, a relatively high concentration of

[20]Schwetz, K.A. 1999. Boron carbide, boron nitride, and metal boride. In Ullmann's Encyclopedia of Industrial Chemistry, Sixth Edition (electronic release). T. Kellersohn, ed. Weinheim, Germany: Wiley-VCH Verlag.

[21]Werheit, H., H.W. Rotter, S. Shalamberidze, A. Leithe-Jasper, and T. Tanaka. 2010. Gap-state related photoluminescence in boron carbide. Physica Status Solidi B 1-5. Available online at http://onlinelibrary.wiley.com/doi/10.1002/pssb.201046342/pdf. Last accessed March 31, 2011.

[22]Kwei, G.H., and B. Morosin. 1996. Structures of the boron-rich boron carbides from neutron powder diffraction: Implications for the nature of the inter-icosahedral chains. Journal of Physical Chemistry 100(19): 8031-8039.

[23]Ibid.

[24]Emin, D. 1988. Structure and single-phase regime of boron carbides. Physical Review B 38(9): 6041-6055.

[25]Thevenot, F. 1990. Boron carbide: A comprehensive review. Journal of the European Ceramic Society 6(4): 205-225.

[26]Lundberg, P., R. Renstrom, and L. Westerling. 2002. Transition between interface defeat and penetration for a given combination of projectile and ceramic material. Ceramic Transactions 134: 173-181.

[27]Bourne, N.K. 2002. Shock–induced brittle failure of boron carbide. Proceedings of the Royal Society A: Mathematical, Physical & Engineering Sciences 458(2024): 1999-2006.

[28]Johnson, G.R., and T.J. Holmquist. 1999. Response of boron carbide subjected to large strains, high strain rates, and high pressures. Journal of Applied Physics 85(12): 8060-8073.

[29]Thevenot, F. 1990. Boron carbide: A comprehensive review. Journal of the European Ceramic Society 6(4): 205-225.

[30]Bourne, N.K. 2002. Shock–induced brittle failure of boron carbide. Proceedings of the Royal Society A: Mathematical, Physical & Engineering Sciences 458(2024): 1999-2006.

[31]Dandekar, D.P. 2001. Shock Response of Boron Carbide, ARL-TR-2456. Available online at http://www.arl.army.mil/arlreports/2001/ARL-TR-2456.pdf. Last accessed April 7, 2011.

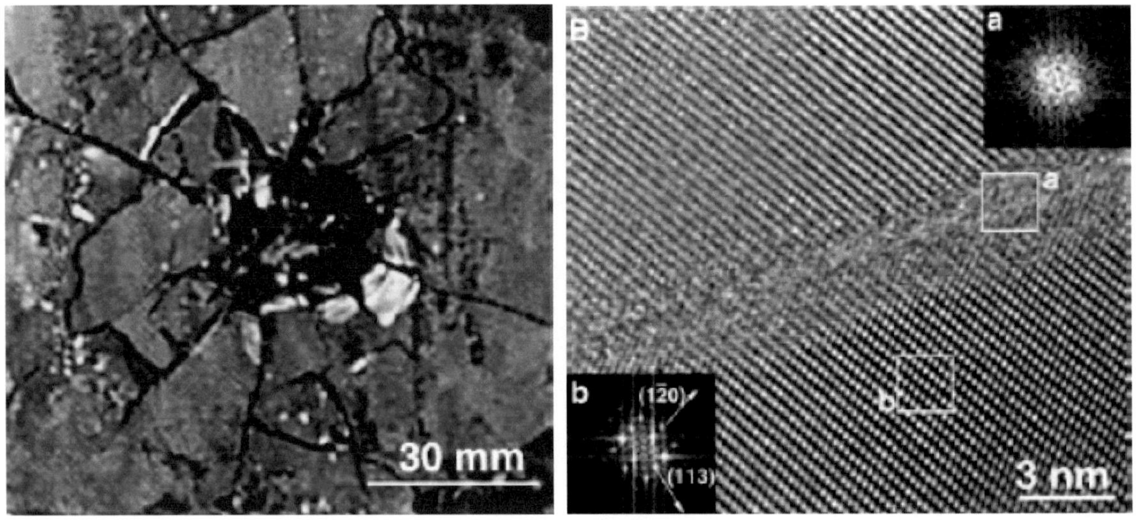

FIGURE 5-4 A boron carbide ballistic target that comminuted during impact (left) and a high-resolution TEM image of a fragment produced by a ballistic test at an impact pressure of 23.3 GPa (right). The lattice images on either side of the band correspond to the [-101] direction of crystalline B_4C, and the loss of lattice fringes in the band indicates localized amorphization in a band within the grain. SOURCE: Chen, M., J. McCauley, and K. Hemker. 2003. Shock-induced localized amorphization in boron carbide. Science 299(5612): 1563-1566.

stacking faults and microtwins was observed, suggestive of plastic deformation of the material under shock loading.[32]

Understanding the pressure dependence of boron carbide phases would shed light on the issue of the pressure-induced, crystal-to-amorphous transformation. Yan et al.[33] used in situ Raman spectroscopy to monitor the quasihydrostatic and nonhydrostatic compression of a boron carbide single crystal up to 50 GPa, followed by depressurization to ambient pressure. Under quasihydrostatic compression, Raman analysis did not detect any signs of amorphization during either loading or unloading, and the material remained a perfect single crystal without any visible surface relief features or cracks. However, under highly nonhydrostatic compressive conditions (i.e., uniaxial compression), the results were significantly different, the pressurized sample having broken into a number of small fragments. In situ Raman spectroscopy detected the formation of the amorphous phase, indicating that a nonhydrostatic high-pressure state can make boron carbide unstable.

This compressive stress transformation has been investigated by simulating molecular dynamics.[34,35] Work by Yan et al. indicated a significant decrease in volume of the B_4C unit cell owing to the bending of the C–B–C chain at a destabilization pressure of 19 GPa for uniaxial compression, consistent with the HEL of 15-20 GPa. At higher pressures, the C–B–C chain bends until the central B atom bonds with neighboring B atoms in the surrounding icosahedra, forming a stable higher energy structure. It has been suggested that the release of this energy during depressurization is responsible for breaking the covalent bonds and for the collapse of the B_4C structure, with the formation of a local amorphous region.[36] A computational study of the phase stability of various boron carbide polytypes at elevated pressures was conducted by Fanchini et al.[37] under increasing purely hydrostatic pressure at room temperature. The results indicated that the energetic barrier for pressure-induced amorphization of boron carbide is lowest for the B_{12}(C–C–C) polytype, which was found to be unstable at 6-7 GPa during hydrostatic loading; however, no such collapse has been observed experimentally up to 40 GPa (see Yan et al.[38] and the references therein). Further, the model of chain bending under uniaxial

[32]Chen, M.W., J.W. McCauley, and K.J. Hemker. 2003. Shock-induced localized amorphization in boron carbide. Science 299(5612): 1563-1566.

[33]Yan, X.Q., Z. Tang, L. Zhang, J.J. Guo, C.Q. Jin, Y. Zhang, T. Goto, J.W. McCauley, and M.W. Chen. 2009. Depressurization amorphization of single-crystal boron carbide. Physical Review Letters 102(7): Article number 075505.

[34]Ibid.

[35]Fanchini, G., J.W. McCauley, and M. Chhowalla. 2006. Behavior of disordered boron carbide under stress. Physical Review Letters 97(6): Article number 035502.

[36]Yan, X.Q., Z. Tang, L. Zhang, J.J. Guo, C.Q. Jin, Y. Zhang, T. Goto, J.W. McCauley, and M.W. Chen. 2009. Depressurization amorphization of single-crystal boron carbide. Physical Review Letters 102(7): Article number 075505.

[37]Fanchini, G., J.W. McCauley, and M. Chhowalla. 2006. Behavior of disordered boron carbide under stress. Physical Review Letters 97(6): Article number 035502.

[38]Yan, X.Q., Z. Tang, L. Zhang, J.J. Guo, C.Q. Jin, Y. Zhang, T. Goto, J.W. McCauley, and M.W. Chen. 2009. Depressurization amorphization of single-crystal boron carbide. Physical Review Letters 102(7): Article number 075505.

compression proposed by Yan et al.[39] assumes transformation to another crystal structure in the loading stage, whereas in situ Raman analysis does not show any sign of such a crystal-crystal transformation.

One way to avoid amorphorization may be to avoid forming the B_{12}(C–C–C) polytype, which occurs as a minority phase during normal processing and sintering. This may be accomplished by doping. Al and Si are both able to substitute for C in B_4C. These dopants occupy sites in the diagonal chain in the rhombohedral B_4C structure. Moreover, it is known that Si addition strongly promotes the sp^3-C content in amorphous carbon materials, which may prevent the segregation of C into two-dimensional graphitic (sp^2) layers. Hence, the notion of significant Si doping to inhibit amorphorization depends on the ability to synthesize a material with stable $B_{12}SiC_2$ polytypes, avoiding B_{12} (C–C–C) formation. Unfortunately, the solubility of Si in boron carbide is quite low (~2.5 at% Si). There are some studies on the B–C–Si system that have explored higher Si concentrations (>20 at%) with the goal of developing useful SiC–B_4C composite materials for potential armor use. Thermodynamic calculations suggest the difference between the Gibbs free energy of the $B_{11}C_{1-\gamma}$, p-Si_γ, p(C–B–C) polytype and that of the most energetically favored minority polytype B_{12}(C–Si_γ–$C_{1-\gamma}$–C) increases with increased Si content. This suggests that if a solid solution of B_4C with Si or Al could be made, it might prove resistant to high-pressure amorphization, which could improve the ballistic performance of this important ceramic armor material. Clearly, further experimental and theoretical work is required to more fully understand the structural changes in boron carbide under impact loading.

Amorphization has limited the effectiveness of boron carbide to high-velocity threats. Modification of the crystal structure via the ternary alloying chemistry of boron carbide may inhibit amorphization. This would provide an armor material that is 25 percent less dense than SiC and 40 percent less dense than Al_2O_3.

Findings

Finding 5-1a. Additional ceramic compositions and structures merit investigation as potential new armor materials. For the currently available armor ceramics, the difficulties in powder synthesis, availability, and processing of the powders into dense ceramics mean that many opportunities for performance improvements remain unexplored, including the addition of alloying elements and variations in nanostructure and microstructure.

Finding 5-1b. There is a need for a fundamental understanding of the equilibrium phases and crystal structures of armor ceramics and for the construction of accurate equilibrium-phase diagrams for the B–C system at ambient pressure and at the pressures used for the manufacturing of the ceramics. Additionally, pressures corresponding to those encountered in ballistic and blast events should be explored to understand the nonequilibrium phase aspects of armor ceramics.

Finding 5-1c. Time-temperature-transformation and time-pressure-transformation diagrams need to be drawn using advanced instrumentation to provide a basic understanding of the kinetics of structural transformations of ceramic materials, in particular boron carbide.

Crystalline Structure of Silicon Carbide

Types and Characteristics

SiC is a simple 1:1 compound of two atoms that both prefer sp^3 bonding. Owing to the similarity of the tetrahedral bonding, SiC has a surprisingly wide variety of polytypes. Whereas many materials are polytypic to a limited extent (e.g., α-Al_2O_3, γ-Al_2O_3), the polytypism of SiC is extensive, with over 200 polytypes having been observed.[40,41,42] The basic unit is a tetrahedron of Si_4C or, equivalently, C_4Si; these are joined at the corners to other tetrahedra. The structure can be seen as invariant in the basal plane; the various polytypes are distinguished by the stacking sequence in the direction normal (c-axis) to the basal planes. An essentially infinite number of stacking sequences can be achieved by altering the number of layers before repeating the sequence. A number of notations have been developed; the most common notation, Ramsdell's, labels the polytypes as *n*L, where *n* is a number indicating the periodicity in the stacking of the tetrahedra layers along the c-axis and L is a letter indicating the general crystal symmetry. For example, 3C is indicative of cubic symmetry with a three-layer repeat. This is in fact the only cubic polytype for SiC and is designated as β-SiC. The most common polytypes—2H, 4H, and 6H—all have hexagonal symmetry. There is one common rhombohedral polytype, 15R, and countless other less common and more exotic combinations like 33R or 1,200R. All of the noncubic polytypes, although different, are grouped together and considered as α-SiC. Five common polytypes of SiC are shown in Figure 5-5.

While it is often simple to qualitatively discern the presence of a particular polytype in an x-ray diffraction pattern by finding certain characteristic peaks, overlapping peaks make it not nearly as straightforward to quantitatively determine all of the polytypes present in samples. Many researchers

[39]Ibid.

[40]Shaffer, P.T.B. 1969. A review of the structure of silicon carbide. Acta Crystallographica Section B: Structural Crystallography and Crystal Chemistry 25(3): 477-488.

[41]Mrotek, S.R. 1998. Microstructural Control of Silicon Carbide via Liquid Phase Sintering, Ph.D. Dissertation. Newark, N.J.: Rutgers University.

[42]Kaza, A. 2006. Effect of Gas Phase Composition in Pores During the Initial Stages of Sintering. Ph.D. Dissertation, Newark, N.J.: Rutgers University.

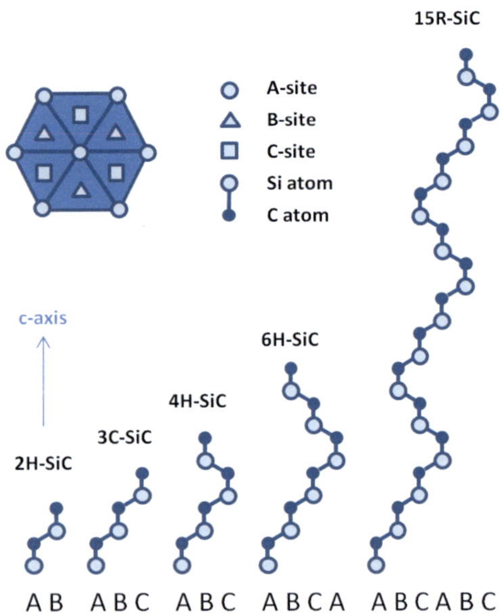

FIGURE 5-5 Schematics of the stacking sequence of layers of Si–C tetrahedra in various SiC polytypes.

measure only the α and β contents of their powder and often fail to be any more specific about the relative amounts of 2H, 4H, 6H, and others because considerable effort would be required. During densification at high temperature, a given polytype can transform into a more stable one, and this can be accompanied by desirable or undesirable grain growth, along with changes in porosity, which influence various properties and ballistic performance.[43,44,45,46] For example, it is common to improve the fracture toughness of SiC by exploiting the anisotropic grain growth that occurs when polytypes transform. The high sintering temperatures required for densifying SiC promote the transformation of β grains to α grains, which can become large, elongated platelet grains. By purposefully seeding an α-SiC powder with β grains before sintering, microstructures with improved fracture toughness can be designed by taking advantage of the increased crack paths around the elongated α grains. In other cases, the large α grains can act as detrimental flaws and decrease other mechanical properties.[47,48]

Impurities and intentional additives to SiC play an important role in the development and transformation of polytypes. As far back as 1948, Lundqvist[49] had observed that different polytypes were often associated with SiC crystals of varying colors in certain powders: 6H were green, 15R were yellow, and 4H grains or samples with mixtures of polytypes appeared black. Through careful x-ray examination of over 200 powders from a variety of locations, accompanied by spectrochemical analysis, large variations in aluminum content and smaller variations in iron content were observed. At very low Al contents, the 6H polytype appeared to be favored, whereas 0.05-0.06 wt percent Al promoted the formation of 15R, with a transition to 4H above 0.10 wt percent Al. Lundqvist also observed inclusions in the grains, most of which were unreacted graphite, noting few inclusions in the clearest to light green samples. In the darker and black samples, large inclusions, found to be compounds of aluminum and iron, were often present along with changes in the nearby crystal structure. In present practice, a wide range of other impurity elements and sintering-aid additions also exert considerable influence over the temperature at which the polytype transformations occur and the exact sequence of the transformations.[50,51,52,53,54]

As mentioned, densifying SiC at temperatures above 1900°C will cause any β grains to transform into various α polytypes, accompanied by rapid anisotropic grain growth.[55] However, if the initial material is instead an α powder, sintering at or above 1900°C will result in a fine, equiaxed α microstructure. Careful control over powder purity, sintering aids,

[43]Shaffer, P.T.B. 1969. A review of the structure of silicon carbide. Acta Crystallographica Section B: Structural Crystallography and Crystal Chemistry 25(3): 477-488.

[44]Pezoldt, J. 1995. Are polytype transitions possible during boron diffusion? Materials Science and Engineering B 29(1-3): 99-104.

[45]Jepps, N.W., and T.F. Page. 1981. The 6H→ 3C reverse transformation in silicon carbide compacts. Journal of the American Ceramic Society 64(12): C-177-178.

[46]Irmscher, K., M. Albrecht, M. Rossberg, H.-J. Rost, D. Siche, and G. Wagner. 2006. Formation and properties of stacking faults in nitrogen-doped 4H-SiC. Physica B: Condensed Matter: 338-341.

[47]Zhan, G.D., M. Mitomo, H. Tanaka, and Y.-W. Kim. 2000. Effect of annealing conditions on microstructural development and phase transformation in silicon carbide. Journal of the American Ceramic Society 83(6): 1369-1374.

[48]Zhan, G.-D., R.-J. Xie, M. Mitomo, Y.-K. Kim, and N.P. Padture. 2001. Effect of beta-to-alpha phase transformation on the microstructural development and mechanical properties of fine-grained silicon carbide ceramics. Journal of the American Ceramic Society 84(5): 945-950.

[49]Lundqvist, D. 1948. On the Crystal Structure of Silicon Carbide and Its Content of Impurities. Acta Chemica Scandinavica 2: 177-191.

[50]Rixecker, G., K. Biswas, A. Rosinus, S. Sharma, I. Wiedmann, and F. Aldinger. 2002. Fracture properties of SiC ceramics with oxynitride additives. Journal of the European Ceramic Society 22(14-15): 2669-2675.

[51]Biswas, K., G. Rixecker, and F. Aldinger. 2003. Improved high temperature properties of SiC-ceramics sintered with Lu_2O_3-containing additives. Journal of the European Ceramic Society 23(7): 1099-1104.

[52]Kim, J., A. Rosenflanz, and I.W. Chen. 2000. Microstructure control of in-situ-toughened α-SiAlON ceramics. Journal of the American Ceramic Society 83(7): 1819-1821.

[53]Kim, Y.-W., Y.-S. Chun, T. Nishimura, M. Mitomo, and Y.-H. Lee. 2007. High-temperature strength of silicon carbide ceramics sintered with rare-earth oxide and aluminum nitride. Acta Materialia 55(2): 727-736.

[54]Kim, W., Y.-W. Kim, and D.-H. Cho. 1998. Texture and fracture toughness anisotropy in silicon carbide. Journal of the American Ceramic Society 81(6): 1669-1672.

[55]Pezoldt, J. 1995. Are polytype transitions possible during boron diffusion? Materials Science and Engineering B 29(1-3): 99-104.

and processing is clearly required in order to systematically modify the microstructure and polytypes in silicon carbide.

Stacking Faults

In addition to the various long repeat sequences that constitute a particular polytype, a localized change in stacking sequence within any specific grain is a type of stacking fault. For example, a specific grain of 6H could contain local regions where the stacking sequence has changed to 4H for a few layers and then back to 6H. An understanding of stacking faults and their connection with plastic deformation behavior has come about in the study of metals over many years. The process of slip on a close-packed plane can produce the same shift in stacking sequence for a number of layers in a crystal; this shifted region is known as a deformation stacking fault but is structurally identical to a growth stacking fault.

The stacking fault can be described as an extended dislocation that is bounded by two partial dislocations. Like all imperfections, the stacking fault has an energy associated with its creation that can differ greatly between materials. Materials with low stacking-fault energy readily form many stacking faults and have large separations between the bounding partial dislocations. Materials with high stacking-fault energy require more energy for their creation and therefore form fewer and narrower, smaller faults. Silicon carbide has low stacking-fault energy, and it is not uncommon to find many growth stacking faults present throughout the crystals. Fragments from ballistic impact experiments do indeed show a considerable amount of stacking faults and twins,[56,57] suggesting that materials with low stacking-fault energy twin readily under shock loading also, because the presence of large numbers of stacking faults provides locations at which twins form easily.[58]

There are a very large number of crystal structures for SiC differing by the stacking sequence of tetrahedral Si_4C or C_4Si units, and the identification and characterization of the polytypes is laborious. The phase content depends on variations in chemical impurities and sample process history.

Because well-defined SiC single crystals are available from the electronics industry, an improved understanding of the deformation of a particular polytype can be conducted. Additionally, the effect of the amount of each polytype and its spatial and size distribution within model polycrystalline materials merits investigation, especially the effect on high-rate behavior. Reducing the activation energy for stacking fault glide by purposeful alloying may provide an opportunity to enhance plasticity and energy absorption.

Availability of Ceramic Powders

Synthesis and processing of armor ceramics begins with ceramic powders, which are compacted and processed using a variety of techniques. The important issue of powder availability is discussed in this subsection. Appendix D further characterizes the current understanding of powder production for the protection materials of interest, including SiC, B_4C, Al_2O_3, AlN, AlON, and spinel, and suggests opportunities to improve the situation.

It is difficult, if not impossible, to fabricate high-quality ceramic components without having control of the powders comprising them. The U.S.-based companies that supply many strategic ceramic components have seen a loss of domestic powder suppliers over the past two decades. Moreover, many critical armor systems rely on unique, highly specified powders for the hard ceramics. Applications ranging from armor for personnel or vehicles to high-intensity mirrors to missile radomes to rocket nozzles rely on powders coming from India, China, and Russia.

There is no powder manufacturer in the United States capable of producing the armor-grade ceramic powders needed by armor manufacturers. Nearly all oxide and carbide powders on the market have been engineered to satisfy the requirements of applications other than armor. As a consequence, ceramic armor manufacturers and university researchers are forced to employ powders that are almost certainly not optimal for armor applications. Beyond impeding research and development generally and, particularly, the development of better protection materials, the precariousness of domestic supply poses a risk for DoD should a need arise for surge production of ceramic armor materials.

The consequence of this eroded domestic supply base has been the inability of component manufacturers to design powders for a specific application. Instead, domestic producers sort or modify highly variable commodity powders of non-U.S. origin to impart the requisite "uniqueness" for an application. This is a problem for a host of powders: those for opaque armor (SiC, B_4C, AlN) and those for transparent armor (MgO-Al_2O_3 [spinel] and AlON). In many cases, lower cost, less highly specified end uses, such as abrasive grain, have given rise to a proliferation of new powder suppliers in the emerging nations. In most cases, the foreign supply chain links many small powder producers with a handful of brokers, virtually eliminating the production of tailored powders and lowering quality.

Furnace reactors were once large-scale operations; now, small producers can introduce highly variable product into a distribution stream. Precursor raw materials are also a problem. For example, for silicon carbide and boron carbide, carbon used to be obtained from high-grade, petroleum-derived coke. However, in China it is not uncommon to

[56]Shih, C.J., M.A. Meyers, V.F. Nesterenko, and S.J. Chen. 2000. Damage evolution in dynamic deformation of silicon carbide. Acta Materialia 48(9): 2399-2420.

[57]Chen, M.W., J.W. McCauley, D.P. Dandekar, and N.K. Bourne. 2006. Dynamic plasticity and failure of high-purity alumina under shock loading. Nature Materials 5(8): 614–618.

[58]Murr, L.E. 1987. Metallurgical effects of shock and high-strain-rate loading. Pp. 1-45 in Materials at High Strain Rates. T.Z. Blazynski, ed. New York, N.Y.: Elsevier Science.

see anthracite coal or low-purity petroleum coke used. The consequence is an end product whose chemistry is highly variable. Component suppliers are now faced with how to make a consistent product meeting today's armor specifications. Improving ceramic performance can no longer entail simply changing the initial powder since the production and supply of powders are no longer within domestic control.

By losing control of powder processing, U.S. armor makers have reached a point at which variability in powders is expected, tolerated, and, in many cases, ignored. While processing treatments have been developed to improve the overall uniformity of powders, this results in dense components whose microscale variability reflects the intrinsic variability of the parent powder. From a simple business or logistical point of view, manufacturers can no longer assure that the powders used in highly specified components will meet strict testing requirements.

The Defense Production Act Title III program gives DoD special authority to issue purchase commitments, loan guarantees, capital investment, or research and development investment to provide an assured domestic supply for critical materials. The business case analysis to support a Title III program in ceramic armor materials is beyond the scope of this study. The committee recommends DoD undertake such an analysis to determine whether domestic production of ceramic armor precursor materials would be a good candidate for Title III.

Finding 5-2a. The goal for future armor systems is not only to maintain current performance but to dramatically increase it as well. As such, it is critical that the United States regain and maintain control of the armor raw material supply chain. There is a need for a strategic powder production infrastructure within the United States to bring about the next generation of opaque and transparent armors. This will not only permit a consistent and reliable supply but also allow for the design of powders whose intrinsic properties are optimized for armor applications.

Finding 5-2b. Powder processing affects the intrinsic properties of many armor ceramics. There is little work on how the powders can be designed and manipulated at the atomic, nano, and micro levels in ways to maximize their potential as raw armor materials.

Finding 5-2c. There are no powders produced specifically for armor applications. The oxide and carbide powders that are commercially available have been designed for other applications. Most powder processes are energy-intensive processes with large carbon emission footprints, and U.S. environmental regulation costs have reduced the competiveness of U.S. producers, with foreign powder producers benefiting from low-cost but environmentally questionable operations. There is no domestic feedback on powder characterization to assist ceramic producers in researching or producing new prototype powders.

Finding 5-2d. Although the availability of high-quality ceramic powders for protection materials is critical to national defense, there is currently no domestic source of ceramic powders to meet DoD needs.

PROCESSING AND FABRICATION TECHNIQUES FOR ARMOR CERAMICS

A variety of fabrication techniques have been employed in the processing of armor ceramic materials. There are two broad classes of forming operations: (1) cold methods—slip casting, extrusion, and die pressing—and (2) high-temperature pressureless and pressure-assisted sintering methods—hot pressing, hot isostatic pressing, and spark plasma sintering (see Box 5-1).

Since armor materials are mostly strongly covalently bonded solids, high-temperature densification, often with pressure-assisted techniques, is required. The goal of densification is to optimize bonding and eliminate porosity in the compacted powder so that full theoretical densities, along with homogeneous microstructures, can be achieved in the final sintered materials. Near-net-shape fabrication that minimizes machining and finishing operations is also desired for cost savings.

"Green" Compaction

The starting point in ceramic forming is the compaction of powders. Die pressing is the predominant forming method for symmetrical shapes such as hexagonal and square tiles. High-pressure compaction methods can be divided into

BOX 5-1
Processing of Ceramic Powders

Hot-pressed SiC and B4C powders yield uniform full-density products with homogeneous microstructures and good ballistic performance. Near-net-shape ceramic processing is of great interest, although powders with additives and sintering aids compacted by means of lower cost conventional pressing methods into "green," or unfired, form and then pressureless-sintered (that is, at atmospheric pressure) yield materials with nonuniform density distribution and microstructure. Their ballistic performance is inferior for higher threat levels compared to that of hot-pressed material.

static and dynamic techniques.[59] With static compaction, a constant pressure is applied onto a sample for a certain period of time, typically a few seconds. Dynamic compaction uses a pressure pulse with a pulse duration of less than a few milliseconds, resulting in a pressure wave that travels through the sample. In both static and dynamic compaction the pressure can be applied in a uniaxial, biaxial, radial, or isostatic/isodynamic mode.

The choice of a process for compacting powders for the fabrication of ceramics depends on the complexity of the shape of the ceramic part.[60] The most widely used forming method for armor production is uniaxial die pressing, whereby uniaxial pressure is applied to the powder placed in a die between two rigid punches. Binder and/or lubricants are added to the powders to reduce the friction and facilitate extracting the formed part from the die. This formed part, often termed a "green" compact because it is unfired, is subsequently heat treated ("sintered") to densify it.

The typical density of the parts achieved after uniaxial die pressing is 50-55 percent of the theoretical density. Density gradients occur depending on the part's shape, aspect ratio, and size. These gradients are a likely source of voids and undesirable porosity in sintered armor tiles. Other defects in laminar character can appear oriented normal to the pressing axis. After die pressing, the part will have shrunk by 20 to 40 percent or so, and the final part dimensions are achieved by machining and grinding.

Uniaxial die pressing is widely used for the low-cost mass production of simple parts. In certain cases, cold isostatic pressing (CIP) is used to further increase the density (up to 73 percent) after die compaction. CIP is conducted as wet bag isostatic pressing in pressure vessels, and parts can be produced as large as a few meters in height and a meter or more in diameter, with large parts having substantially higher costs. Hydrostatic pressures of 100-700 MPa can be achieved with suitable CIP systems.[61]

Dynamic compaction approaches are potential alternatives for making near-net-shape parts with very high "green" densities (up to 95-100 percent). Dynamic compaction depends on the way the pressure waves needed to densify the sample are generated and how the reflected waves are absorbed. One of the best-known methods is compaction using explosives. However, this method would be problematic as an industrial manufacturing process in a factory environment. The alternative dynamic magnetic compaction (DMC) technique uses magnetic pulse pressures and is suitable for a factory environment. The DMC rapid consolidation technique developed by IAP Research, Inc., is based on a magnetic pulse that launches a pressure wave that travels at 100-300 m/s through the powders, giving rise to stress gradients; the technique is designed so as to absorb reflected waves.[62] The stress gradients cause particle motion, particle deformation, and particle fracture, especially in brittle powder materials at high pressure; accordingly, they bring about a higher degree of consolidation than static pressing.[63] Very high green densities of the compacts can be realized—in fact, they approach theoretical densities even before sintering. Because DMC samples have higher compact densities they can be sintered at lower temperatures or for shorter periods of time to obtain close-to-full-density materials. The dynamically pressed samples exhibit rather homogeneous microstructures after pressureless sintering and properties similar to those of hot-pressed material. In addition, dynamic processing techniques allow retention of special powder microstructures (including nano grain size) after sintering owing to the short sintering time and lower sintering temperatures. In light of its advantages, dynamic compaction needs to be seriously investigated for armor production methods.

Sintering

Appendix E characterizes commonly used ceramic sintering processes and discusses issues surrounding their application to opaque armor materials. The advantages and disadvantages of these processes are summarized in Table 5-1.

The effect of specially designed powder microstructures, such as nano grain sizes, on the controlled fracture to enhance ballistic performance is being investigated by the Army Research Laboratory (ARL) and other laboratories. Ceramic manufacturers are also exploring ways to improve performance through modifications to the front surface of a ceramic armor plate—in one case by molding multiple nodes with conical or rounded shapes. By modifying the impact angle of the projectile, the ballistic performance of ceramics could be improved. Ceramic nodes, spheres, or hollow ceramic spheres give the structure a multiplicity of surfaces for a multiplicity of crack initiation sites.[64] These nodes cause part of the energy of the projectile to initiate a multiplicity of cracks at the node surface; however, spherical nodes arrest cracks. Other candidates for exploring the improvement of performance include novel alloying and doping methods.

[59]Jak, Michiel J.G. 2004. Dynamic compaction of nano-structured ceramics. Nanocomposites. Volume 10 in Electronics Materials Science & Technology. Springer.

[60]Tressler, R.E. 2004. An assessment of low cost manufacturing technology for advanced structural ceramics and its impact on ceramic armor. Pp. 451-462 in Progress in Ceramic Armor. New York, N.Y.: John Wiley & Sons.

[61]Nishimura, T., K. Jinbo, Y. Matsuo, and S. Kimura. 1990. Forming of ceramic powders by cyclic-CIP: Effect of bias pressure. Journal of the Ceramic Society of Japan 98(7): 735-738.

[62]Chelluri, B., E. Knoth, E. Schumacher, and L.P. Franks. 2010. Method for Producing SiC armor tiles of higher performance at lower cost. Pp. 199-205 in Advances in Ceramic Armor VI: Ceramic Engineering and Science Proceedings, Volume 31, Issue 5. J.J. Swab, ed. Hoboken, N.J.: John Wiley & Sons.

[63]National Research Council. 1983. Dynamic Compaction of Metal and Ceramic Powders. Washington, D.C.: National Academy Press.

[64]Medvedovski, E. 2010. Ballistic performance of armor ceramics: Influence of design and structure. Part 2. Ceramics International 36(7): 2117-2127.

TABLE 5-1 Manufacturing Processes for Opaque Ceramic Armor Materials

Process	Material	Advantages	Disadvantages
Hot pressing	Ceralloy B_4C Norbide B_4C Ceralloy SiC, SiC–N, TiB_2	Lower temperature, lowest porosity	Shape limitation
Solid-state sintering (SSS) or pressureless sintering	Hexoloy SiC Purbide SiC (MCT/SSS) SiC	No grain boundary phase, low porosity	Higher temperature, grain coarsening
Liquid-phase sintering (LPS)	Ekasic-T (MCT LPS) SiC	Lower temperature, fine grains, low porosity	Oxide grain boundary phase
Reaction bonding	Si/SiC, Si/B_4C (MCT/RBSC, RBBC)	Low temperature, excellent complex shape capability	Residual silicon

NOTE: MCT, M Cubed Technologies, Inc., RBBC, reaction-bonded boron carbide; RBSC, reaction-bonded SiC.
SOURCE: Karandikar, P.G., G. Evans, S. Wong, M.K. Aghajanian, and M. Sennett. 2009. A Review of Ceramics for Armor Applications. Ceramic Engineering and Science Proceedings 29(6): 163-175.

Finding 5-3. Refractory ceramics such as SiC and B_4C require very high sintering temperatures (>2150°C-2200°C) for long periods (more than 2 hours of dwell time) and the use of sintering aids to obtain full density by solid-state sintering or liquid-phase sintering. Neither are the hot pressing or the high-temperature, pressureless sintering methods satisfactory for processing powders with special microstructures (including nano grain sizes) because they induce grain growth. Fast, high-density compaction techniques, coupled with low-temperature sintering methods (including spark plasma sintering), are therefore needed to permit the retention of specially designed initial powder microstructures.

TRANSPARENT ARMOR

Infrared domes, lenses, reconnaissance windows, and windows on military vehicles must be transparent to radiation of certain wavelengths and must also resist damage from airborne debris and penetration from projectiles. They should retain a degree of transparency after a hit and should withstand multiple hits.

Glass is the traditional window material. Glass windows can be produced in large sizes by the relatively inexpensive float glass process. Most windows consist of multiple layers of glass, usually made of soda lime or borosilicate, adhesively bonded to one another and backed by a polycarbonate layer. Care is taken to avoid flaws on the layer surfaces. Glasses can be toughened with thermal or chemical treatments that induce compressive stresses on their surfaces.

Laminated glass windows can resist penetration by certain threats, but at an areal density of 50 to 55 lb/ft^2 they impose a severe weight penalty. Current Humvee windows are about 4 in. thick and weigh about 90 lb each. The six windows on a typical vehicle thus weigh as much as several soldiers. Window weight contributes to worn-out transmissions and suspension systems, which in turn cause military vehicles to be taken out of service for repair. This underlines the motivation to innovate lighter-weight window materials and window structures.

Glass ceramics are glass-based materials in which a dense population of ceramic nanocrystallites is embedded in the amorphous silica matrix (Figure 5-6). These materials were originally developed as zero-coefficient-of-expansion materials for use in cook tops and fireplace screen windows. Because they were found to have ballistic performance tens of percent better than the baseline glasses, these glass ceramics are beginning to replace traditional window glass on military vehicles. Many chemistries and processing routes have been explored to achieve glass ceramics with even better ballistic performance while maintaining transparency (nanocrystallite size must be kept at less than ~0.1 × the wavelength of light). Laminated glass ceramic windows are currently being installed in military vehicles.

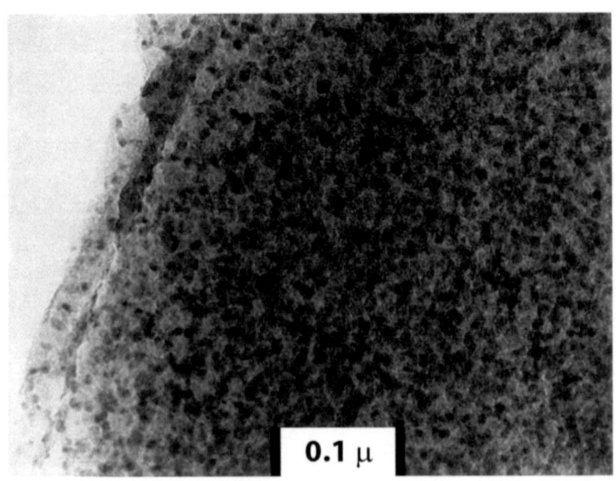

FIGURE 5-6 Scanning TEM micrograph of the microstructure of spinel glass ceramic. Shown is the uniform dispersion of the dark 10-20 nm spinel crystals throughout the lighter continuous, highly siliceous glass matrix. SOURCE: Pinckney, L.R., and G.H. Beall. 2008. Microstructural evolution in some silicate glass-ceramics: A review. Journal of the American Ceramic Society 91(3): 773-779.

TRANSPARENT CRYSTALLINE CERAMICS

Three candidates for transparent armor—aluminum oxynitride ($AlN_x{\cdot}(Al_2O_3)_{1-x}$), known as AlON; spinel (MgAl2O4); and sapphire (Al_2O_3)—are harder, stronger, and tougher than soda lime and borosilicate glass, and they have been shown to provide protection against armor-piercing rounds at roughly half the weight and thickness of conventional glass laminates. However, the materials are quite expensive compared to glass and they are not available in large quantities. Their high cost and low production volume prevent their widespread use in armor material and currently limit their application to strike-face materials.

AlON is a polycrystalline, large-grained (200 μ) ceramic material formed from a solid solution of Al_2O_3 and AlN. This solution is stable over a wide range of mixture ratios centered at $9Al_2O_3$–5AlN (35.7 mol percent AlN), has the chemical formula $Al_{23}O_{27}N_5$, and exhibits an ambient density of 3.67 g/cm^3. Because of its cubic crystalline structure, AlON is optically isotropic and therefore transparent, even in polycrystalline form. Conventional powder processing techniques have been used to produce large (17 × 34-in.) plates of high optical quality. A tiling approach to the production of larger transparent armor windows using AlON tiles with dimensions between 12 × 12 in. and 14 × 20 in. is in development.[65,66,67]

Producing a nanocrystalline transparent ceramic may offer improved mechanical behavior. Several nontransition metal oxides and oxynitrides, including Al_2O_3 and AlON, can be produced as nanocrystalline ceramics. If the ceramic is transparent as a micron-scale polycrystal or single crystal, it will maintain that transparency when it is produced in the nanocrystalline form, provided there is no porosity at the 0.5 μ scale and above. However, B_4C, SiC, and Si_3N_4 often contain carbon impurities, which can lead to loss of transparency.

Spinel is a stoichiometric compound of magnesium and aluminum oxides, $MgAl_2O_4$. Fine-grained (micron or submicron grains) spinel has high transparency and hardness and good ballistic resistance (Figure 5-7). Exceptionally fine-grained material (grain size as small as 0.6 μ) can be obtained by hot pressing and subsequent heat treatments. Novel processing methods, such as spark plasma sintering, are also being investigated to achieve high mechanical strength and hardness without sacrificing optical transparency. Prototype 11- × 14-in. armor windows have been successfully fabricated and delivered to the Army. Larger mosaic windows with minimum detectability of the seam have also been produced.[68,69]

Sapphire is a single-crystal alumina (Al_2O_3). Although sapphire has a density nearly double that of conventional window glass, its superior performance allows equivalent ballistic protection at a reduction in system weight and thickness of about 40 percent. Saint-Gobain Crystals, the sole supplier of integrated ceramic transparent armor in the world, currently produces plates up to 9 × 26 in. and 12 × 24 in.[70,71,72] Since sapphire is grown from the melt in individual crystal growers, plates are produced one by one and are expensive. Smaller plates, which are less expensive to produce, can be seamed together to achieve larger windows.

Finding 5-4a. Transparent crystalline ceramics are harder, stronger, and tougher than glasses and glass ceramics and have much better penetration resistance. Transparent ceramics could most likely meet the Army's requirements for lightweight protective windows. However, they are expensive and are not available in the quantities and sizes needed to replace existing vehicle windows. The cost of transparent ceramics might be reduced by identifying less expensive sources of powders; improving powder processing procedures, fabrication, and finishing; increasing the production volume by identifying and developing secondary markets; and advancing seaming technologies that enable large windows to be produced by joining smaller tiles.

Finding 5-4b. It could be productive to explore transparent crystalline ceramics with different chemistries and processing methods and microstructures, including nanocrystalline ceramics, to achieve control over fracture and fragmentation behavior while maintaining transparency.

Finding 5-4c. Composite windows made of three materials—glasses, glass ceramics, and transparent crystalline ceramics—represent a viable trade-off between cost and performance. Superior laminated armor configurations made of transparent glasses, glass ceramics, crystalline ceramics, and polymers for the front (strike), back, and intermediate

[65]McCauley, J.W., P. Patel, M. Chen, G. Glide, E. Strassburger, B. Paliwal, K.T. Ramesh, and D.P. Dandekar. 2009. AlON: A brief history of its emergence and evolution. Journal of the European Ceramic Society 29(2): 223-236.

[66]Goldman, L.M., R. Foti, M. Smith, U. Kashalikar, and S. Sastri. 2009. AlON transparent armor. Pp. 225-232 in Advances in Ceramic Armor V, Volume 30, Issue 5. J. Swab, ed. Hoboken, N.J.: John Wiley & Sons.

[67]Goldman, L.M., R. Twedt, R. Foti, M. Smith, and S.A. Sastri. 2009. Large area AlON windows for reconnaissance and armor applications. Paper 7302 06 in Window and Dome Technologies and Materials XI, Proceedings of SPIE Volume 7302. R.W. Tustison, ed. Bellingham, Wash.: Society of Photo-Optical Instrumentation Engineers.

[68]Krell, A., J. Klimke, and T. Hutzler. 2009. Advanced spinel and submicron Al_2O_3 for transparent armor applications. Journal of the European Ceramic Society 29(2): 275-281.

[69]Krell, A. 2009. Ballistic strength of opaque and transparent armor. American Ceramic Society Bulletin 86(4): 9201-9207.

[70]Rioux, J., C. Jones, M. Mandelartz, and V. Pluen. 2007. Transparent armor. Advanced Materials and Processes 165(10): 31-33.

[71]Jones, C.D., J.B. Rioux, J.W. Locher, V. Pluen, and M. Mandelartz. 2009. Ballistic performance of commercially available Saint-Gobain sapphire transparent armor composites. Pp. 113-125 in Advances in Ceramic Armor III: Ceramic and Engineering Science Proceedings, Volume 28, Issue 5. L.P. Franks, J. Salem, and D. Zhu, eds. Hoboken, N.J.: John Wiley & Sons.

[72]See the Saint-Gobain Crystals's sapphire substrates Web site, http://www.photonic.saint-gobain.com/sapphire-substrates.aspx.

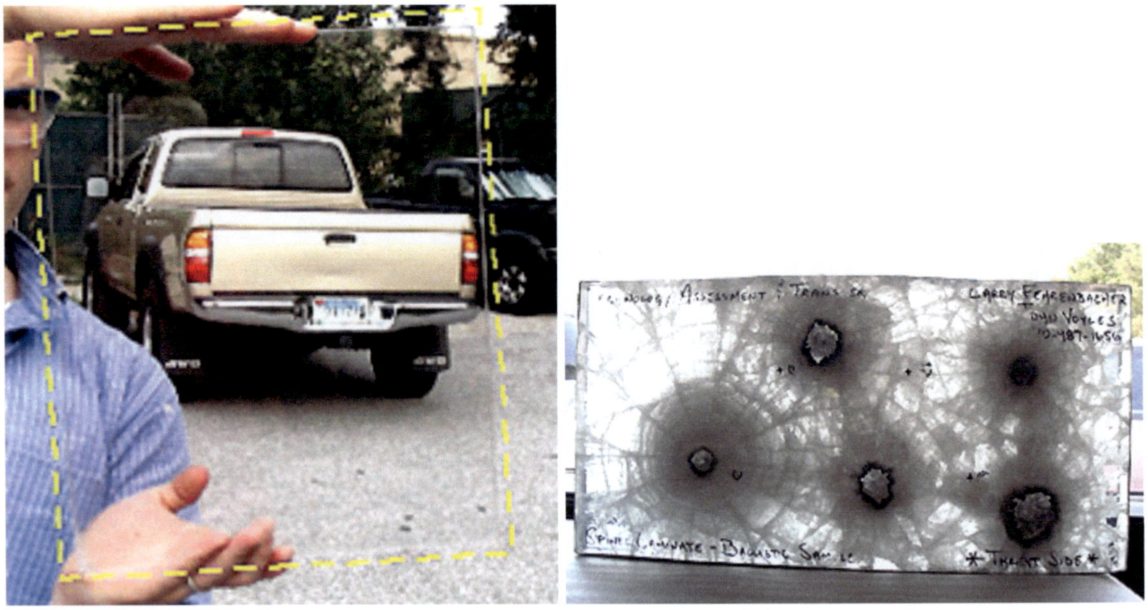

FIGURE 5-7 Photo showing the transparency (left) and multi-hit performance (right) of spinel. SOURCE: Spinel and Optical Ceramics-Armor. Undated. Available online at www.techassess.com/tech/spinel/spinel_armor.htm. Last accessed April 8, 2011.

plates should be identified by computationally simulating projectile impacts in which the number, thickness, location, order, and so forth of the plates are varied.

FIBERS

The field of high-performance fibers is only about 50 years old. This section briefly reviews the history of their development and current production technology and then discusses opportunities for technological innovations relevant to protection materials. Nylon and silk fibers had been used to make armor vests for soldiers, but with very limited success. Nylon was invented at DuPont and commercialized in 1939. In the 1960s, DuPont developed polyparaphenylene terephthalamide (PPTA), a much stiffer semirigid rod molecule that resulted in a liquid crystalline spinning solution and produced revolutionary structured fibers of very high crystallinity. When these fibers were woven into yarns and the yarns into a flexible multi-ply fabric, the resulting material, now known as Kevlar, was able to stop a bullet.

Typical properties of selected fibers and some high-performance fibers are given in Table 5-2; their specific strength is plotted against specific stiffness in Figure 5-8. Figure 5-9 schematically depicts the molecular structures of a typical textile fiber (e.g., polyethylene terephthalate or nylon with ~10 µ diameter or other larger diameter fibers); current high-performance polymeric fibers such as gel-spun polyethylene, with folding and entanglement, and semirigid-rod polymers like PPTA; and the ideal fiber made up of either polymer molecules or carbon nanotubes (CNT).

Commodity textile fibers contain significant amounts of amorphous-phase (50 percent) and chain-folded crystals, resulting in typical strengths of 0.5-1.0 GPa. The ability to highly orient macromolecules and to form extended chain crystals with a high degree of crystallinity creates high-performance fibers, whose strength is typically an order of magnitude larger (up to 6 GPa). Realizing that the ideal fiber structure is as shown in the right-hand panel of Figure 5-9 allows straightforward estimation of its theoretical properties based on the strength of the bonds in the chain and the cross-sectional area of the molecule.[73,74,75] The predicted tensile strength of a perfect, fully extended polymeric fiber such as polyethylene is about 30 GPa and that of a perfectly packed, single-wall CNT fiber of ideal diameter should be about 150 GPa. Achieving fibers that approach the predicted theoretical strength will require the removal of any voids, foreign particles, and chain entanglements. In addition, a polymer's molecular weight—that is, the polymer chain or CNT length—also plays a role in governing tensile strength since the number of chain end defects is inversely proportional to the molecular weight. Thus, synthetic methods that make it possible to also increase the polymer chain or CNT length will also have to be developed to narrow the gap between the current tensile strength and the theoretical limit.

Appendix F presents a brief review of high-performance fibers, including the following:

[73]Elices, M., and J. Llorca. 2002. Fiber Fracture. Amsterdam: Elsevier.

[74]Dumitrica, T., M. Hua, and B.I. Yakobson. 2006. Symmetry-, time-, and temperature-dependent strength of carbon nanotubes. Proceedings of the National Academy of Sciences of the United States of America 103(16): 6105-6109.

[75]Kelly, A., and N.H. MacMillan. 1986. Strong Solids, third edition. Oxford, England: Clarendon Press.

TABLE 5-2 Typical Properties of Selected Fibers

Fiber	Density (g/cm³)	Modulus (GPa)	Tensile strength (GPa)	Compressive strength (GPa)	Strain-to-failure (percent)
Polymeric fibers					
Nylon 66	1.14	4.3,[a] 1.2,[b] 0.4[c]			25
Silk	1.36	30-60	1.1-2.9		7-12
Kevlar 49	1.45	125,[a] 2.5,[b] 1.4[c]	3.5[a]	0.4,[a] 0.06[b]	2.6-4.2
Kevlar 149	1.47	185,[a] 2.5,[b] 1.2[c]	3.4[a]	0.4,[a] 0.07[b]	
Spectra 1000		89,[a] 1.2,[b] 0.18[c]	2.4-3.4	0.2[a]	2.8-3.0
Zylon HM	1.56	270,[a] 1.0[c]	5.8[a]	0.3[a]	2.5
M5 (PIPD)	1.70	270[a]	>4.0[a]		>1.4
Vectran	1.47	65[a]	2.9[a]		3.3
Carbon fibers					
Pitch based (P-100)	2.15	758,[a] 4.1,[b] 4.7[c]	2.41[a]	0.5,[a] 0.13[b]	0.3
Pitch based K-1100	2.2	965[a]	3.10[a]		
PAN based (T-300)	1.79	230,[a] 6.0,[b] 15.0[c]	3.75[a]	~3.0,[a] 2.7[b]	
PAN based (T-800)	1.8	300[a]	5.6[a]	~3.0[a]	
Ceramic and glass fibers					
Alumina (Al_2O_3)	3.7	350,[a] 12-26[b]	1.7[a]	6.9,[a] 2.3[b]	
Boron	2.5	415[a]	3.5[a]	5.0[a]	
SiC (Nicalon)	2.8	200[a]	2.8[a]	3.1[a]	
SiC (CVD)	3.0	400[a]	3.4[a]		
E glass	2.58	76,[a] 68,[b] 38[c]	3.4[a]	4.2,[a] 2.7[b]	2
S glass	2.46	90[a]	4.5[a]		
Alumina borosilicate (Nextel 440)	3.05	186[a]	2.1[a]		
Steel	7.8	200[a]	2.8[a]		1.4

NOTE: HM, high-modulus; PIPD, poly[2,6-diimidazo(4,5-b-4′,5′-e)pyridinylene-1,4(2,5-dihydroxy)phenylene]; PAN, polyacrylonitrile; CVD, chemical vapor deposition.

[a]Longitudinal.

[b]Transversal.

[c]Shear.

SOURCE: Warner, S.B. 1995. Fiber Science. Upper Saddle River, N.J.: Prentice-Hall; Minus, M., and S. Kumar. 2005. The processing, properties, and structure of carbon. JOM 57(2): 52-59; Kozey, V.V., H. Jiang, V.R. Mehta, and S. Kumar. 1995. Compressive behavior of materials 2: High-performance fibers. Journal of Materials Research 10(4): 1044-1061.

- Semirigid-rod PPTA, polybenzoxazole, and poly(pyridobisimidazole) fibers (e.g., Kevlar, Twaron, Technora, Zylon, and M5),
- Polyethylene (Spectra, Dyneema),
- Thermotropic liquid crystalline polymeric fibers (Vectran),
- Carbon fibers,[76,77,78]
- CNT fibers, and
- Alumina, boron, silicon carbide, glass, and alumina borosilicate ceramic fibers.[79,80,81]

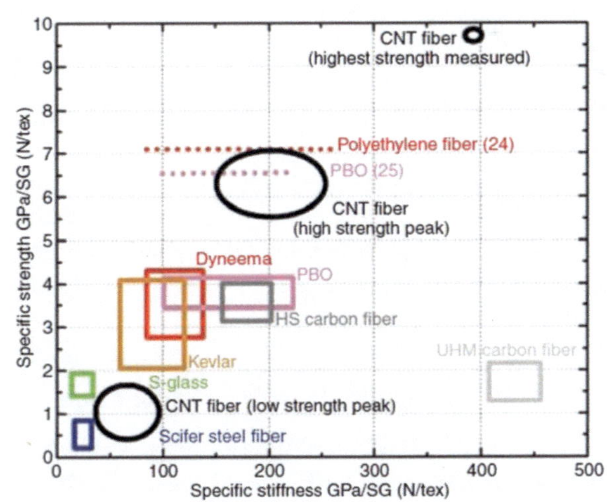

FIGURE 5-8 Strength and stiffness of the strongest fiber sample and of fibers typical of the high-strength and low-strength peaks in the 1-mm gauge length distribution versus the properties of other commercially available, high-performance fibers. Two laboratory observations of higher strengths in commercialized systems are also included (reference numbers are shown). SOURCE: Koziol, K., J. Vilatela, A. Moisaa, M. Motta, P. Cunniff, M. Sennett, and A. Windlel. 2007. High-performance carbon nanotube fiber. Science 318(5858): 1892-1895.

[76]Donnet, J.-B., T.K., Wang, S. Rebouillat, and J.C.M. Peng, editors. 1998. Carbon Fibers, third edition. New York, N.Y.: Marcel Dekker.

[77]Peebles, L.H.. 1995. Carbon Fibers: Formation, Structure, and Properties. Boca Raton, Fla.: CRC Press.

[78]Minus, M., S. Kumar. 2005. The processing, properties, and structure of carbon. JOM 57(2): 52-59.

[79]Elices, M., and J. Llorca. 2002. Fiber Fracture. Oxford, England: Elsevier Science, Ltd.

[80]Chawla, K.K. 1998. Fibrous Materials. Cambridge, England: Cambridge University Press.

[81]Watt, W.W., and B.V. Perov. 1985. Strong Fibers. Amsterdam: Elsevier Science Publishers.

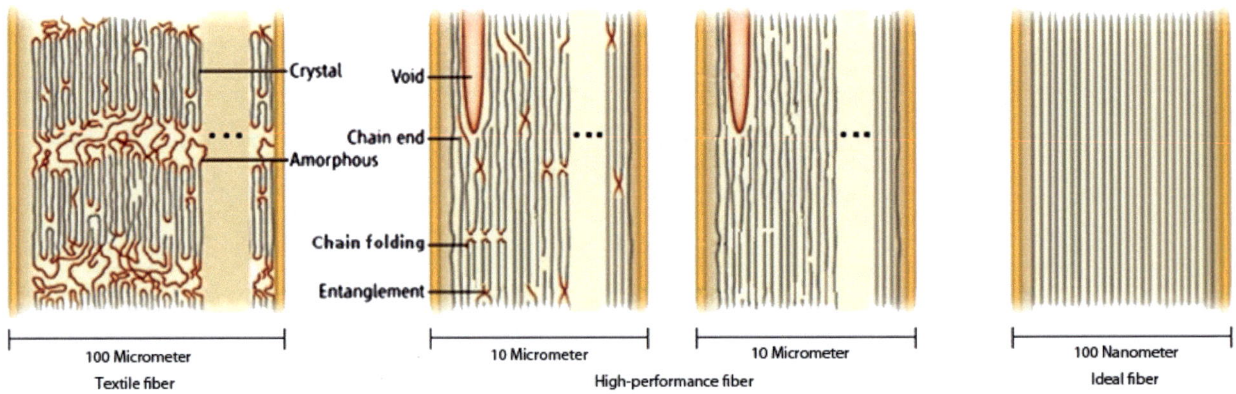

FIGURE 5-9 Schematic of transverse sections of fibers. Textile fibers are large diameter (~100 μ) with a partial crystalline structure (left); high-performance fibers are around 10 μ in diameter and feature more extended chains, leading to higher strength and modulus, but still contain many defects (center), whereas the ideal fiber would have a much smaller diameter (~100 nm) and be essentially defect free (right). SOURCE: Modified from Chae, H.G., and S. Kumar. 2008. Materials science: Making strong fibers. Science 319(5865): 908-909. Reprinted with permission from AAAS.

Effect of Fiber Diameter on Strength in High-Performance Fibers

Fiber tensile strength increases with decreasing fiber diameter. This has been demonstrated for polymeric fibers, carbon fibers, and ceramic and glass fibers. The current commercial carbon fibers range in diameter from 4 μ to 10 μ; for polymeric and most ceramic and glass fibers, diameters are in the range of 10 μ to 15 μ. Fibers processed by chemical vapor deposition, such as boron fibers, tend to have much larger diameters, typically 100-150 μ. The probability of finding defects decreases with decreasing fiber diameter. Developing new processing technologies for the economical production of smaller diameter fibers (1 μ or less) that also provides good control of fiber drawability is expected to significantly improve fiber tensile strength. Additionally, processing to achieve hollow fibers with a relatively thin wall (less than 1 μ) may reduce tensile strength just as it reduces the overall fiber diameter. Processing techniques for this new fiber class are actively under investigation.

Relating Tensile Properties to Ballistic Performance

The development of a new fiber material and a process to manufacture it typically costs several hundred million dollars. This presents a very high barrier for new materials and new process entry in the marketplace. It would be beneficial to be able to forecast ballistic performance from simple fiber tensile strength data obtainable from spinning small amounts of fiber. Further development of the Cuniff equation/model could be a step in that direction. Rigorous testing, evaluation, and refinement of this and other fiber ballistic models[82] could allow predicting fabric ballistic performance at a much earlier stage.

Finding 5-5. Rigorous testing, evaluation, and refinement of fiber ballistic models could allow predicting fabric ballistic performance at a much earlier stage.

Approaching the Theoretical Tensile Strength and Theoretical Tensile Modulus

Between 75 and 90 percent of the theoretical fiber modulus can now be achieved in some commercially produced fibers. This has been demonstrated in carbon fiber (K-1100), in Kevlar 149, and in SiC fiber processed by the chemical vapor deposition method. However, the situation for the theoretical tensile strength of most commercial fibers is quite different: Current best fibers still exhibit 10 percent or so less strength than their theoretical strengths. For example, the theoretical tensile strength of polyethylene fiber is predicted to be 33 GPa,[83] and that of carbon fibers, including CNT fibers, is predicted to be between 100 and 150 GPa.[84,85] The tensile strength of polymeric fibers is limited by the presence of chain ends, chain entanglements, chain misorientation, voids, and impurities.[86] Similarly, various types of defects limit the tensile strength of carbon and

[82]Phoenix, S.L., and P.K. Porwal. 2003. A new membrane model for the ballistic impact response and V_{50} performance of multi-ply fibrous systems. International Journal of Solids and Structures 40(24): 6723-6765.

[83]Elices, M., and J. Llorca. 2002. Fiber Fracture. Oxford, England: Elsevier Science.

[84]Dumitrica, T., M. Hua, and B.I. Yakobson. 2006. Symmetry-, time-, and temperature-dependent strength of carbon nanotubes. Proceedings of the National Academy of Sciences of the United States of America 103(16): 6105-6109.

[85]Kelly, A., and N.H. MacMillan. 1986. Strong Solids, third edition. Oxford, England: Clarendon Press.

[86]Chae, H.G., and S. Kumar. 2008. Materials science: Making strong fibers. Science 319(5865): 908-909.

ceramic fibers. Eliminating these defects has the potential to improve both the strength and the strain-to-failure and, potentially, the ballistic performance, by a factor of as much as 10 to 50. Achieving such strength would require making a perfectly oriented extended polymer chain or nanotube fiber free of any type of defect (last section in Figure 5-9).

Is the goal of making a near perfect and continuous fiber achievable? Arguably the highest perfection in materials processing has so far been achieved by the microelectronics industry. The electronic devices made for communication, data storage, and data processing contain materials with near perfection at the nanometer scale. This material perfection can be repeated not in just a few devices but in devices that are mass produced and perform with a high degree of reliability. If the investment in the field of high-performance fibers was at even a small fraction of the investment in the microelectronics industry, it might be possible to achieve fibers approaching a nearly ideal structure and to thus attain theoretical strength values. Continued developments in nanotechnology, both the characterization and processing tools, will help in the production of enhanced fibers. For example, a gel-drawn polyethylene nanofiber was recently reported by Massachusetts Institute of Technology scientists to have an axial thermal conductivity value exceeding 100 W/m/K.[87] This makes polyethylene, normally considered a thermal insulator, into a thermal conductor, with an axial thermal conductivity exceeding that of a number of metals. This extraordinary thermal conductivity is a result of the near-perfect extended chains in this experimental polyethylene fiber that has not yet been achieved in commercial polyethylene fibers such as Spectra and Dyneema.

The Need for Mechanical Tests at High Strain Rates

The tensile properties of elastic fibers such as Kevlar and carbon exhibit very low or no strain-rate dependence, while fibers that have some viscoelastic behavior, such as current high-performance polyethylene fibers (Spectra and Dyneema), exhibit a relatively high degree of strain-rate dependence due to the many rate-dependent dissipative processes. In general, as the structural perfection of the fibers improves, there should be a decrease in the strain-rate sensitivity. The data in Table 5-2 are generally from quasi-static tests; however, a great need exists for mechanical property tests on fibers at high strain rates (greater than 10,000/s) for assessing their true potential as ballistic materials.

Fundamental questions abound: What is the optimum tow size—that is, the number of filaments—for the fiber assembly for a given application? Are there systems in which having individual filaments in a given layer, resulting in a layer that is about as thick as the fiber diameter, would be preferable to a yarn layer, or fabrics that are thousands of filaments thick? What is the effect of factors such as interfiber friction in a one-dimensional fiber assembly like yarn; a two-dimensional fiber assembly, such as a woven fabric; and other systems in which fibers are used?

Similarly, the role of interphase and interface properties in fiber/matrix systems needs to be understood. In this context it should be noted that the effect of fiber friction on fiber assemblies has historically been studied and understood at relatively low and moderate strain rates. Likewise, the role of interfaces and interphases has been characterized and studied at low to moderate strain rates.

On the basis of predicted strain-to-failure,[88] ideal CNT fibers—that is, with no defects or entanglements—will have a specific tensile strength of 70 N/tex,[89] for a single-wall CNT fiber with a CNT diameter of 2 nm, this equates to a tensile strength of 70 GPa.[90]

Finding 5-6a. Near-term opportunities (5 to 15 years) are very promising, especially for achieving advances in fiber strength that would allow tensile strengths to reach 8 to 15 GPa from the present 6 GPa. Even without a change in modulus, a twofold increase in strength increases the work-to-fracture by a factor of four owing to the improved mechanical behavior of small-diameter fibers that result from enhanced crystal perfection and orientation. Specific materials with good potential for further development include the following:

- *Ultrahigh-molecular-weight polyethylene (UHMWPE) fiber.* Polyethylene fiber with a tensile strength of 7 GPa has been reported in the laboratory.[91]
- *Glass fiber.* Glass fibers with a tensile strength of >10 GPa have recently been reported by Corning.[92]
- *Novel carbon fibers.* Combining gel spinning and incorporating CNTs having smaller diameters than the current state-of-the-art fibers may align the polymer chains surrounding the CNTs during processing.[93,94]

---

87 Shen, S., A. Henry, J. Tong, R. Zheng, and G. Chen. 2010. Polyethylene nanofibres with very high thermal conductivities. Nature Nanotechnology 5(4): 251-255.

88 See Dumitrica, T., M. Hua, and B.I. Yakobson. 2006. Symmetry-, time-, and temperature-dependent strength of carbon nanotubes. Proceedings of the National Academy of Sciences of the United States of America 103(16): 6105-6109.

89 "Tex" is the mass of a 1,000-meter length of fiber in grams.

90 Chae, H.G., and S. Kumar. 2008. Materials science: Making strong fibers. Science 319(5865): 908-909.

91 Judah M. Goldwasser, Program Manager, Defense Advanced Research Projects Agency, "DARPA's role in armor," presentation to the committee on June 8, 2010.

92 J. Jay Zhang, Program Manager, Dow Corning, presentation to the committee on June 8, 2010.

93 Chae, H.G., Y.H. Choi, M.L. Minus, and S. Kumar. 2009. Carbon nanotube reinforced small diameter polyacrylonitrile based carbon fiber. Composites Science and Technology 69(3-4): 406-413.

94 Chae, H.G., M.L. Minus, A. Rasheed, and S. Kumar. 2007. Stabilization and carbonization of gel spun polyacrylonitrile/single wall carbon nanotube composite fibers. Polymer 48(13): 3781-3789.

Postprocessing tensioned heat treatment could lead to a novel composite graphitic carbon fiber.

- *CNT fiber.* Pure CNT fibers with a tensile strength of ~10 GPa have been made in the laboratory. Nano graphene ribbon may also be useful for producing fibers with high tensile strength and high tensile modulus.
- *Poly(pyridobisimidazole) fiber.* Based on early laboratory developments, M5 fiber may have potential as a new ballistic fiber.

Finding 5-6b. Longer-term opportunities (25+ years) point to continued improvements in fiber strength. The theoretical strengths of polymeric, carbon, and ceramic fibers are 30 to 150 GPa.[95,96,97] Advances in nanotechnology—in both characterization and processing tools—will aid in the production of fibers that approach theoretical tensile strength values, which will, in turn, have a strong impact on weight savings for body armor.

Finding 5-6c. A better understanding of the role of fiber friction in fiber assemblies and of the interface and interphase in composites and in nanocomposites at high strain rates (typically greater than 10^4 per second) is needed to predict their ballistic performance.

BALLISTIC FABRICS

The ability to create fabrics from fibers by weaving affords the creation of low-density materials that can withstand both ballistic and blast events. Most ballistic fabrics have two-dimensional plain weave yarns in two orthogonal directions, although some work is being done on three-dimensional weaves and on nonwoven fabrics. Organic polymers or inorganic glass or ceramic fibers have the requisite high stiffness and high-strength-to-weight ratios to produce lightweight high-performance fabrics. Leading organic fibers were discussed previously. Alumina (Nextel) fibers are an important ceramic fiber. Nextel and Kevlar fabrics are used for shielding against hypervelocity impacts such as meteorites in the International Space Station.[98] Fabrics are also used for blast containment—Kevlar fabrics, for example, have been developed for blastproof cargo containers for airplanes.[99] Many factors affect the response of fabrics to ballistic impact. These include material properties of the fiber, the yarn, the weave architecture, the far-field boundary conditions, interyarn friction, friction between the projectile and the yarn, and projectile geometry and velocity.

Ballistic Testing and Experimental Work on Fabrics

For typical ballistic testing, fabric samples are clamped on the warp yarns and the incoming projectile is caught by the yarn network. Kinetic energy is transferred to the fabric as the stress wave spreads outward from the point of impact. The energy is partially dissipated by fiber deformation and interfiber friction caused by interfiber slippage. A projectile with sufficiently high mass and velocity may penetrate the fabric and cause it to fail.

Inspection of the impact area can help in understanding the failure mechanism of fibers under ballistic impact; however, such postfailure analysis provides little information on other energy dissipation paths. For insight into the dynamics of the material's response to ballistic impact, Wilde et al.[100] observed a single layer of nylon fabric by high-speed photography and found that the majority of the energy was absorbed by breaking of the orthogonal yarns. Starratt et al.[101] designed a simple and cost-effective system for continuously measuring ballistic impact. They used an enhanced laser velocity system to monitor the continuous motion of the projectile and thus determine the impact force and energy loss. Schmidt et al.[102] proposed an advanced deformation and strain analysis method based on three-dimensional image correlation photogrammetry, which can provide information on full-field dynamic deformation.

The study of ballistic impact of fabrics includes work by Shockey et al.[103] on projectile impact, residual velocity, load-stroke response, energy absorption, and tensile properties of yarns. A high-speed camera provided information on initial and residual velocity and on fragment orientation under different viewpoints and resolutions. When the specific energy absorptions of aluminum fuselage skin and of Kevlar, Spectra, and Zylon fabrics were compared, the organic-based fabrics were found to be superior to the Al skin. A finite-element model was developed for woven fabrics; inputs included the shape and geometry of the yarns (taken from high-resolution images of actual yarns) and other property data taken from laboratory tests on woven fabrics and individual yarns.

[95]Elices, M., and J. Llorca. 2002. Fiber Fracture. Oxford, England: Elsevier Science.

[96]Dumitrica, T., M. Hua, and B.I. Yakobson. 2006. Symmetry-, time-, and temperature-dependent strength of carbon nanotubes. Proceedings of the National Academy of Sciences of the United States of America 103(16): 6105-6109.

[97]Kelly, A., and N.H. MacMillan. 1986. Strong Solids, third edition. Oxford, England: Clarendon Press.

[98]Eric Christiansen, NASA Johnson Space Center, "Hypervelocity shields," presentation to the committee on June 10, 2010.

[99]See, for example, http://www.telair.com/02-01News/index.html.

[100]Wilde, A.F., D.K. Roylance, and J.M. Rogers. 1973. Photographic investigation of high-speed missile impact upon nylon fabric-1: Energy absorption and cone radial velocity in fabric. Textile Research Journal 43(12): 753-761.

[101]Starratt, D., T. Sanders, E. Cepus, A. Poursartip, and R. Vaziri. 2000. Efficient method for continuous measurement of projectile motion in ballistic impact experiments. International Journal of Impact Engineering 24(2): 155-170.

[102]Schmidt, T., J. Tyson, and K. Galanulis. 2003. Full-field dynamic displacement and strain measurement using advanced 3D image correlation photogrammetry: Part 1. Experimental Techniques 27(3): 47-50.

[103]Shockey, D.A., J.W. Simons, and D.C. Erlich. 1999. Improved Barriers to Turbine Engine Fragments: Interim Report I, DOT/FAA/AR-99/8, I. Menlo Park, Calif.: SRI International.

Failure Mechanisms of Fabrics

When a projectile hits the individual fiber or yarn,[104,105] longitudinal and transverse waves propagate from the impact point. Most of the kinetic energy transfers from the projectile to the principal yarns (those that come directly into contact with the projectile); the orthogonal yarns, which intersect the principal yarns, absorb less energy. The transient deformation within the fabric was simulated by Grujicic et al.[106] The transverse deflection continuously increases until it reaches the breaking strain of the fibers and causes failure. Specific failure mechanisms are reviewed in Appendix G, including these:

- Breakage of fiber bonds and yarns,
- Yarn pullout,
- Remote yarn failure,
- Wedge-through phenomenon (hole smaller than the diameter of projectile),
- Fibrillation, and
- Effects of friction between the projectile and the fabric, yarns, and fibers.

Appendix G also reviews some of the concepts that have been proposed for improving the performance of ballistic fabrics, such as the addition of shear thickening fluids and other coatings.

Important Issues for Ballistic Performance of Fabrics

As discussed, the ballistic performance of fabrics depends on many factors, including the structure of the fabrics, the projectile, friction, temperature, and moisture. This section discusses the main factors and reviews related studies.

Fiber Properties

Although the tensile properties of fibers, including tensile strength, modulus, and strain at failure, are important to the ballistic performance of fibers, single-fiber properties do not determine it. For example, Kevlar yarn is less tough than nylon, but its ballistic performance is better; high-strength polypropylene is approximately 50 percent stronger than nylon, but its ballistic performance is worse.

To understand relative ballistic fabric performance based on single-fiber mechanical properties, Cunniff[107] developed a parameter U^* to evaluate the ballistic performance of fibers. U^* is the product of fiber-specific toughness and strain wave velocity and is given by

$$U^* = \frac{\sigma\varepsilon}{2\rho}\sqrt{\frac{E}{\rho}F1}$$

where E is Young's modulus, σ is fiber ultimate tensile strength, ε is ultimate strain, and ρ is density. U^* can be used to predict V_{50} rankings of fibers.

The mechanical properties of some high-performance fibers (e.g., UHMWPE) are strain-rate dependent while those of other fabrics (e.g., carbon and PPTA fibers) are much less so, which is not accounted for in the expression for the parameter U^*.

Test methods that can provide fiber tensile property at strain rates greater than 10^3 s^{-1} are needed. Since the mechanical behavior of polymers is pressure sensitive—for example, UHMWPE has a relatively low melting point (140°C)—the effects of pressure and temperature on materials behavior at high rates also need consideration.

Fabric Architecture

Normally fibers are twisted to form yarn. Farris et al.[108] investigated the influence of twist on the strength and modulus and found that all the fiber yarns exhibit the best tensile strength at an optimum twist angle of about 7°. In ballistic applications, the most common weave patterns are plain and basket weaves. Cunniff et al.[109] observed that loosely woven fabric or unbalanced weave led to poor ballistic performance. Shockey et al.[110] studied single-ply Zylon fabrics and observed that absorbed energy was proportional to fabric areal density but that ballistic effectiveness was not strongly dependent on mesh density or weave tightness. Chitrangad[111] observed that the cover factor (the ratio of the area covered by the yarns to the whole area of the fabric) of fabrics in the range of 0.60 to 0.95 is suitable for ballistic applications. Lower value fabrics become too loose, and at higher cover factor values, degradation occurs during weaving. The V_{50}

[104]Cheeseman, B.A., and T.A. Bogetti. 2003. Ballistic impact into fabric and compliant composite laminates. Composite Structures 61(1-2): 161-173.

[105]Cunniff, P.M. 1992. An analysis of the system effects in woven fabrics under ballistic impact. Textile Research Journal 62(9): 495-509.

[106]Grujicic, M., W.C. Bell, G. Arakere, T. He, X. Tie, and B.A. Cheeseman. 2010. Development of a meso-scale material model for ballistic fabric and its use in flexible-armor protection systems. Journal of Materials Engineering and Performance 19(1): 22-39.

[107]Cunniff, P. 1999. Dimensionless parameters for optimization of textile-based body armor systems. Pp. 1303-1310 in Proceedings of the 18th International Symposium on Ballistics, San Antonio, Texas. W.G. Reinecke, ed. Lancaster, Penn.: Technomic.

[108]Rao, Y., and R.J. Farris. 2000. Modeling and experimental study of the influence of twist on the mechanical properties of high-performance fiber yarns. Journal of Applied Polymer Science 77(9): 1938-1949.

[109]Cunniff, P.M. 1992. An analysis of the system effects in woven fabrics under ballistic impact. Textile Research Journal 62(9): 495-509.

[110]Shockey, D.A., D.C. Elrich, and J.W. Simmons. 2001. Improved Barriers to Turbine Engine Fragments: Interim Report III, DOT/FAA/AR-99/8, III. Menlo Park, Calif.: SRI International.

[111]Chitrangad, I., 1993. Hybrid Ballistic Fabric, Patent No. 5,187,003. Available online at http://www.freepatentsonline.com/5187003.pdf. Last accessed April 12, 2011.

of composite fabrics with higher elongation in weft yarns and lower elongation-to-break in warp yarns was greater than that of fabrics made from a single material, which may be due to the lesser influence of yarn crimp. By considering yarn crimp in modeling, Tan et al.[112] obtained more accurate results. The number of fabric plies also affects the ballistic performance (note that typically there may be 20-50 plies). Shockey et al.[113] observed increased specific energy absorbed for multi-ply targets owing to the friction forces between layers. The influence of interply distance on ballistic performance has also been investigated.[114,115] The influence of projectile geometry also becomes less important with the increased number of plies.[116,117,118] A three-dimensional woven structure was studied in a fabric composite[119] designed to provide greater through-thickness direction reinforcement than in conventional two-dimensional woven fabrics; this structure showed higher ballistic performance and led to fewer penetrated layers under impact.

Projectile Characteristics and Fabric Damage

The geometry of a projectile will strongly affect its penetration ability. A sharp-edged or pointed projectile perforates the fabric more easily than a blunt-faced projectile, shearing yarns across their thickness direction and leading to a smaller specific energy absorbed.[120,121,122] Tan et al.[123] investigated the effects of projectile shape, including ogival, conical, hemispherical, and flat-headed, on the ballistic performance of single-ply Twaron fabrics; they observed the sequence hemispherical>flat-headed>ogival≥conical when projectile velocity is 100-600 m/s. Conical and ogival projectiles caused the least yarn pullout, which suggests that they were able to wedge through the fabrics.

The velocity of the projectile will also affect the performance of fabrics. In low-velocity impact, the transverse wave has a longer time to propagate and more fabric area is involved, which leads to higher energy absorption. Also, yarn pullout becomes the predominant failure mode. At high-velocity impact, some types of fibers become stiffer and stronger owing to their viscoelastic properties, and primary bond failure becomes the predominant failure mechanism.[124]

Fabric Boundary Conditions

When fabrics are impacted by a projectile, the size of the target and gripping conditions are important. For instance, a longer yarn can absorb more deformational energy than a shorter one before failure; thus a larger target area will lead to higher energy dissipation. However, this is not true when the velocity of the projectile is very high compared to the velocity of the shock wave in the fibers since then only a small portion of the target can dissipate the kinetic energy of the projectile. The boundary conditions of the target also play an important role. Shockey et al.[125] observed that a two-edge gripped fabric absorbs more energy than a four-edge gripped fabric, and fabrics with free boundaries absorb the least energy. Chitrangad[126] observed that when pre-tension is applied on aramid fabrics, their ballistic performance is improved. Zeng et al.[127] observed that for four-edge gripped fabrics, energy absorbed is improved if the yarns are oriented at 45° relative to the edge.

Friction

Frictional effects between a projectile and a fabric are observed at low-velocity impact but diminish at a higher velocity.[128] A quantitative study on Kevlar yarn friction

---

[112]Tan, V.B.C., V.P.W. Shim, and X. Zeng. 2005. Modelling crimp in woven fabrics subjected to ballistic impact. International Journal of Impact Engineering 32(1-4): 561-574.

[113]Shockey, D.A., D.C. Elrich, and J.W. Simmons. 2001. Improved Barriers to Turbine Engine Fragments: Interim Report III, DOT/FAA/AR-99/8, III. Menlo Park, Calif.: SRI International.

[114]Cunniff, P.M. 1992. An analysis of the system effects in woven fabrics under ballistic impact. Textile Research Journal 62(9): 495-509.

[115]Lim, C.T., V.B.C. Tan, and C.H. Cheong. 2002. Perforation of high-strength double-ply fabric system by varying shaped projectiles. International Journal of Impact Engineering 27(6): 577-591.

[116]Ibid.

[117]Montgomery, T.G., P.L. Grady, and C. Tomasino. 1982. Effects of projectile geometry on the performance of ballistic fabrics. Textile Research Journal 52(7): 442-450.

[118]Prosser, R A., S.H. Cohen, and R.A. Segars. 2000. Heat as a factor in the penetration of cloth ballistic panels by 0.22 caliber projectiles. Textile Research Journal 70(8): 709-722.

[119]Grogan, J., S.A. Tekalur, A. Shukla, A. Bogdanovich, and R.A. Coffelt. 2007. Ballistic resistance of 2D and 3D woven sandwich composites. Journal of Sandwich Structures & Materials 9(3): 283-302.

[120]Lim, C.T., V.B.C. Tan, and C.H. Cheong. 2002. Perforation of high-strength double-ply fabric system by varying shaped projectiles. International Journal of Impact Engineering 27(6): 577-591.

[121]Montgomery, T.G., P.L. Grady, and C. Tomasino. 1982. Effects of projectile geometry on the performance of ballistic fabrics. Textile Research Journal 52(7): 442-450.

[122]Prosser, R.A., S.H Cohen, and R.A. Segars. 2000. Heat as a factor in the penetration of cloth ballistic panels by 0.22 caliber projectiles. Textile Research Journal 70: 709-722.

[123]Tan, V.B.C., C.T. Lim, and C.H. Cheong. 2003. Perforation of high-strength fabric by projectiles of different geometry. International Journal of Impact Engineering 28(2): 207-222.

[124]Shim, V.P.W., V.B.C Tan, and T.E Tay. 1995. Modelling deformation and damage characteristics of woven fabric under small projectile impact. International Journal of Impact Engineering 16(4): 585-605.

[125]Shockey, D.A., J.W. Simons, and D.C. Erlich. 1999. Improved Barriers to Turbine Engine Fragments: Interim Report I, DOT/FAA/AR-99/8, I. Menlo Park, Calif.: SRI International.

[126]Chitrangad, I., 1993. Hybrid Ballistic Fabric. U.S. Patent 5,187,003. Available online at http://www.freepatentsonline.com/5187003.pdf. Last accessed April 12, 2011.

[127]Zeng, X.S., V.P.W. Shim, and V.B.C. Tan. 2005. Influence of boundary conditions on the ballistic performance of high-strength fabric targets. International Journal of Impact Engineering 32(1-4): 631-642.

[128]Tan, V.B.C., C.T. Lim, and C.H. Cheong. 2003. Perforation of high-strength fabric by projectiles of different geometry. International Journal of Impact Engineering 28(2): 207-222.

was conducted by Briscoe et al.:[129] The yarn pullout force increases with an increase in interyarn friction, and the increase in effective yarn modulus is attributed to the increase in interfilament friction. Fabrics with high friction and lower effective modulus can dissipate more energy than those with lower friction. Duan et al.[130] modeled the effects of interyarn friction and found that it accounts for only a small portion of energy dissipation during impact. Friction does help maintain the integrity of local fabrics in the impact region by allowing more yarns to be involved in the impact, and it increases energy absorption by increasing yarn strain and kinetic energy. Dischler[131] applied a thin polymeric film on Kevlar (20-ply), which increased the coefficient of friction from 0.19 to 0.27; he observed a 19 percent improvement in ballistic performance in stopping a flechette.

Environmental Degradation

Environmental factors such as temperature, moisture, residual spinning solvents, and UV radiation may cause high-performance fabrics to degrade, reducing their ballistic performance over time. In particular, Zylon (PBO) ballistic fabrics exhibited loss of performance when exposed to UV radiation or moisture.[132] See Appendix F for details on the environmental effects on fibers. In addition, the effect of cyclical deformation/fatigue on the ballistic performance of fibers, fabrics, and composites needs to be investigated.

Ultimately, high-performance polymer fibers are used as fabrics or as fabric panels, which are reinforced with resin in helmets. Laboratory-scale work has enabled fiber microstructures to increasingly approach the ideal structure, and corresponding single-fiber properties (strength, modulus, strain-to-failure) have recently reached impressive levels. Prospects for further improvements appear promising: They are expected to attain the fully extended and aligned state, which optimizes fiber tensile properties. Ballistic and blast performance of fabrics depends, however, on a host of parameters beyond single-fiber tensile properties, including yarn friction, yarn pullout, and others. More sophisticated modeling and simulation efforts that examine important influences such as environmental factors need to be performed.

[129]Briscoe, B.J., and F. Motamedi. 1992. The ballistic impact characteristics of aramid fabrics: The influence of interface friction. Wear 158 (1-2): 229-247.

[130]Duan, Y., M. Keefe, T.A. Bogetti, and B.A. Cheeseman. 2005. Modeling friction effects on the ballistic impact behavior of a single-ply high-strength fabric. International Journal of Impact Engineering 31(8): 996-1012.

[131]Dischler, L. 2001. Bullet Resistant Fabric and Method of Manufacture. U.S. Patent 6,248,676. Available online at http://www.google.com/patents/about?id=nGsIAAAAEBAJ&dq=Martin-Electronics&ie=ISO-8859-1. Last accessed April 12, 2011.

[132]Holmes, G.A., K. Rice, and C.R. Snyder. 2006. Ballistic fibers: A review of the thermal, ultraviolet and hydrolytic stability of the benzoxazole ring structure. Journal of Materials Science 41(13): 4105-4116.

Finding 5-7a. Environmental factors can lead to degradation of fiber and fabric properties and hence ballistic performance, particularly when exposed to extreme temperatures, ultraviolet radiation, cyclical deformation, and humidity over long times. Reliable methods need to be developed for predicting the effect of these factors on the mechanical properties at high strain rates over the useful life of fibers and fabrics.

Conclusion

The ideal microstructure for fibers is known and has been experimentally approached. Further emphasis on processing to eliminate molecular-level irregularities in chain packing and to reduce residual solvents should provide severalfold improvements in fiber properties.

Finding 5-7b. A combination of high-rate experimental measurements and computational modeling and simulation is needed to more deeply understand the dynamic deformation and failure mechanisms of ballistic fabrics and to provide insight into the most desirable high-level organization of fibers into yarns and yarns into plies and fabrics. In situ imaging of impact events and post-test assessment of fibers and fabrics need to be undertaken to reveal damage and failure mechanisms and to improve multi-hit performance.

METALS AND METAL-MATRIX COMPOSITES

Metals have been the defining armor materials for more than 2,000 years, and steel has been the armor material of choice for most of the world's armed forces. Steel technology is sophisticated, cheap, and has a very large installed industrial base. The modern army is a very heavy user of steel as a protection material, particularly in the form of rolled homogeneous armor steel, also known as rolled homogeneous armor (RHA). The substitution of lighter (nonferrous) metals for steel has always been of interest to armies, and such substitution became increasingly important in the early 20th century as armed forces became more mechanized. In the current security climate, the global reach of our nation is intimately tied to its ability to rapidly deploy mechanized armored forces. This continues to drive the development of lighter armored systems and thus the use of lighter metals for armor.

The largest fraction of protection materials in currently deployed vehicular fleets is metallic, primarily in the form of steel and aluminum alloys and particularly in the rolled condition. Some of the reasons for the large role of metals include the fact that they are relatively cheap to make, easily weldable, and able to play dual roles as structural materials and as armor materials. Because these materials have a significant commercial market, a large industrial base has grown up, along with downward pressures on the costs associated with extraction, processing, and metalworking.

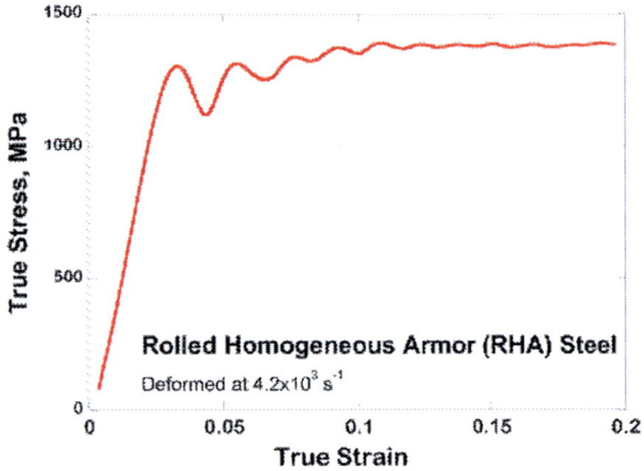

FIGURE 5-10 Stress-strain curve for RHA steel deformed in compression at a high strain rate. The oscillations in the curve are an artifact of the experimental technique and do not represent material behavior. The elastic response is also not captured accurately in such experiments. SOURCE: Zhang, H., J. Ye, S.P. Joshi, J.M. Schoenung, E.S.C. Chin, G.A. Gazonas, and K.T. Ramesh. 2007. Superlightweight nanoengineered aluminum for strength under impact. Advanced Engineering Materials 9(5): 355-359. Copyright Wiley-VCH Verlag GmbH &Co KGaA. Reproduced with permission.

These nontechnical factors surrounding the dual use potential and the economics of processing, metalworking, and joining are likely to continue to make metallic materials strong candidates for major components of robust and affordable armor systems. While the development of increasingly intense threats makes it less likely that an all-metal structure for armor will prevail, metals will probably continue to play an important role in cost-effective armor packages that can represent an optimal solution to an array of potential threats.

RHA continues to be the benchmark with respect to which most protection materials are judged: A typical objective is framed in relative terms—for example, "at least the performance of RHA at a lower areal density." Although the performance of an RHA-based armor system is measured in terms of a specific threat, the fundamental stress-strain response is a good initial benchmark for materials design. A compressive stress-strain curve for RHA at high strain rates is shown in Figure 5-10. The dynamic strength is well over 1 GPa, and there is a small but distinct strain-hardening domain. A collection of such experiments over a range of strain rates provides an estimate of the strain-rate sensitivity of the flow stress, and similar experiments performed over a range of temperatures provides an estimate for the thermal softening of the flow stress. The constitutive behavior of metals such as steels and aluminum alloys can be relatively easily incorporated within a J_2-flow type plasticity theory by associating the stress and strain mentioned above with the equivalent stresses and the equivalent strains. This kind of model is typically sufficient to describe the constitutive response of the metal at high strain rates. The parameters that define the overall behavior include the modulus, the yield strength in uniaxial tension or compression, the strain hardening, the rate sensitivity, the ultimate tensile strength in uniaxial tension, and the strain-to-failure in a uniaxial tensile test.

Desirable Attributes of Metals as Protective Materials

There is general agreement on some of the key features of good metal protection materials: high-strength, good ductility, some strain hardening, and some increase in strength with an increasing rate of deformation ("rate-sensitivity"). Other characteristics that are desirable include good formability so that the material can be formed into structures of the appropriate shapes, good long-term performance in the operating environment (e.g., corrosion and fatigue resistance), and weldability for ease of joining.

A commonly asked question about potential substitutes for RHA is whether their yield strength is on the order of 1 GPa. This strength-driven approach can be misleading, because in addition to a protection material's basic constitutive behavior, its dynamic failure processes have a major influence on its performance in the face of a specific threat. The primary failure mechanisms consist of void growth under largely tensile conditions (typically defined by the spall strength[133]) and adiabatic shear localization[134] under conditions of superimposed pressure and shear (both mechanisms are described in Chapter 3). The resistance of a metal to dynamic failure is essentially its resistance to spall and its susceptibility to adiabatic shear localization (the conventional properties of ductile fracture, while also relevant, are relatively well understood). The spall strength of a metal can be effectively bounded (at the lower end) using analyses such as those of Wu et al.,[135] which incorporate the constitutive response, and the approaches of Molinari and Wright,[136] which account for the internal defect distribution. The susceptibility to adiabatic shear localization is dependent primarily on the rate of thermal softening of the material and the strain-rate sensitivity.[137] Thus the hardening and softening mechanisms within the material must be considered in addition to the mechanisms that simply raise the initial yield strength. For instance, many high-strength metals have very low rate sensitivity and may therefore be susceptible to adiabatic

[133]Antoun, T., L. Seaman, D.R. Curran, G.I. Kanel, S.V. Razorenov, and A.V. Utkin. 2003. Spall Fracture. New York, N.Y.: Springer.

[134]Wright, T.W. 2002. The Physics and Mathematics of Adiabatic Shear Bands. New York, N.Y.: Cambridge University Press.

[135]Wu, X.Y., K.T. Ramesh, and T.W. Wright. 2003. The effects of thermal softening and heat conduction on the dynamic growth of voids. International Journal of Solids and Structures 40(17): 4461-4478.

[136]Molinari, A., and T.W. Wright. 2005. A physical model for nucleation and early growth of voids in ductile materials under dynamic loading. Journal of the Mechanics and Physics of Solids 53(7): 1476-1504.

[137]Wright, T.W. 2002. The Physics and Mathematics of Adiabatic Shear Bands. New York, N.Y.: Cambridge University Press.

shear localization. The degree to which each of these failure mechanisms is important for a specific material depends on the specific geometry of the armor structure as well as the specific threat; this means a detailed understanding of the connection between the microstructure and the failure mechanisms in the metal is important in the design of new metallic protection materials.

The mechanical properties of metals can be changed substantially by controlling the microstructure by chemical and thermomechanical means. Typical strengthening mechanisms include solid solution hardening, precipitation and dispersoid hardening, and grain boundary strengthening. In addition, many metals and metal alloys can be strengthened substantially by increasing the internal dislocation density through processes such as work hardening. This is one of the advantages of metallic materials: that the processing routes associated with metalworking can often be optimized to increase the strength and the ductility. An example of such a useful work-hardening route is rolling, which is typically used to produce plate geometries. Rolled metal can be much stronger than the metal before rolling, and indeed the largest tonnages of metallic armor materials are rolled alloys (such as RHA). Materials with submicron structural features are known to have higher yield strengths. Indeed, controlling not only grain size but also feature size—for example, in metals and bicontinuous composites like ceramic/polymer or metal/ceramic—can improve mechanical behavior.[138,139]

Finding 5-8a. Although metal alloys have been in use for many years, only a small fraction of the alloys in use have actually been characterized at the high strain rates relevant to ballistic problems. As a result, much of the modeling and simulation that is performed using these alloys has become heuristic rather than based on fundamental experimental data. This makes it very difficult to design with these alloys when new threats are presented. A sustained effort to develop a database of high-strain-rate material properties for metallic materials would benefit armor designers.

Nonferrous Metal Alternatives

Appendix H provides a more detailed review of the main nonferrous metals that may (and sometimes do) compete substantially with steel: titanium and titanium alloys, aluminum and aluminum alloys, magnesium and magnesium alloys, and metal-matrix composites. A brief discussion on aluminum and magnesium and their alloys follows.

Aluminum and Aluminum Alloys

Aluminum and aluminum alloys were developed early in the twentieth century, and beginning around the time of World War II, they were pressed into service, beginning with armor for aircraft, to reduce weight. The introduction in the late 1950s of the T113 (later M113) vehicle type built of an aluminum alloy was followed by the deployment of significant quantities of aluminum alloys in the armored fleet. While pure aluminum is very soft, conventional aluminum alloys can have yield strengths that easily compete with those of the simpler steels. Specific approaches such as solid solution strengthening and age-hardening have been developed to strengthen Al alloys.

The trade-offs between weight, structural performance, ballistic performance, ease of production, and ease of maintenance, including resistance to corrosion, play a significant role in the choice of alloy for vehicular applications. Most of these aluminum alloys are used as rolled plate, and work-hardening alloys such as the 5000 series (5083 being the prime example) have some advantages. Aluminum alloys used as armor in Army vehicles include 2024, 2519, 5083, 5059, 6061, 7039, and 7075. Promising new commercial alloys include 2139 Al, a commercial alloy with significant strength (around 600 MPa at high strain rates) and reasonable ductility.

There is significant potential for the development of novel aluminum-based materials with very high strengths through alloying approaches and by the development of nanostructured systems and aluminum-based composites.

Finding 5-8b. There is a substantial potential for the development of new and improved aluminum alloys that can substitute for steel in military vehicles. Efforts to increase both strength and ductility of aluminum alloys at high strain rates are likely to bring significant benefits.

Magnesium and Magnesium Alloys

Magnesium has a remarkably low density of 1,700 kg/m^3 (in comparison, Al is 2,800 kg/m^3, Ti is 4,950 kg/m^3, and steels are 7,800 kg/m^3). Its density approach es that of polymers. Magnesium and magnesium alloys are thus among the lightest structural metals, and they are becoming increasingly important in the automotive and hand-tool industries. The rapid growth in the commercial use of magnesium is intimately tied to the increasing cost of energy. Their low density makes these materials very attractive for defense applications, but magnesium alloys have historically had relatively low strengths (~250-300 MPa) in comparison to aluminum alloys. There has also been lingering, albeit somewhat exaggerated, concern about the flammability of magnesium and about the relative ease with which these alloys corrode in severe environments. However, these concerns are relatively easily mitigated by proper design and appropriate protocols for maintenance.

---

[138] Kraft, O., P.A. Gruber, R. Monig, and D. Weygand. 2010. Plasticity in confined dimensions. Annual Review of Materials Research 40: 293-317.

[139] Lee, J.-H., L. Wang, S. Kooi, M.C. Boyce, and E.L. Thomas. 2010 Enhanced energy dissipation in periodic epoxy nanoframes. Nano Letters 10(7): 2592-2597.

A substantial effort has begun over the last decade to generate high-strength magnesium alloys using a variety of approaches, including solid solution strengthening and precipitation strengthening. Commercial magnesium alloys that can substitute for some aluminum alloys include AZ31[140] and ZK60, and several alloys containing rare earths show promise. Most of the innovation in this area is occurring abroad, particularly in China and Japan, and this may represent a long-term risk for the United States. A recent workshop at Johns Hopkins on the potential of magnesium and magnesium alloys as protection materials highlighted a variety of opportunities. One of the more promising strengthening approaches appears to be the development of ultra-fine-grained or nanostructured magnesium alloys through severe plastic deformation. A major research effort to gain a fundamental understanding of the strengthening mechanisms in magnesium alloys is likely to be very fruitful, and the opportunities presented by the metal's low density should not be missed.

Finding 5-8c. A fundamental research effort to improve magnesium alloys could have a big impact on the weight of the armored vehicular fleet (a nascent effort exists at this time, driven by the ARL). There is a need to intensify research into other lightweight protective metals as well, which would help maintain the U.S. infrastructure for critical materials associated with protection systems for the warfighter. The outstanding performance of metals, the ease of fabricating and joining them, and the well-established industrial base ensure that these materials will be significant components of protection material systems for the foreseeable future. As the efforts in aluminum have demonstrated, there is a substantial potential for dramatic improvements even within the classes of metal alloys currently available.

ADHESIVES FOR ARMOR AND FOR TRANSPARENT ARMOR

Adhesive interlayers are key components of both ballistic glass and composite armor. Understanding, testing, and modeling of the adhesive interlayers in composite armor are crucial for their future design and improvement. Studies of nontransparent armor[141] as well as transparent armor indicate that the adhesives significantly influence the ballistic behavior of the composite structure. While results and material information are generally kept secret, published information on the subject provides insight into which adhesives are commonly used, the state of their development, and their modeling capabilities. Most interlayer materials are polymers, including thermoplastics and thermosets. The role of the adhesive is to hold the armor together before and after impact and to both deform and delaminate to absorb energy. These functions require an intermediate level of bonding between the adhesive and the hard components (glass, ceramic, metal)—neither too weak an interface nor too strong. Adhesives must absorb little to no light, because any absorption will lower the overall transmission of the transparent composite material. This applies not only to the visible wavelengths of light but also to the near-infrared as some applications are used with night vision goggles and sensors. As well as giving an overview of commonly used adhesives, this section also reports on design criteria for composites and their testing, both experimental and computational.

General Considerations for the Selection of an Adhesive Interlayer

Many material properties are important for the adhesives that are used as interlayers in armor and transparent armor applications. These include the strength of the adhesive bonds across the various interfaces, which is highly dependent on chemistry but also on surface roughness, environmental stability, mechanical impedances, mechanical properties over a very large range of strain rates, and transparency in the visible and near-infrared spectrum to name but a few. Figure 5-11 shows a cross-sectional view of a typical ballistic-resistant glass composite made up of a ceramic strike face, an outer region of plies of thick glass and alternating adhesive interlayers plus a transition section of a thick plastic (e.g., polyurethane) and an absorbing section, usually made of polycarbonate.

Important Issues Surrounding Adhesives for Lightweight Armor Applications

In automotive safety glass the strength of adhesion between the adhesive interlayer and the polymer, glass, and ceramic layers has been tailored with good effect.[142,143,144] Interlayers with bonds that can delaminate from the sub- and superstrates may better absorb a projectile's kinetic energy. At the same time, however, they must retain enough integrity

[140]Mukai, T., M. Yamanoi, H. Watanabe, and K. Higashi. 2001. Ductility enhancement in AZ31 magnesium alloy by controlling its grain structure. Scripta Materialia 45(1): 89-94.

[141]Zaera, R., S. Sánchez-Sáeza, J.L. Pérez-Castellanos, and C. Navarro. 2000. Modelling of the adhesive layer in mixed ceramic/metal armours subjected to impact. Composites Part A: Applied Science and Manufacturing 31: 823-833.

[142]Fock, K., H.D. Hermann, K. Fabian, and J. Ebigt. 1987. Reduction in the Adhesion to Glass of Thermoplastic, Plasticized Polyvinylbutyral Molding Compositions. U.S. Patent 4,663,235.Available online at http://www.patents.com/us-4663235.html. Last accessed April 12, 2011.

[143]Hermann, H.D., K. Fabian, and J. Ebigt. 1985. Polyvinylbutyral Films Which Contain Plasticizer and Have a Reduced Adhesive Power on Glass. U.S. Patent 4,533,601. Available online at http://www.patents.com/us-4533601.html. Last accessed April 12, 2011.

[144]Garrison, W.E. 1969. Glass Laminate. U.S. Patent 3,434,915. Available online at http://www.freepatentsonline.com/3434915.pdf. Last accessed April 12, 2011.

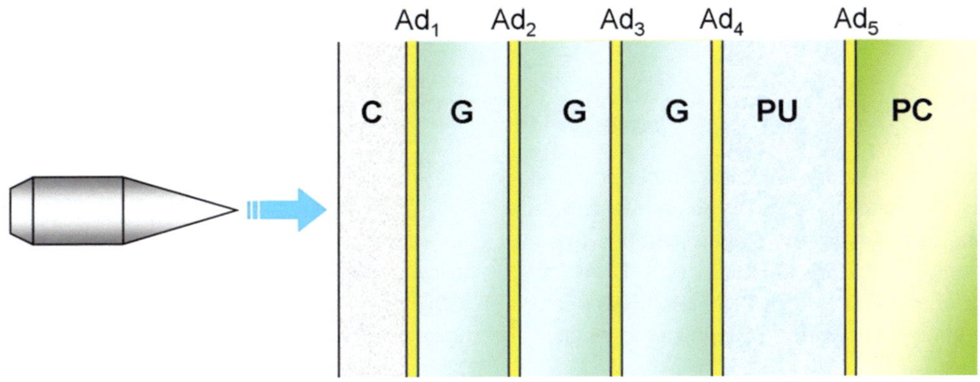

FIGURE 5-11 Composite stack of transparent layers: a ceramic strike face (C), adhesive interlayers (Ad), glass (G), polyurethane (PU), and polycarbonate (PC).

to minimize flying debris (spall).[145] Additionally, the adhesion of the interlayer affects multi-hit performance; retaining and/or confining the hard ceramic and/or glass is necessary and so must be considered when designing interlayers.

Ultraviolet Radiation/Humidity/Temperature (Environmental) Stability

Service temperatures for armor can vary widely depending on where in the world it is deployed. Additionally, environmental degradation as a result of ultraviolet (UV) radiation or oxidation of the polymeric interlayer can affect both the transparency and the adhesive strength of the interlayer.

Strain-Rate Dependence

Most adhesive interlayers are for protection against ballistic threats. Thus, the properties of the adhesive at relevant strain rates must be known. Polymeric materials, of which most of the interlayers are made, typically have mechanical properties that depend greatly on strain rate and pressure.

Mechanical Impedance

Waves traveling through composite armor can be reflected or transmitted depending on the impedance mismatch between consecutive layers. Control of the mechanical impedance of the interlayer is therefore important in the design of the armor. The impedance of many polymers is only about 0.05 to 0.005 of that of ceramics, so that most (90 percent or more) of the incident energy is reflected from the ceramic-interlayer interface.

[145]Hou, S., and H. Reis. 2009. Adhesive bond evaluation in laminated safety glass using guided wave attenuation measurements. Pp. 33-44 in Advances in Ceramic Armor IV: Ceramic Engineering and Science Proceedings, Volume 29, Issue 6. L.P. Franks, ed. Hoboken, N.J.: John Wiley & Sons.

Thermal Expansion Coefficient

The coefficient of thermal expansion (CTE) of the adhesive interlayer is important for applications that will see a wide range of temperatures. Transparent armors are usually constructed with both high CTE (plastics such as polycarbonate) and low-CTE (ceramic and glass) materials. When directly bonded and exposed to a change in temperature, materials with much different CTEs will change in dimension by different amounts, resulting in stresses and deformation, including shape change and/or delamination. The CTE of the adhesive interlayer, along with its mechanical properties, can mitigate the effect of CTE mismatch in these systems.

Index of Refraction

Reflection from interfaces reduces the amount of light transmitted through a composite material. Ideally the refractive index of the interlayer material should be chosen according to the relationship $n_{adhesive} = \sqrt{(n_{mat1} \times n_{mat2})}$ with the thickness equal to a quarter wavelength so as to minimize reflection.[146]

Cost

Generally the cost of the adhesive is insignificant compared to that of armor materials such as AlON and sapphire.

Additional Functionality

The implantation of metal wires or conductive materials for resistive heating (for defogging, for instance) relies on the ability of the soft interlayer material to act as a host matrix. Implementation of heads-up displays or other electronics

[146]Hecht, E. 2001. Optics, 4th edition. Old Tappan, N.J.: Addison-Wesley Longman.

are sometimes required, which places additional demands on the adhesive.

Glass Transition Temperature and Mechanical Loss Peaks

The transition temperature is key for processing and for providing a flexible material at the service temperature. Polymeric materials can absorb energy due to their many molecular motions. Such motions are temperature and frequency dependent and the associated mechanical loss peaks can be tailored for energy absorption in high-rate events.

Types of Adhesive Interlayers

Thermoplastic Polyvinyl Butyral

Developed in the late 1930s and commonly used in automotive glass applications, thermoplastic polyvinyl butyral (PVB), which is generally plasticized, has been the workhorse of polymeric adhesive interlayers. Examples are Saflex (Solutia, Inc.), Butacite (DuPont), Trosifol (Kuraray Europe), S-LEC (Sekisui Chemical), and KB (GlasNovations). Positive features of PVB include good optical transparency when bonded to glass, controllable adhesion to glass, resistance to elongation when struck with a projectile, and good UV stability.[147]

Thermoplastic Polyurethanes

Thermoplastic polyurethanes (TPUs) come in two broad categories, aliphatic or aromatic, depending on the precursor from which they are synthesized. Examples are Dureflex (Bayer Material Science), IM800 (and others from Inter Materials), Deerfield 4700 (and others from Deerfield Urethane), and Huntsman 399. Aliphatic TPUs are generally preferred for transparent armor applications because of their superior clarity compared to aromatic TPUs. TPUs are sometimes preferred to PVB since they do not contain plasticizer, which can chemically attack other polymers such as acrylics and polycarbonate.[148] TPUs are typically extruded and rolled in sheet form. The composite is formed by layering the materials, which are then sealed in a bag that is then evacuated of air and autoclaved to consolidate the layers.

Thermosets

Other cross-linkable polyurethanes may be used for adhesive interlayer materials. One such example uses a poly(urethane urea) elastomer.[149] Blends of mercaptans with epoxies have shown improved performance.[150] Epoxies for use as adhesives in nontransparent composite armor have also been studied.[151]

Other Materials

Thermoplastic poly(ethylene vinyl acetate), and low-temperature flowing glass or glass ceramics have been used as an interlayer for bonding alumina and sapphire or other high-temperature materials.[152] Other hybrid materials specifically engineered for combining the adhesive and the rear panel are available from some manufacturers (e.g., FAENAC, a transparent plastic composite from Saint-Gobain Sully[153,154]).

Testing, Simulation, and Modeling of Adhesives

Adhesives are generally tested as part of a composite. It is the combined properties of the system that matter, and the interplay between the various hard components of the armor is transmitted via the adhesive interlayers. Chapter 2 of *Advances in Ceramic Armor*[155] provides a good overview of destructive testing methods for adhesives. The simplest composite armor can be considered to be a ceramic plate adhesively bonded to a metal plate. An obvious question beyond the choice of material for the adhesive concerns the impedance and thickness of the adhesive. As for deformation and failure of the ceramic, an interlayer with a higher impedance is better since less of the energy would be reflected back from it as a tensile wave, and a thinner interlayer likewise is better. This is because more of the incident energy is more quickly transmitted to the metal layer, and when a compressive wave reflected from the interlayer-metal interface arrives at the ceramic, it helps to prevent bending and subsequent cracking of the ceramic. As for the metal, however, a thicker interlayer is better since it would spread the deformation

---

[147] Freeguard, G.F., and D. Marshall. 1980. Bullet-resistant glass: A review of product and process technology. Composites 11(1): 25-32.

[148] See http://www.bayerfilms.com/tpu/content.php?p=security-glaze for more information.

[149] Sarva, S.S., and A.J. Hsieh. 2009. The effect of microstructure on the rate-dependent stress-strain behavior of poly(urethane urea) elastomers. Polymer 50(13): 3007-3015.

[150] Uram, Jr., J.R. 1984. Moisture-Resistant Transparent Mercaptan Compositions. U.S. Patent 4,555,450. Available online at http://www.patents.com/us-4555450.html. Last accessed April 13, 2011.

[151] Zaera, R., S. Sánchez-Sáeza, J.L. Pérez-Castellanosa, and C. Navarro. 2000. Modelling of the adhesive layer in mixed ceramic/metal armours subjected to impact. Composites Part A 31(8): 823-833.

[152] Patel, P.J., G.A. Gilde, P.G. Dehmer, and J.W. McCauley. 2000. Transparent armor. The AMPTIAC Newsletter 4(3): 1-2.

[153] Saint-Gobain Sully. Undated. Technical Datas FAENAC® Film. Available online at http://www.saint-gobain-sully.com/GB/quality/tech/FICHE%20TECHNIQUE%20FILM%20FAENAC%20A.pdf. Last accessed April 2011.

[154] Jones, C.D., J.B. Rioux, J.W. Locher, E.S. Carlson, K.R. Farrell, B.C. Furchner, V. Pluen, and M. Mandelartz. 2009. Transparent Ceramic Composite Armor. U.S. Patent 7,584,689. Available online at http://www.freepatentsonline.com/7584689.pdf. Last accessed April 13, 2011.

[155] Sun, X., K.C. Lai, T. Gorsich, and D.W. Templeton. 2009. Optimizing transparent armor design subject to projectile impact conditions. Pp. 15-22 in Advances in Ceramic Armor IV: Ceramic Engineering and Science Proceedings, Volume 29, Issue 6. L.P. Franks, ed. Hoboken, N.J.: John Wiley & Sons.

over a larger volume of the metal.[156,157] Clearly, detailed modeling and simulation can provide optimized solutions for armor design. Studies elucidating the influence of defects in plate materials[158] have been conducted. The ARL is actively modeling transparent composite armor.[159,160]

Finding 5-9. There is need for an improved understanding of the dynamic behavior of the adhesive by itself and of the adhesive placed between dissimilar hard materials as part of an armor system.

JOINING

Armor systems use different classes of materials—ceramics, metals, polymers, and composites—to meet defined ballistic threat requirements. The quality of the joints between dissimilar materials therefore plays an important role in the final armor performance, since these linkage sites have to withstand the dynamic loads under ballistic and blast conditions. With the increased use of diverse advanced materials, the number of armor joints is increasing and greater demands are being placed on them for better performance. Current methods of combining dissimilar armor material are somewhat empirical and based on experience with other products so that systematic research efforts are needed for understanding the dependence of ballistic performance on bond characteristics.

Joining different materials is often not an easy task.[161] In general, bonding methods are chosen based on the particular materials to be combined, their geometrical configuration, and the performance requirements. In joining different materials careful attention has to be paid to minimize mismatches in properties and structural discontinuities. The key to a successful joint between dissimilar materials with different bonding characteristics and properties is the design of a buffer interface capable of accommodating the dissimilarities.[162,163]

The fundamentals that need attention when joining surfaces are surface roughness and surface contamination. When two surfaces are brought into contact, the true area of contact is less than the apparent area of contact owing to inherent surface roughness and the nonplanarity on an atomic scale of any surface. This inherent roughness is overcome by deformation, diffusion of surfaces (direct bonding), or infiltration of bonding filler between the two surfaces (indirect bonding). In armor systems, materials such as ceramics (for example, alumina, silicon carbide, or boron carbide) are bonded to metals (steels, aluminum, titanium, or magnesium), which in turn are joined to fibers, woven fibers, or polymer structures, mostly using indirect or mechanical bonding processes.

The selection of a technique for manufacturing a particular component will be based on a number of factors:

- Types of materials to be joined,
- Desired component function—for example, strength,
- Operational temperature,
- Applied mechanical stresses (static and dynamic) on the joint,
- Required level of joint airtightness,
- Component design, and
- Cost.

Mechanical joints typically have poor joint strength (10-50 MPa) and create areas of stress in ceramics, limiting design flexibility. Their use in armor applications[164] is thus restricted except where such conditions can be tolerated. In both indirect and direct bonding, charge or mass transfer can occur between surfaces.[165] In indirect joining, an intermediate layer of filler alloy is used for bonding different surfaces. Examples of indirect bonding include soldering, brazing, adhesive bonding, and other processes that provide contact between the surfaces through the intervening filler materials. In contrast, direct bonding uses no fillers, and the bonding occurs by means of the solid-state processes that depend on deformation and diffusion between surfaces.

Bonding via solid-state diffusion requires the application of heat and long exposure times, while deformation requires relative sliding of surfaces with substantial applied stresses. Thus, while solid-state bonding methods yield

[156] Ibid.

[157] Zaera, R., S. Sánchez-Sáeza, J.L. Pérez-Castellanosa, and C. Navarro. 2000. Modelling of the adhesive layer in mixed ceramic/metal armours subjected to impact. Composites Part A 31(8): 823-833.

[158] Fountzoulas, C.G., J.M. Sands, G.A. Gilde, and P.J. Patel. 2009. Applying modeling tools to predict performance of transparent ceramic laminate armors. Pp. 45-53 in Advances in Ceramic Armor IV: Ceramic Engineering and Science Proceedings, Volume 29, Issue 6. L.P. Franks, ed. Hoboken, N.J.: John Wiley & Sons.

[159] Fountzoulas, C.G., B.A. Cheeseman, P.G. Dehmer, and J.M. Sands. 2009. A Computational Study of Laminate Transparent Armor Impacted by FSP, ARL-RP-249, June. Available online at http://www.arl.army.mil/arlreports/2009/ARL-RP-249.pdf. Last accessed April 13, 2011.

[160] MacAloney, N., A. Bujanda, R. Jensen, and N. Goulbourne. 2007. Viscoelastic Characterization of Aliphatic Polyurethane Interlayers, ARL-TR-4296, October. Available online at http://www.dtic.mil/cgi-bin/GetTRDoc?AD=ADA474714&Location=U2&doc=GetTRDoc.pdf. Last accessed April 13, 2011.

[161] do Nascimento, R.M., A.E. Martinelli, and A.J.A. Buschinelli. 2003. Review article: Recent advances in metal-ceramic brazing. Cerâmica 49(312): 178-198.

[162] Paiva, O.C., and M.A. Barbosa. Brazing parameters determine the degradation and mechanical behaviour of alumina/titanium brazed joints. Journal of Materials Science 35(5): 1165-1175.

[163] Howe, J.M. Bonding, structure, and properties of metal/ceramic interfaces: Part 2 interface fracture behaviour and property measurement. International Materials Reviews 38(5): 257-271.

[164] Klomp, J.T., and G. de With. 1993. Strong metal-ceramic joints. Materials and Manufacturing Processes 8(2): 129-157.

[165] Martinelli, A.E. 1995. Diffusion Bonding of Silicon Carbide and Silicon Nitride to Molybdenum. Ph.D. dissertation. Montral, Canada: McGill University.

strong joints, they may not be suitable for most of the armor applications because dissimilar materials have different temperature tolerances and deformation characteristics. The role of temperature in joining dissimilar materials for a given set of surfaces is also an important operational parameter in selecting bonding media. The vast majority of joining processes involve heating surfaces that, upon cooling, develop residual stresses owing to mismatches in elastic modulus and CTE.

More often, the ceramic-metal joint in armor applications is achieved through indirect bonding processes. Adhesive joining is widely used. Even though adhesive joining with epoxy is executed under ambient conditions and is applicable to most materials, the resulting bond strength is relatively weak relative to brazing or soldering. Further, the low modulus of epoxy joints leads to a large elastic impedance mismatch with ceramic and metal, which could lead to poor ballistic performance.[166] Very few adhesive materials exist with impedance close to that of metals and ceramics, because wave velocity depends on the elastic modulus and density of the material. The class of adhesives whose impedance most closely matches that of ceramics and metals is high-temperature ceramic adhesives. However, such ceramic adhesives are not as strong as polymer glues, and they are often used as matching layers in mechanically bonded systems. By combining ceramic adhesives with polymeric and other glues, performance could be considerably improved. Multilayer adhesives with better impedance match have demonstrated[167] improved multi-hit ballistic performance and structural integrity.

Bonding options such as brazing and soldering typically result in higher modulus interfaces and thereby decrease (compared to adhesives) the elastic impedance mismatch with ceramic and metal substrates. Brazing or soldering ceramics to metal relies on wetting the ceramic surface with a suitable metal or alloy that will react with both the metal and the ceramic to form a joint. However, heating the surfaces to high temperatures develops residual stresses on cooling due to mismatches in elastic modulus and CTE. The heating temperatures for braze alloys are above 450°C and for soldering below 450°C. Achieving a superior brazed or soldered bond while minimizing residual stresses is important.

Mizuhara et al.[168] developed an active brazing method in which the active component, such as titanium, is incorporated into silver-copper eutectic brazing alloys to enhance the wetting of ceramic and metal surfaces. This one-step vacuum brazing process wets most armor materials (ceramics, titanium alloys, and steels) and forms a superior metallurgical bond. However, the high processing temperature required for "active" brazing results in a large buildup of stress upon cooling owing to the different CTEs of metals and ceramics. Intermediate layers to alleviate expansion coefficient mismatches are being continuously developed, and the quality of the bond depends on filler layers that mediate the joining with minimum stress buildup.

Active soldering is an emerging technology similar to active brazing but performed at lower temperatures (<450°C) to reduce mismatch stresses during heating and cooling cycles. Here, reactive elemental titanium is added to the solder alloy as it is to a brazing alloy to enable direct wetting and bonding. The lower joining temperatures offered by active soldering minimize thermal stresses while yielding reasonable elastic impedance matching. Solder joint strengths are similar to those of epoxy joints. The tensile strength of an epoxy joint for bonding a hot-pressed SiC surface to an annealed Ti-6Al-4V surface is greater (73 MPa) than that of an active solder joint (43 MPa), while the elastic impedance of a solder joint is 10 times better than that of an epoxy joint, thus approaching the elastic impedance of ceramic and metal surfaces.

Finding 5-10a. Reliable methods for manufacturing dissimilar materials are in a nascent stage. Systematic studies to understand the relationships between ballistic performance, bond adherence, key filler material characteristics, and elastic impedance matching are needed to enable the manufacture of armor systems containing dissimilar advanced materials.

Finding 5-10b. Investment is needed in research and development in active brazing and soldering materials, adhesives, and processing methods for joining armor material to produce joints with minimal thermal mismatch stresses during the heating and cooling cycle of the bonding method.

OTHER ISSUES IN LIGHTWEIGHT MATERIALS

Nondestructive Evaluation Techniques

Nondestructive evaluation (NDE) techniques have been employed for the characterization of armor and armor materials for several decades.[169] These techniques are preferred over destructive ones since they leave the material intact and ready for use. NDE tests of entire lots of materials can identify specific pieces that do not meet the appropriate criteria without having to rely on statistical interpolations of destructive test results carried out on a few select samples. Different evaluation techniques are applied to garner different kinds of information from the armor material. NDE is applied at various stages in the testing of armor to assess the performance capabilities of armor materials, to ensure the integrity of assembled arrays of tiles, and to understand how materials become damaged when introduced to various threats.

[166]James, B. 2004. Practical issues in ceramic armor design. Pp. 33-44 in Progress in Ceramic Armor. New York, N.Y.: John Wiley & Sons.

[167]Ibid.

[168]Mizuhara, H., and K. Mally. 1985. Ceramic to metal joining with active brazing filler metal. Welding Journal 64(10): 27-32.

[169]See the American Society for Nondestructive Testing Web site at http://www.asnt.org/.

NDE testing focuses on determining whether materials for compiled armor assemblies will perform adequately when they are used in the field. This can be as basic as a simple go/no-go test or as complicated as a three-dimensional representation of internal flaws and density gradients. A variety of nondestructive methods has historically been used to rapidly locate and identify anomalous internal flaws in dense armor materials. These methods, which include resonant ultrasound spectroscopy, high-frequency ultrasound scans, infrared thermography, and microfocus x-ray computed tomography, are discussed in Appendix I.

Fiber-Reinforced Polymer Matrix Composites

Polymer matrix composites (PMCs), discussed further in Appendix J, consist of a polymer resin reinforced with fibers. One application is the combat helmet. PMCs can be subdivided into two categories based on whether the fiber reinforcement is continuous or discontinuous. PMCs with discontinuous fibers (less than 100 mm long) are made with thermoplastic or thermosetting resins, whereas PMCs with continuous fibers usually employ thermosetting resins.

The most common design for PMCs is a laminate structure made of woven fabrics held together by a polymer resin. Fabrics are incorporated to take advantage of their high strength and stiffness and to improve energy absorption and distribute the kinetic energy laterally. Owing to their highly engineered structures, PMCs are lightweight with high specific strength and high specific stiffness.

Common reinforcement materials are carbon, glass, aramid, and polyethylene fibers. PMCs can be manufactured by wet and hand lay-up, molding (compression, injection, and transfer), vacuum bag molding, infusion molding, vacuum-assisted resin transfer molding, prepreg[170] molding, and other common techniques. Unlike the usual structural composites, which typically contain up to about 60 vol percent fibers, ballistic PMCs contain a higher volume fraction of fibers or fabrics, up to about 80 vol percent, although the effect of this variation in structure on the ballistic protection properties of PMCs has not been thoroughly investigated.

Because PMCs respond to ballistic impact in ways that depend on their particular structure, they are different from other protective materials. Unlike fabric materials, the PMC material responds only in the neighborhood of the impact position; thus the response is completely governed by the local behavior of the material and unaffected by the boundary conditions. Additionally, the penetration mechanism is dependent on the thickness of the composite. For thin composites, the deformation across the thickness direction does not vary with depth, while for thick composites it does.[171] Ballistic performance initially increases linearly with the increased thickness; however, as the composite becomes thicker, the marginal protective gain from increasing the thickness is less,[172,173] while the rate at which the weight increases is maintained.

[170]Semifinished fiber products preimpregnated with epoxy resin (prepregs).

[171]Naik, N.K., and A.V. Doshi. 2008. Ballistic impact behaviour of thick composites: Parametric studies. Composite Structures 82(3): 447-464.

OVERALL FINDINGS

The overall findings of Chapter 5 are summarized below and are addressed by the recommendations presented in Chapter 6.

Finding 5-11. A sustained effort is needed to develop a database of high-strain-rate material behavior for the ceramic, polymeric, and metallic materials in use today and to expand the database as new materials are developed.

Finding 5-12. The intrinsic properties of opaque and transparent ceramics and ceramic powders are underrealized in armor systems. There is a need for an atomic, nano, and micron-size understanding of how powders and processing can be designed and manipulated to realize the benefits of dense and porous ceramic armor.

Finding 5-13. A need exists to build a production infrastructure for strategic ceramic powders within the United States for the next generation of opaque and transparent ceramic armor.

Finding 5-14. Current opportunities include the development of finer diameter and more ideal polymeric and carbon fibers with a two- to fivefold improvement in specific tensile strength over the current state-of-the-art fibers. Such improvements would lead to significant reduction in the weight of body armor.

Finding 5-15. Since polymers are often parts of systems (e.g., fabrics, matrixes, and reinforcing elements in composites), a fundamental understanding of how to model the deformation mechanisms and failure processes of polymers is critical to the successful large-scale modeling of complex multicomponent armor systems.

Finding 5-16. Advances are needed in test methods for determining the high-strain-rate (10^3 to 10^6 s^{-1}) properties of fibers, polymers, and ceramics and their dynamic failure processes. Results could be used to develop a comprehensive database of strain-rate behavior for such materials.

Finding 5-17. The very low density of magnesium, including magnesium alloy fibers, could lead to the development of very lightweight alternatives to traditional metallic materials

[172]Ibid.

[173]Faur-Csukat, G. 2006. A study on the ballistic performance of composites. Macromolecular Symposia 239(1): 217-226.

in protection material systems. A better basic understanding of the strengthening mechanisms in magnesium, especially the development of ultra-fine-grained magnesium alloys through severe plastic deformation, could be highly beneficial. Magnesium-based fibers are also worthy of exploration.

Finding 5-18. The development of bonding materials (adhesives, brazes, and solders) whose elastic impedances and thermal expansion coefficients match those of the materials to be bonded will improve the ballistic and blast performance of laminated armor.

6

The Path Forward

A NEW PARADIGM

The need for rapid advances in the effectiveness and affordability of lightweight protection materials and systems is compelling and will continue for the foreseeable future. The experience with body armor and vehicle armor in Iraq and Afghanistan has shown that the weight penalty of today's materials exacts a significant toll on U.S. forces, both in human terms and in increased costs for equipment, maintenance, and fuel. Escalating threats have greatly accentuated the need for continued rapid development of lightweight armor.

The ideal situation is to have new materials available to meet these challenges. However, while new materials are the subject of research efforts, their introduction into military systems is very slow. As shown in Chapter 1 (Figure 1-3), the advances indicated by the areal density plot of lightweight protection materials have slowed in recent years. The inability to rapidly transition materials with the properties and behavior needed for armor systems is due not to a lack of excellent materials research, but rather to the approach by which protection materials research is accomplished.

As described in Chapter 2 (see also Figure 6-1), armor

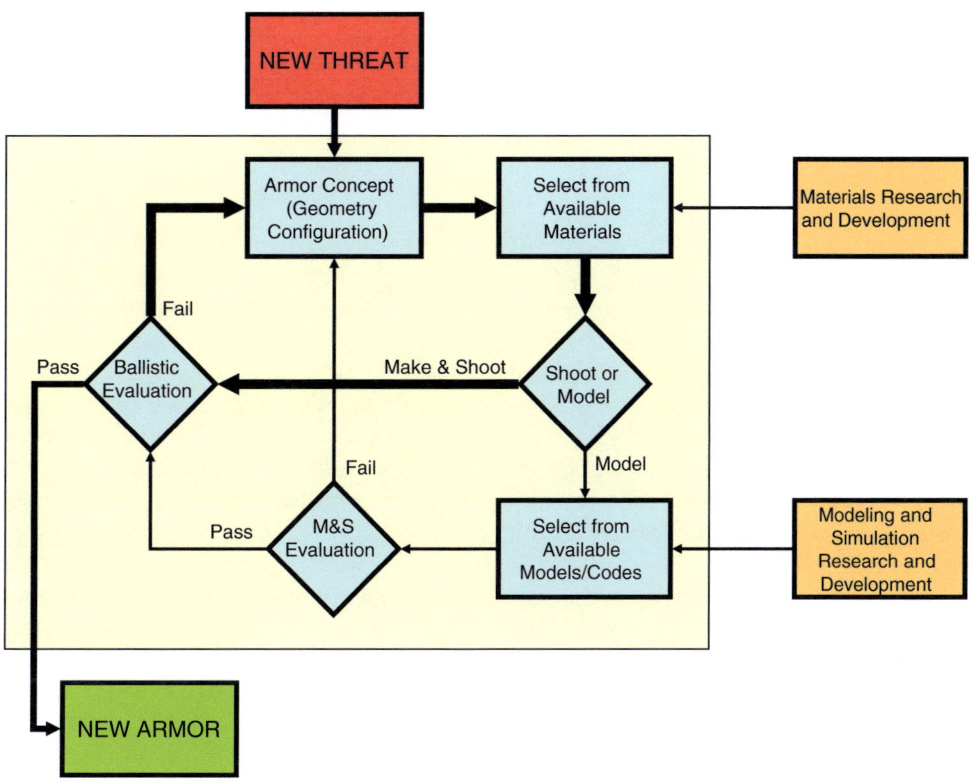

FIGURE 6-1 Current paradigm for armor design. As mentioned in Chapter 2, a shoot-and-look approach is much more prevalent than a modeling approach.

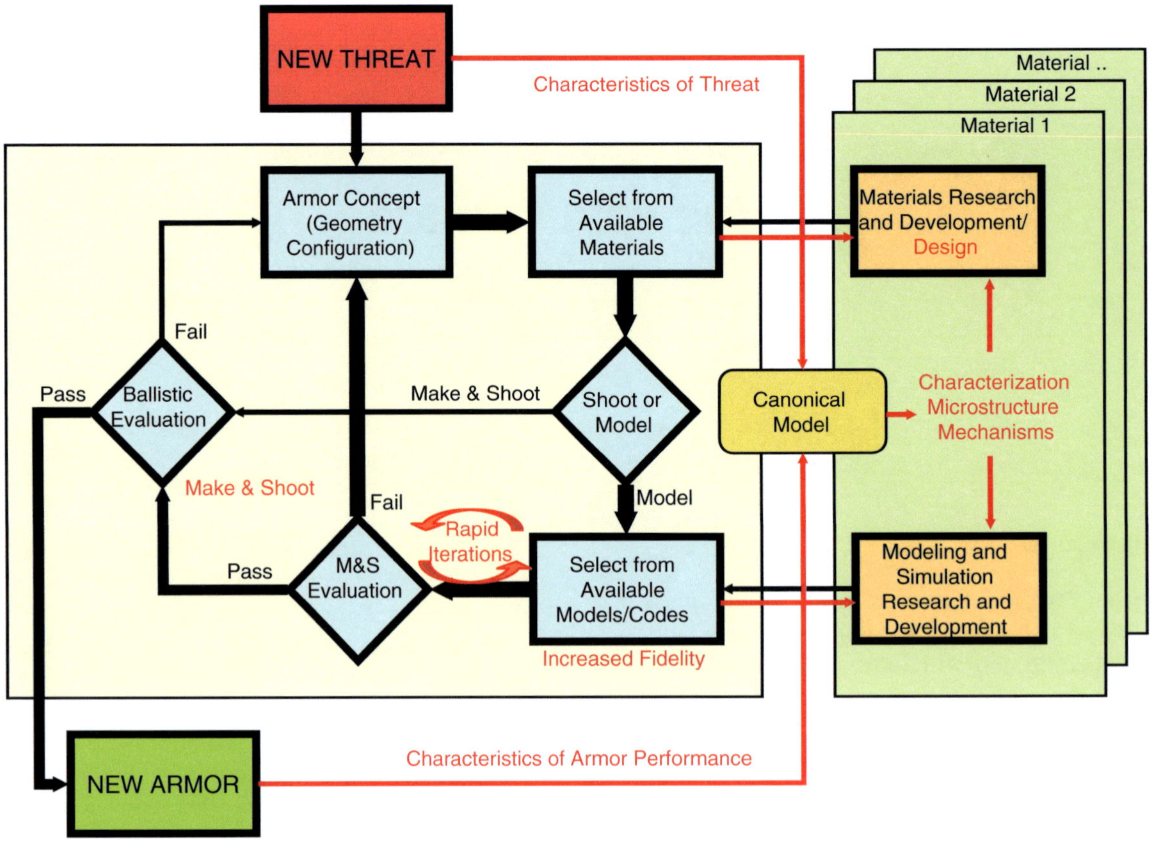

FIGURE 6-2 New paradigm for armor development. The new design path for armor provides enhanced and closer coupling of the materials research and development community and the modeling and simulation community, resulting in significantly reduced time for development of new armor. This path connects the armor design process to the materials research and development community through canonical models to deal with the restricted information problem. The elements of armor system design are not themselves new, but the emphasis shifts from design-make-shoot-redesign to rapid simulation iterations, and from designing with off-the-shelf-materials to designing that explores materials for their protective properties. The feedback loop between armor system design and material design contrasts with current practice, in which a one-way flow puts new materials on the shelf to be tried in the make-shoot-look process.

systems in operational use today are the product of years of heuristic-based advances. Development of the protection materials used in these systems is coupled only loosely to armor system design, with the coupling taking the form of inferred desired properties. The current paradigm of material and system development can be characterized as a design-make-test-redesign-repeat … iterative loop. The time and expense involved in such an approach limit the number of optimization iterations and slow the advance of new material systems that could provide the needed protection with reduced areal density.

The current paradigm and the research programs and organizations that support it are not sufficient to accelerate advances in lightweight protection materials. New research initiatives, organizational structures, and implementation approaches will be needed to increase the rate of progress.

The committee concludes that the ability to design and optimize protection material systems can be accelerated and made more cost effective by operating in a new paradigm for lightweight protection material development (Figure 6-2). In this new paradigm, the current armor system design practice is replaced by rapid iterations of modeling and simulation, with ballistic evaluation used selectively to verify satisfactory designs. Strong coupling with the materials research and development community is accomplished through canonical models that translate armor system requirements (which are often classified) into characterizations, microstructures, behaviors, and deformation mechanisms that an open research community can use. The principal objective of this new paradigm is to enable the design of superior materials and to accelerate their implementation in armor systems. The new paradigm will build on the multidisciplinary collaboration concepts and lessons from other applications documented in *Integrated Computational Materials Engineering* (ICME),[1] which cites many advances and several examples of successful implementation. It advocates pushing the large body

[1]NRC. 2008. Integrated Computational Systems Engineering: A Transformational Discipline for Improved Competitiveness and National Security. Washington, D.C.: The National Academies Press.

of existing computational materials science to the next step. Unfortunately, while the optimization of the materials, manufacturing processes, and component design is well described in the ICME report, the path forward for protection materials is far more complicated, since designs must deal with highly nonlinear and large deformations typically not encountered in commercial products, where applied stresses are kept well below the elastic limit in the linear regime.

The new paradigm can be focused on the most promising opportunities in lightweight protection materials, bringing such current products as ceramic plates and polymer fiber materials well beyond their present state of performance and opening the possibility for radically new armor system solutions to be explored and optimized in tens of months rather than tens of years.

The added features (indicated in red in Figure 6-2) of the new paradigm compared to the current paradigm are these:

- Canonical models explicitly link armor system design, which is typically done in a restricted setting, to protection materials research and design, which is typically done in an unclassified setting. A particular canonical model puts some aspect(s) of the dynamic protection problem into a standard form to be used as the basis for material system experiments and simulations. Each canonical model abstracts the key features of a threat and an armor configuration and expresses them in unclassified terms of dynamic material properties and behaviors needed to meet protection requirements. Such canonical models would be defined by an individual or group with the appropriate expertise and an intimate knowledge of both restricted and open research and development activity. A particular canonical model provides both material developers and model developers with a benchmark to use to evaluate potential improvements. Benefit: controlled linkage between open and restricted environments and a better match between armor system needs and potential material solutions.
- Design of the material or material system is based on an understanding of failure mechanisms invoked by projectiles or blast loads and uses physics-based modeling and simulation of the material or material system's behavior or performance within the dynamic. The design modeling and simulation of the material take place prior to the longer iterations that involve physical testing. The rapidity of model iterations makes it possible to explore more alternatives and optimize the material to provide the desired behavior. Benefit: faster development of higher-performance armor materials.
- A feedback loop to the armor system design flow better defines the required material behaviors. Benefit: faster iterations than today's make-and-shoot process and better requirements for achievable material behavior and dynamic properties.
- Simulation can model the consequences of specific process flows on the microstructure and hence the subsequent dynamic behavior and other important attributes (such as cost) before physically making the material. Benefit: higher yields, faster deliveries, lower costs.

Successful implementation of the new paradigm can, by dint of the insights gained from modeling and simulation, give armor system designers the freedom to work with novel as well as established materials to meet performance requirements. It can identify more rapidly than in the past how newly envisioned and to-be-developed materials and systems could create new opportunities for the protection afforded personnel, vehicles, ships, aircraft, and structures at lower weight and cost. The new approach would enable the reliable identification of materials that could be advantageous in protection applications, establish their merits and limitations, drive research and development to exploit the protective capacity of the new materials and systems, and, most importantly, bring about their rapid insertion into the field.

To realize the vision of this new paradigm and achieve these benefits, advances are needed on multiple fronts, including these:

- Better fundamental understanding of the mechanisms by which ballistic penetrators and blast loads interact with material systems at multiple scales, including insights into (1) dynamic properties at large strains, pressures, and high rates that go far beyond the usual quasi-static measures and (2) the resulting material behaviors that affect protection performance (see Chapter 3);
- Better computational approaches (physics-based models and codes for the evolution of failure) coupled with new experimental approaches allowing improved spatial and temporal measurement of damage evolution (see Chapters 3 and 4);
- The ability to design material compositions, crystal structures, microstructures, and composites over length scales from the atomic to the mesoscale to achieve behaviors that are important for protection performance and innovative processes that can synthesize and process these materials affordably (see Chapter 5); and
- An organizational structure and method of dealing with security constraints that will facilitate interaction and information sharing and enable successful basic and applied research to accelerate the development of these improved lightweight protection materials (detailed later in this chapter).

RECOMMENDATIONS FOR PROTECTION MATERIALS BY DESIGN

The recommendations in this section point out a way forward that will address the challenges outlined above by bringing together the efforts of university researchers, government labs, and industry to engage collaboratively in a long-term program of use-inspired fundamental research.

Overarching Recommendation. Given the long-term importance of lightweight protection materials to the Department of Defense (DoD) mission, DoD should establish a defense initiative for protection materials by design (PMD), with associated funding lines for basic and applied research. Responsibility for this new initiative should be assigned to one of the Services, with participation by other DoD components whose missions also require advances in protection materials. The PMD initiative should include a combination of computational, experimental, and materials testing, characterization, and processing research conducted by government, industry, and academia. The program director of the initiative should be given the authority and resources to collaborate with the national laboratories and other institutions in the use of unique facilities and capabilities and to invest in DoD infrastructure where needed.

This overarching recommendation requires actions in four important elements of the PMD initiative:

Element 1—Fundamental Understanding of Mechanisms of Deformation and Failure Due to Ballistic and Blast Threats

The first element of the PMD initiative would be to develop better fundamental understanding of the mechanisms of high-rate[2] material deformation and failure in various protection materials, discussed in Chapter 3. As part of the new paradigm, armor development should be considered not from the viewpoint of conventional bulk material properties but from the viewpoint of mechanisms. The deeper fundamental understanding could lead to the development of more failure-resistant material compositions, crystal structures, and microstructures and to protective materials with better performance. Moreover, by identifying the operative mechanisms and quantifying their activity, mathematical damage models can be written that may allow computational armor design. Chapter 3 discusses failure mechanisms for the several classes of materials.

Recommendation 6-1. The Department of Defense should establish a program of sustained investment in basic and applied research that would facilitate a fundamental understanding of the mechanisms of deformation and failure due to ballistic and blast events. This program should be established under a director for protection materials by design, with particular emphasis on the following:

- Relating material performance to deformation and failure mechanisms. Developing models and data for choosing materials based on their ability to inhibit or avoid failure mechanisms as opposed to choosing them based on bulk properties as measured in quasi-static and dynamic tests.
- Developing superior armor materials by identifying compositions, crystalline structures, and microstructures that counteract observed failure mechanisms and by establishing processing routes to the synthesis of these materials.
- Reducing the cost of production of protection materials by improving the processes and yields and by enhancing the ability to manufacture small lots.

Element 2—Advanced Computational and Experimental Methods

The second element of the PMD initiative would be to advance and exploit the capabilities of the emerging computational and experimental methods discussed in Chapter 4. The first objective is to predict the ballistic and blast performance of candidate materials and materials systems as a prelude to the armor design process. The second objective is to define requirements that will guide the synthesis, processing, fabrication, and evaluation of protection materials. The PMD initiative would develop the next generation of

- DoD advanced protection codes that incorporate experimentally validated, high-fidelity, physics-based models of material deformation and failure, as well as the necessary high-performance computing infrastructure;
- Experimental facilities and capabilities to assess and certify the performance of new protection materials and system designs, as well as provide insight into fundamental material behaviors under relevant conditions with unprecedented simultaneous high spatial and temporal resolution; and
- Collaborative infrastructure for encouraging direct communication and improved cooperation between modelers and experimenters, through both (1) the establishment of collaborative environments and (2) requirements in proposals when the specific research topic is well served by such collaboration.

The high-priority opportunities identified in Chapter 4 will need sustained investment and program direction to advance computational and experimental capabilities. The envisioned computational capabilities must be devel-

[2]Ballistic velocities typically range from several hundred to several thousand meters per second and can lead to strain rates of up to $10^5\ s^{-1}$.

oped in partnership with a strong experimental effort that identifies the dynamic mechanisms of material behavior. These mechanisms must be understood and modeled for the activity to be successful, the material characteristics and properties must be known for the simulations to be carried out, and the outcomes of the computational modeling must be validated.

Recommendation 6-2. The Department of Defense should establish a program of sustained investment in basic and applied research in advanced computational and experimental methods under the director of the protection materials by design (PMD) initiative, with particular emphasis on the following:

- *Dynamic mechanism characterization.* Identify and characterize (1) the failure mechanisms underlying damage to a material caused by projectiles from weapons and detonations and (2) the compositional and microstructural features of each constituent of the material, as well as the material's overall structure. An enhanced experimental infrastructure will be needed to make progress in high-resolution (time and space) experiments on material deformation and failure characterization.
- *Code validation and verification.* Focus on multiscale, multiphysics material models, integrated simulation/experimental protocols, prediction with quantified uncertainties, and simulation-based qualification to help advance the predictive science for protection systems.
- *Challenges and canonical models.* Periodically propose open challenges comprising design, simulation, and experimental validation that will convincingly demonstrate the PMD. Each challenge problem must address the corresponding canonical model and must result in quantifiable improvements in performance within that framework.

Element 3—Development of New Materials and Material Systems

The third element of the PMD initiative is the development and production of new materials and material systems whose characteristics and performance can achieve the behavior validated in modeling and simulation of the new armor system. The recommendations in this element target the most promising opportunities identified in Chapter 5.

Recommendation 6-3. The Department of Defense should establish a program of sustained investment in basic and applied research in advanced materials and processing, under the director of the PMD initiative program, with particular emphasis on the following:

- *A sustained effort to develop a database of high-strain-rate materials for armor.* Material behavior and dynamic properties must be measured and characterized over the range of strains, strain rates, and stress states in the context of penetration and blast events. Develop a comprehensive database of materials that exhibit high-strain-rate behavior and consider them as materials of interest. The PMD director should designate a custodian for this database and arrange for experimental results of the PMD program to be provided to the database and shared with the research community. The database should include ceramics, polymers, metals, glasses, and composite materials in use today and should be expanded as new materials are developed.
  —*Opaque and transparent ceramics and ceramic powders.* The intrinsic properties of opaque and transparent ceramics and ceramic powders are not yet fully realized in armor systems. There is need for understanding at the atomic, nano-, and micron levels of how powders and processing can be designed and manipulated to maximize the intrinsic benefits of dense ceramic armor and reduce production costs.
  —*Polymeric, carbon, glass, and ceramic fibers.* There is an opportunity to develop finer diameter and more ideally microstructured polymeric and carbon fibers with potentially a two- to fivefold improvement in specific tensile strength over the current state of the art. Such improvements would significantly reduce the weight of body armor.
  —*Polymers.* In addition to polymer fibers, thermoplastic and thermoset polymers are used as monolithic components and also serve as matrixes in various composites. Improved measurements of and models for the deformation mechanisms and failure processes are needed for thermoplastic- and thermoset-based protection materials.
  —*Magnesium alloys.* The very low density of magnesium provides potential for the development of very lightweight alternatives to traditional metallic materials in protection material systems. The basic understanding of strengthening mechanisms in magnesium should be advanced, especially the development of ultra-fine-grained magnesium alloys through severe plastic deformation. Magnesium-based fibers are also worthy of exploration.
- *Adhesives and active brazing/soldering materials.* Development of adhesives and active brazing/soldering materials and their processing methods to match the elastic impedance of current materials while minimizing the thermal stresses will improve the ballistic and blast performance of panels made of bonded armor, including transparent armor.

- *Test methods.* Advances are needed in test methods for determining the high strain rates (10^3 to 10^6 s^{-1}) and dynamic failure processes of (especially) fibers, polymers, and ceramics. Results should be passed on to the designated database of materials with high-strain-rate behavior.
- *Material characterization.* The characterization of, composition, crystalline structure, and microstructure at appropriate length scales is a key task that will need more attention to take advantage of the improved experimental tools for quantifying initial and deformed microstructures.
- *Cost reduction.* Advances are needed to reduce the cost of producing protection materials by improving their processing and yield and by improving small-lot manufacturing capability.
- *Processing science and intelligent manufacturing.* Advances are needed in basic understanding of and ability to model the consequences of material processing for performance and other characteristics of interest. Intelligent manufacturing sensing and control capabilities are needed that can maintain low variance and produce affordable protection materials, even in relatively low volumes.

Element 4—Organizational Approach

The fourth element of the PMD initiative is an organizational construct for multidisciplinary collaboration among academic researchers, government laboratories, and industry, in both restricted-access and open settings. The PMD initiative will need strong top-level leadership with insight into both the open and restricted research environments and the authority to direct funding and set PMD priorities. The program will require committed funding to ensure long-term success and should be subject to periodic external reviews to ensure that high standards of achievement are established and maintained. To meet these requirements, the committee considered several organizational alternatives, described in the sections below, and concluded that the notional DoD organizational approach depicted in Figure 6-3 includes the features necessary for success.

Recommendation 6-4. In order to make the major advances needed for the development of protection materials, the Department of Defense should appoint a PMD program director, with authority and resources to accomplish the following:

- Plan and execute the PMD initiative and coordinate PMD activities across the DoD;

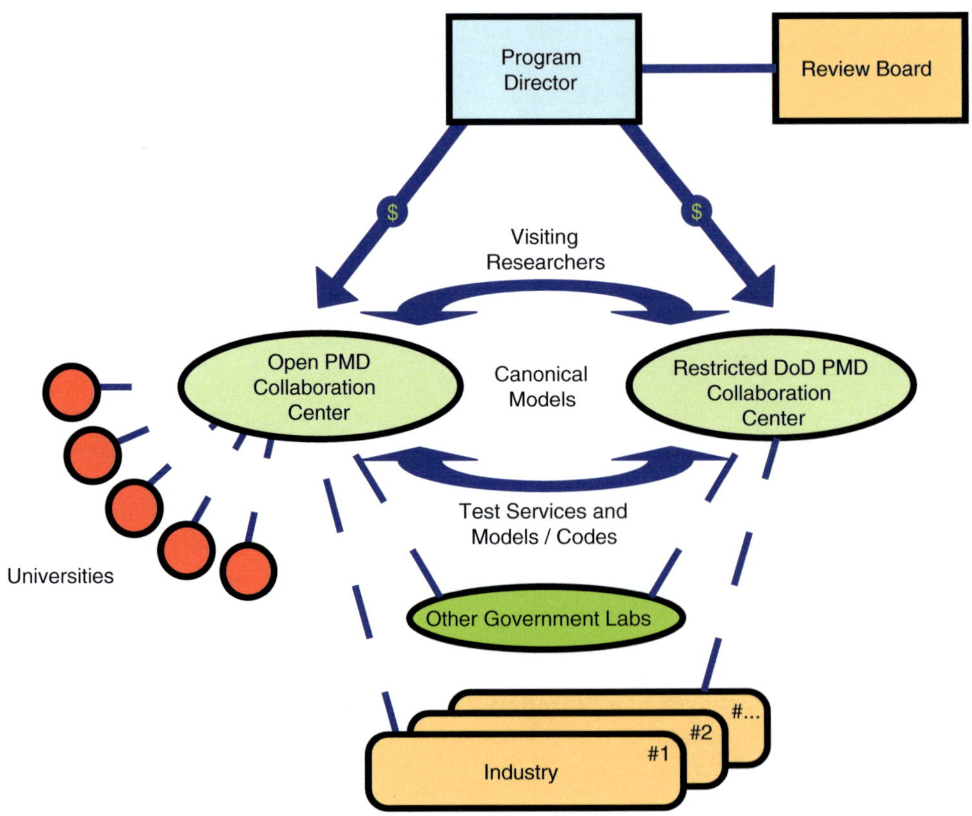

FIGURE 6-3 PMD initiative organizational structure involving academic researchers, government laboratories, and industry.

- Select an existing facility to be the DoD center for PMD and fund a research director and the staff, equipment, and programs needed by the PMD initiative;
- Award a competitive contract for an open access PMD center whose mission would be to host and foster open collaboration in research and development of protection materials;
- Establish an external review board to conduct periodic reviews of programs in both centers; and
- Provide liaison with the Department of Energy, the National Institute of Standards and Technology, and other government laboratories on matters related to PMD.

CRITICAL SUCCESS FACTORS FOR THE RECOMMENDED NEW ORGANIZATIONS

DoD Center for the PMD Initiative

The essential features of the recommended organization are as follows:

- A program director for the PMD initiative who is responsible for planning and overseeing the execution of basic and applied research in both classified and unclassified settings. The director might be organizationally located within the lead service for protection materials, the Department of the Army, but in any event would be responsible for coordinating research and development in lightweight protection materials across the rest of DoD.
- Selection by the above-mentioned program director of an existing DoD organization to become the DoD center for the PMD initiative. This program director would appoint a research director for the center and would organize funding for it. The center would be staffed and equipped for classified research in materials synthesis, processing, and testing, as well as computational facilities to enable materials design and armor design. The DoD center for the PMD initiative would have both internal capabilities and access to external capabilities as needed for advanced modeling and simulation, protection materials synthesis, processing, characterization, fabrication, ballistic and blast testing, and evaluation of protection material systems. This organization would accommodate visiting researchers who would be granted security clearances for the duration of their rotating assignments. It would also provide testing services and evaluation results at both classified and unclassified levels to qualified external researchers.
- An open PMD collaboration center, with a physical experimentation center and virtual collaboration links to distributed academic, government, and industry researchers. This organization and its research director would promote collaboration among modelers and experimentalists. The center could facilitate multidisciplinary information exchange and enforce appropriate boundaries for restricted information. It would have both internal facilities and remote access to facilities for experimentation and material characterization. It would also maintain awareness of the open literature and global technology developments and actively enhance them and would provide input to the program director and the staff of the classified center at DoD.
- An external review board, duly constituted to review programs and advise the director in the planning and conduct of research in protection materials, both restricted and open.
- Links to Department of Energy labs, the National Institute of Standards and Technology labs, and other government labs whose research and capabilities are relevant to protection materials research.

The proposed research and development program would require the collaboration of scientists and engineers from DoD research laboratories, other national laboratories, universities, independent research institutes, and commercial companies in settings that can foster collaboration while maintaining boundaries for unclassified, proprietary, export controlled, and classified information.

Given the constraints of current classification guidelines, research and development involving specific threats and vulnerabilities will require access to a facility where restricted research and testing—that is, research that is either classified or otherwise not available for public release—can be conducted. The committee believes designation of an existing DoD organization as the lead laboratory for the PMD initiative would be the best way to meet this need. This open collaboration center would need to have capabilities for the following:

- Materials characterization,
- Model development and simulation against real threats,
- Armor system prototyping, and
- Ballistic and blast testing and evaluation against real threats.

Such capabilities exist at the Army Research Laboratory facility in Aberdeen, Maryland, and at other DoD facilities.

To tap the sources of innovation in academia and industry, an environment for collaboration outside restricted governmental facilities would be needed. This open research community would have the following capabilities:

- Experimental facilities for dynamic measurements of material behavior, including in situ visualization of high-rate deformation and failure processes.
- Modeling and simulation capabilities.

- Materials design and processing design capabilities.
- Collaboration between modelers and experimentalists, supported by information sharing and virtual collaboration environments.
- An enclave where classified or restricted information could be exchanged among researchers with the appropriate clearances.
- Physical meeting facilities.
- Proximity or easy transportation access to the DoD PMD center.

The key to success would be to link these two research environments through formal organizational relationships, personnel exchanges, funding and program direction, and processes to translate classified information on threats and materials into canonical models suitable for academic research topics. Procedures would be needed to adapt data from the classified center for use by the open environment.

Open PMD Collaboration Center

Of the organizational elements, the newest and most far-reaching area for investment would be the open PMD collaboration center. This center would be a vibrant intellectual engine that attracts the best academic researchers across multiple organizations to address well-defined problems in material design, high-strain-rate experimentation, analytical and computational modeling across the length and timescales, and materials processing for protection applications. It could foster precompetitive collaboration with industry for both fundamental research and technology transfer.

The key features of the open center would include these:

- Academic opportunity for interesting problems, funding, access to state-of-the-art facilities, up-to-date data, workshops, and publications.
- Canonical models that support unclassified research objectives stated in terms of material behaviors or other fundamental phenomena.
- Open competition for new research awards from both the PMD initiative and other basic and applied research sponsors.
- Multidiscipline and multiuniversity collaboration.
- University-industry-government collaboration.
- Means to host visiting researchers.

The committee considered several organizational alternatives that might have the desired attributes of this new entity, including the following:

- *Multidisciplinary University Research Initiative (MURI).* DoD currently uses MURI awards to support university research that intersects two or more traditional science and engineering disciplines. Teaming in multidisciplinary research helps to transition basic research findings to practical application and accelerates research progress by cross-fertilization of ideas. In supporting these team efforts, MURI complements other DoD programs that support university research via single-investigator awards. Typically, awards cover a period of 3 years; 2 additional years are possible. This model is strong on multidiscipline and multiuniversity collaboration but is not typically used for collaboration with industry.[3]
- *University Affiliated Research Center (UARC).* This appellation is given to university laboratories that maintain critical competencies in technology and systems that support national defense. University Affiliated Research Centers (UARCs) are awarded as noncompetitive DoD contracts through a provision of the Competition in Contracting Act (CICA) of 1984, as codified in 10U.S.C. 2304(c)(3)(B), which authorized noncompetitive contracts with educational institutions where necessary for DoD to establish or maintain essential engineering, research, and developmental capabilities. UARCs support DoD through a special strategic relationship, wherein they serve as trusted technical advisors free from commercial conflicts of interest. The requirement for maintaining a UARC and its associated funding is driven by the specific needs of sponsoring DoD RDT&E programs. These special needs are manifest in core competencies, specified by the sponsors, which define the scope of services to be provided by the UARC. The current ARL Material Centers of Excellence in Ceramics, Metals, Polymers and Composites are not UARCs but have similar characteristics. The UARC model has to be broadened to provide the type of open competition and academic opportunity envisioned for an open PMD collaboration center.[4]
- *Collaborative Technology Alliance (CTA).* CTAs are collaborations between academia, Army labs or centers, and private industry. Their goal is to rapidly transition new technologies to warfighters, thereby enabling the Army's Future Force. This partnership is at the core of the CTA concept, wherein each partner brings a unique approach to research. Academia, for example, brings cutting-edge innovation; ARL researchers maintain the focus on solving complex Army technology problems; and the industrial partners are able to solve technology bottlenecks and to leverage existing research results. In this way, multidisciplinary research teams bring about the complex technology needed to solve the Army's complex problems. The program brings world-class research and development talent to bear on meeting Army

[3]For more information, see http://www.wpafb.af.mil/library/factsheets/factsheet.asp?id=9327.

[4]For more information, see http://www.hawaii.edu/uhmfs/uarc/Attach_003.UARCMgmt.pdf.

needs for technology. The CTA model has most of the desired characteristics but has traditionally been used when technology advances are driven more by market forces than by government needs, which is not the case in protection materials.[5]

- *Fraunhofer-like institute.* Fraunhofer institutes originated in Germany. Affiliated centers have been established in the United States to perform applied research under contract to government and industry for such customers as federal and state governments, multinational corporations, and small- to medium-sized companies. Each center is partnered with a major research university. These partnerships serve as bridges between academic research and industrial needs. Such bridges would fill some but not all the government needs for the open PMD collaboration center.[6]
- *Engineering Research Centers (ERCs) such as those sponsored by the National Science Foundation (NSF).* Located at universities throughout the United States, ERCs are a group of interdisciplinary centers that partner closely with industry. Each ERC provides an environment in which academe and industry can collaborate in pursuing strategic advances in complex engineered systems and systems-level technologies that could spawn whole new industries or radically transform the processing technologies, product lines, or service delivery of current industries. Activity within ERCs lies at the interface between the innovation-driven culture of engineering and the discovery-driven culture of science. The centers provide the intellectual foundation for industry to collaborate with faculty and students on producing the knowledge base needed for steady advances in technology, resolving long-range, generic challenges and rapidly transitioning results to the marketplace. The academic opportunity criterion could be well met by incorporating ERC characteristics, but additional features would be needed to give the government a stronger role in guiding the research to meet government needs.[7]

[5]For more information, see http://www.arl.army.mil/www/default.cfm?page=93.

[6]For more information, see http://www.fraunhofer.org/.

[7]For more information, see http://www.erc-assoc.org/index.htm.

- *University-industry consortium models.* Numerous examples of university research consortia, industry research consortia, and membership organizations fall into this category. The committee considered the NSF Industry/University Cooperative Research Centers (I/UCRCs),[8] in particular the Ceramic, Composite and Optical Materials Center at Rutgers University,[9] to be in this category. It also considered the National Textile Center,[10] the Semiconductor Research Consortium,[11] and the National Warheads and Energetics Consortium[12] as examples. Some of these consortia are federally funded and others operate on industry funding. A common denominator in such consortia is a business case for industry participation. Assuming such a business case can be made, these models, like the CTA model above, could meet most of the desired criteria. The model would need to be tailored to focus on canonical models as the bridge to the government needs.

The committee concluded that none of these models would meet all the needs of an open PMD collaboration center but that the various university and industry consortium models have proven features that the Army could combine to define the contract for such a center.

Time Frame for Anticipated Advances

While it is always problematic to try to predict the future, it is apparent that some areas are ripe for rapid progress and discovery. The committee believes, for example, that increased funding of basic research on high-rate deformation of polymer fibers and ceramics could, within about 10 years (depending on the level of effort), achieve a level of understanding that would rival the current understanding of metals. Progress will be aided as national lab facilities for extremely fast data acquisition during high-rate events become available and as researchers design experiments to take advantage of such facilities.

[8]For more information, see http://www.nsf.gov/eng/iip/about.jsp.

[9]For more information, see http://www.ccmc.rutgers.edu.

[10]For more information, see http://www.ntcresearch.org/mission.htm.

[11]For more information, see http://www.src.org/about/.

[12]For more information, see http://www.nwec-dotc.org/.

Appendixes

Appendix A

Background and Statement of Task

Current operations in Iraq and Afghanistan unambiguously demonstrate the need for threat-specific, ultra-lightweight transparent and opaque armor in many Army systems, constructed facilities, and personnel protection. As the threats have escalated and become more varied, the challenges for rapidly developing optimized threat-specific, passive lightweight armor packages have grown complex because of the interplay of issues involving energy absorption and momentum transfer issues. Critical components for further accelerating the optimization of these material systems are the development of validated predictive-performance computer models, materials design tools, and integrated structural design to take advantage of advanced materials technology. This approach is based on the determination and quantification of the various impact energy absorption mechanisms, including the various deformation modes, damage nucleation and accumulation processes, and the resulting eventual failure of materials at high rates under very high impact stress (shock wave).

Over the past few years there have been major initiatives that bear on this activity. In 2007, the Army Research Office, in conjunction with other Army Research Laboratory groups, convened the workshop "Impact Damage on the Performance of Armor Ceramics." More recently, the National Research Council's National Materials Advisory Board completed a study entitled *Integrated Computational Materials Engineering: A Transformational Discipline for Improved Competitiveness and National Security*. In addition, the Basic Energy Sciences Office of the U.S. Department of Energy convened a high-level study committee on "Directing Matter and Energy: Five Challenges for Science and the Imagination" and another on "Basic Research Needs for Materials under Extreme Environments." In May 2008, the U.S. Army's Engineer Research and Development Center (ERDC) and the Army Research Office (ARO) conducted a workshop to share emerging fundamental discoveries in experimentation, theory, and computational methods for the mechanics of cementitious and ceramic materials. Then on September 22-24, 2008, a major Army Research Laboratory workshop, "Multi-Scale Materials Behavior in Ultra-High Loading Rate Environments," focused on multiscale materials and mechanics for dynamic energy management at the macro- and microscale.

In order to design and produce impact-resistant advanced materials and systems, validated, robust multiscale physics-based models (atomistic to polycrystalline to continuum) are needed to simulate reliably the mechanical response of such materials and systems in extreme environments. It is well known, for example, that variation in material characteristics (phases, microstructure, and defects) including grain boundaries and intergranular films can significantly affect the quasi-static, mechanical behavior of structural ceramics. There are, however, many significant differences between the high-rate and quasi-static stress environments, including differences in the following areas: stressed volume; overstressed condition; propagation and rate of stress waves (compression, tensile, and shear); kinetic effects; mixed, spatially varying macrostress states; activation of new micromechanical mechanisms; and possibility of phase transformations, among others. Ultimate failure is a function of the temporal and spatial interaction of the macrostresses with the ceramic materials at the microstructural and nanostructural scales, including elastic and inelastic (plastic) deformation, damage nucleation, and evolution and resulting failure from the macroscale (top down) or from the nanoscale (bottom up). The macromechanical responses (constitutive equations), assuming homogeneous, defect-free mechanically isotropic bodies, are very well known, but the spatial micromechanical responses and stochastic variability are not nearly as well established. As computing power and speed continue to increase, the ability to simulate the mechanical response at the microstructural and mesostructural level will become much more important. Many existing models and codes, being extrapolations from metal behavior, exclude defects, microcracking, ceramic plasticity, ceramic-specific failure mechanisms, high-pressure phase transformations, sample homogeneity, and sample-to-sample variability. Although this example focuses on the challenges associated with ceramics, similar materials-specific complexities arise in composite materials, fibers, and textiles, in concrete and laminated assemblies of multiple materials, and the associated interfaces contained therein.

STATEMENT OF TASK

An ad hoc committee will conduct a study and prepare a report on protection materials for the Army to explore the possibility of a path forward for these materials. Specifically, the committee will:

1. Review and assess the current theoretical and experimental understanding of the major issues surrounding protection materials.
2. Determine the major challenges and technical gaps for developing the future generation of light weight protection materials for the Army, with the goal of valid multi-scale predictive simulation tools for performance and, conversely, protection materials by design.
3. Suggest a path forward, including approach, organizational structure and teaming, including processing, material characterization (composition and microstructure), quasi-static and dynamic mechanical testing and model development and simulation and likely timeframes for the Army to deliver the next generation protection materials.

In considering these questions, the committee should consider the following:

- Shock wave energy dissipative (elastic, inelastic and failure) and management mechanisms throughout the full materials properties spectrum (nano through macro).
- Experimental approaches and facilities to visualize and characterize the response at nano and mesoscales over short time scales.
- Multi-scale modeling techniques to predict energy dissipative mechanisms (twinning, stacking faults, etc.) from the atomic scales and bulk material response.
- Materials and material systems issues including processing and characterization techniques focusing on intrinsic (single crystal) properties and processing controlled extrinsic characteristics (phases, microstructure, interfaces).

Appendix B

Biographical Sketches of Committee Members

Edwin L. Thomas (NAE), *Chair*, professor and department head of materials science and engineering at Massachusetts Institute of Technology (MIT), carries out research on photonics, phononics, interference lithography and mechanical behavior of microtrusses, polymer physics and engineering of the mechanical and optical properties of block copolymers, liquid-crystalline polymers, and hybrid organic-inorganic nanocomposites. Professor Thomas has a special interest in the area of photonics and the fabrication of polymeric photonic crystals using self-assembly, especially with block copolymers, and holographic interference lithography. For these studies, much emphasis is placed on the understanding of complex relations between the lattice symmetry and optical properties of periodic structures. Another area of particular focus is phononics. Professor Thomas's group is exploring the way that light and sound propagate in quasi-crystalline photonic and phononic structures. Other major topics in his research are structured polymers. His structured-materials research concentrates on enhancing the ability to fabricate complex structures with characteristic length in submicron and nanometer ranges in order to create materials with superior properties that can be tailored to a particular application. Understanding the influence of composition and processing conditions on the resultant microstructure of polymers and how this determines the properties is the central part of his polymer morphology research. Professor Thomas is also the founding director of MIT's Institute for Soldier Nanotechnologies (ISN), where advanced nanotechnology research seeks to improve the survival of the soldier of the future. The ISN was founded in March 2002 with the help of a $50 million contract from the U.S. Army, and now entering its third 5-year contract, its charge is to pursue a long-range vision for how technology can make soldiers less vulnerable to enemy and environmental threats. The ultimate goal is to create a 21st-century battlesuit that combines high-tech capabilities with light weight and comfort.

Michael F. McGrath, *Vice Chair*, is the vice president for systems and operations analysis at Analytic Services Inc. (ANSER), a not-for-profit government services organization. He previously served as the Deputy Assistant Secretary of the Navy for Research, Development, Test and Evaluation, and in that position, he was a strong proponent for improvements in technology transition, modeling and simulation, and testing and evaluation. In prior positions, Dr. McGrath served as the vice president for government business at the Sarnoff Corporation, assistant director for manufacturing at the Defense Advanced Research Projects Agency (DARPA), and director of the Department of Defense's (DoD) Computer-aided Acquisition and Logistics Support (CALS) Program. While at DARPA, Dr. McGrath managed the Affordable Multi-Missile Manufacturing Program and the Agile Manufacturing Program, which developed technologies for distributed engineering and manufacturing processes and teams. He also led DoD's Research and Development (R&D) planning program Technology for Affordability. He has maintained research interests in information systems, supply chains, and manufacturing technologies. He is a member of the National Research Council's (NRC's) Board on Manufacturing and Engineering Design, and he chaired the 2002 NRC study *Equipping Tomorrow's Military Force: Integration of Commercial and Military Manufacturing in 2010 and Beyond*. Dr. McGrath's expertise includes defense R&D programs and organizational management, defense acquisition, systems engineering, manufacturing enterprise systems, and life-cycle support. He holds a B.S. in space science and applied physics (1970) and an M.S. in aerospace engineering (1972) from Catholic University and a doctorate in operations research from George Washington University (1985).

Relva C. Buchanan is a professor and former head of ceramics and materials science in the Department of Chemical and Materials Engineering at the University of Cincinnati. His research focus is on electroceramics materials as components for passive devices and various microelectronics sensing applications. Included are ferroelectric thin-film systems and also core-shell/barrier-layer structures, developed in donor doped $BaTiO_3$ ceramics, with superior dielectric and strain properties, of interest for supercapacitor and sensor applications. Dr. Buchanan's research interests include Ni-ZrO_2 and Ni/NiO composite film structures for fuel cell and capacitive electrode systems and thermistor use. Conductive polymer/carbon composite structures for electromagnetic shielding and thermistor and toxic gas detection, as well as low-temperature glasses for thick-film use, are also areas of his ongoing research. Dr. Buchanan is a fellow of the Graduate College, University of Cincinnati; a fellow of the American Ceramic Society; a fellow of the American Society of Metals (International); and a member of the National Institute of Ceramic Engineers. He has served as trustee of the American Ceramic Society and chair of its Programs and Meetings Committee. He is a member of the Ferroelectrics Program Committee of the Institute of Electrical and Electronics Engineers and currently serves on several international review committees: the International Panel on Evaluation of Portuguese Materials Science Research, the International Advisory Committee of Electroceramics European Conferences V-VIX, the International Conference on Electroceramics, 2003 through 2009, and the U.S.-Japan conference committee on dielectrics. He has served also on several national review committees, including as chair of the Energy Technology Review Committee, University of Chicago/Argonne National Laboratory, and on the Ohio Science and Technology Council. Dr. Buchanan has authored more than 150 technical and review articles (e.g., in the *Journal of Materials Research*, *Applied Physics Letters*, the *Journal of the American Ceramics Society*, *Sensors and Actuators*, and others). He has given more than 120 invited talks and more than 100 technical presentations (with his students) and has co-authored or authored six books. His book *Ceramic Materials for Electronics: Process, Properties, and Applications* (Marcel Dekker, 1991; 3rd ed., 2004) is widely used in the field, as is his book *Materials Crystal Chemistry* (Dekker, 1997). He teaches courses in materials science, ceramic processing, materials crystal chemistry, functional ceramic devices, electrical ceramics, and glass and glass properties.

Bhanumathi Chelluri is a senior research scientist and program manager at IAP Research, Inc. Dr. Chelluri received her M.S. in physics (1974) and Ph.D. in materials science and engineering (1980) from the University of Illinois at Champaign-Urbana. After completing her Ph.D., she worked at the Max-Planck Institute in Germany for 2 years. On returning to the United States, Dr. Chelluri joined AT&T Bell Laboratories in New Jersey in the molecular beam epitaxy and research and development group. In 1989, she joined IAP as program manager of the advanced materials group. She has initiated and worked on a broad range of materials research areas, including metals, ceramics, composites, magnetic materials, thin films, nanomaterials, and semiconductors, with an emphasis on production and production capacity. Her recent focus has been on dynamic processing and production of powder materials using submillisecond-duration dynamic pressures. The process has also been successfully applied to armor-grade materials. Dr. Chelluri is the inventor of the dynamic magnetic powder compaction process. She holds six patents and has four patents pending related to the processing of advanced powder materials. She led numerous development projects as principal investigator, including Applied Technology Programs and Department of Defense and Department of Energy research programs. Dr. Chelluri has authored over 60 publications, of which several are invited feature articles. She has presented numerous invited talks at national and international conferences. Dr. Chelluri is the IAP corporate representative for the Metal Powder Industries Federation and the Edison Welding Institute. She holds professional membership in the following: Metal Powder Industries Federation; the Advanced Particulate Materials Association; the American Society for Metals; the Metals, Minerals and Materials Society; the American Ceramic Society; and the European Powder Metallurgy Association.

Richard A. Haber is a professor of material science and engineering at Rutgers University. Professor Haber is also the director of the Center for Ceramic Research, the oldest active National Science Foundation Industry/University Cooperative Research Center in the United States. Professor Haber also is the manager of the U.S. Army Research Laboratory's Material Center of Excellence for Ceramics in Lightweight Vehicular Armor at Rutgers. He has been on the faculty of Rutgers since 1984. He received his B.S., M.S., and Ph.D. degrees from Rutgers University. He is a fellow and past vice president of the American Ceramic Society and past president of the Ceramic Manufacturers Council. Professor Haber has written more than 90 papers and presented more than 250 lectures worldwide, on a range of topics including the following: ceramic processing, minerals processing, characterization of ceramic materials, strategic mineral and material utilization, nondestructive analysis, and structure-property relations in armor ceramics.

John Woodside Hutchinson (NAS/NAE) is the Lawrence Professor of Engineering, School of Engineering and Applied Sciences, Harvard University. Professor Hutchinson and his collaborators work on problems in solid mechanics concerned with engineering materials and structures. Buckling and structural stability, elasticity, plasticity, fracture, and micromechanics are all relevant research topics. Research activities include efforts to develop a mechanics framework for assessing the durability of thermal barrier coatings (TBCs) and the development of a fracture approach for structures subject to intense dynamic loads. Industrial efforts are under way to exploit these ceramic coatings, which are now widely used in aircraft and power generation turbines to shield engine blades and essentially all metal surfaces from high temperatures, thus enabling even higher operating temperatures. The technological challenge is to enhance the lifetime of the coatings under more severe operating conditions given their tendency to delaminate and spall. The effort involves collaboration with a broad community of engineers and material scientists who are actively exploring all aspects of TBCs. A wide range of efforts are also under way to develop new concepts for metallic structures with enhanced blast resistance (fracture now generally limits the maximum sustainable load). Professor Hutchinson's current work in this area is focused on the development of fracture models that can be employed in structural analysis codes to predict both the onset of failure and its progression.

Gordon R. Johnson is a program director in the Engineering Dynamics Department at Southwest Research Institute. Previously he was a chief fellow at Honeywell/ATK and a senior scientist at Network Computing Services/Army High Performance Computing Research Center. He received a B.S., M.S.C.E., and Ph.D. from the University of Minnesota in 1964, 1966, and 1974, respectively. Dr. Johnson is the originator and principal developer of the EPIC computer code, which has been used extensively by the Department of Defense, the Department of Energy, and industry for computations involving high-velocity impact and explosive detonation. He has developed numerous algorithms for finite elements, meshless particles, contact, and linking of particles to elements. He has also been a developer of the Johnson-Cook strength and failure models for metals, the Johnson-Holmquist models for ceramics (JH-1 and JH-2), the Johnson-Holmquist-Beissel model for ceramics with a phase change, the Holmquist-Johnson-Cook model for concrete, and the Johnson-Beissel-Cunniff models for fabrics and composites. He is the author of numerous publications, served on the National Research Council's Committee on the Safety and Security of Commercial Spent Fuel Storage, and received the H.W. Sweatt Award from Honeywell and the Distinguished Scientist Award from the Hypervelocity Impact Society.

Satish Kumar is professor of materials science and engineering at the Georgia Institute of Technology. He received his M.Sc. degree in physics in 1975 from the University of Roorkee, India (now I.I.T. Roorkee) and his Ph.D. in the field of fiber science in 1979 from the Textile Technology Department at the Indian Institute of Technology, New Delhi. He obtained his post-doctoral experience in polymer science and engineering under the tutelage of Professor R.S. Stein at the University of Massachusetts, Amherst. He conducted research as a foreign collaborator at C.E.N.G. at Grenoble, France, a laboratory of the Atomic Energy Commission of France, using small-angle scattering studies to understand the structure of ion-containing polymers. From 1984 to 1989, Dr. Kumar was associated with the polymer branch at the Air Force Research Laboratory, Wright-Patterson Air Force Base, Ohio, as an onsite contractor through Universal Energy Systems and subsequently through the University of Dayton Research Institute. At the Air Force Research Laboratory, the focus of his research was structure, processing, and properties of rigid-rod polymeric fibers, as well as structural studies of carbon fibers and thermosetting and thermoplastic resins. His current research and teaching interests are in the areas of structure, processing, and properties of polymers, fibers, and composites, with an emphasis on polymer-carbon nanotube nanocomposites. Dr. Kumar has conducted fiber processing and structure-property studies on a broad range of polymers, including synthetic and natural polymers, as well as carbon fibers. His areas of research interest also include the ability of carbon nanotubes to nucleate polymer crystallization as well as their ability to template polymer orientation. He is also conducting research on carbon-based electrochemical supercapacitors, with the objective of enhancing their energy density. He serves on the editorial advisory boards of several journals in the field.

Robert M. McMeeking (NAE), a professor of mechanical engineering and professor of materials, University of California, Santa Barbara (UCSB), earned a B.Sc. (with first class honors) in mechanical engineering at the University of Glasgow, Scotland, in 1972, finishing first in his class of mechanical engineers. He then completed his M.S. and Ph.D. in solid mechanics at Brown University under the supervision of Professor James R. Rice, with dissertations on finite deformation plasticity methods for finite elements and ductile crack tip blunting in metals. He was at Stanford University for 2 years working on metal forming problems with Professor Erastus H. Lee. After 7 years at the University of Illinois at Urbana-Champaign on the faculty of the Theoretical and Applied Mechanics Department, Professor McMeeking went to UCSB in 1985 as a professor of materials and of mechanical and environmental engineering. He was chair of the Department of Mechanical and Environmental Engineering at UCSB in 1992-1995 and again during 1999-2003. He has written more than 220 scientific papers on such subjects as plasticity, fracture mechanics, computational methods, glaciology, tough ceramics, composite materials, materials processing, powder consolidation and sintering, ferroelectrics, structural evolution, nanotribology, actuating structures, blast and fragment protection of structures, fluid structure interactions arising from underwater blast waves, and the mechanics of the cell and its cytoskeleton. In 1983, Professor McMeeking was a Science and Engineering Research Council Senior Visiting Fellow at Cambridge University. In 1995-1996 he was a visiting professor at Cambridge University and was honored as a visiting scholar at Pembroke College. He was Southwest Mechanics Lecturer in 1988, and a plenary lecturer at the Seventh International Congress on Fracture in 1989, and he was honored as a

Midwest Mechanics Lecturer in 1992-1993 and as the Arthur Newell Talbot Lecturer at the University of Illinois at Urbana-Champaign in 2007. In 1998 he was elected a fellow of the American Society of Mechanical Engineers and in 2002 was recognized by the Institute for Scientific Information as a highly cited researcher in the fields of materials science and engineering. He was also promoted to fellow of the American Academy of Mechanics in 2002 and in 2004 was given an Alexander von Humboldt Research Award for senior U.S. scientists. Professor McMeeking was elected to the National Academy of Engineering in 2005 and was given the Brown University Engineering Alumni Medal in 2007. He has served as a reviewer for funding agencies such as the National Science Foundation, the Department of Energy, and the Army Research Office in the United States and for funding agencies of the United Kingdom, Austria, Denmark, Hong Kong, Ireland, and Sweden. He is active in consulting for manufacturers of medical devices and other companies on mechanical stress, fatigue life, fracture, and ferroelectric devices. He was associate editor of the American Society of Mechanical Engineers *Journal of Applied Mechanics*, 1987-1993, and is currently editor for the 2002-2012 term. He is an editorial board member for several journals in the fields of solid mechanics and materials and has reviewed for all the major journals in his field. In addition to his appointment at UCSB, Professor McMeeking is Sixth Century Professor of Engineering Materials (part-time) at the University of Aberdeen, Scotland; visiting professor of materials engineering at the University of the Saarland, Germany; and external member of the Leibniz Institute for New Materials, Saarbrücken, Germany.

Nina A. Orlovskaya is an assistant professor of mechanical, materials, and aerospace engineering at the University of Central Florida (UCF). Her research interests lie in the field of ceramics and ceramics composites for various engineering applications. During her research career she has addressed numerous topics both in the processing of ceramics and ceramic composites and in the characterization of materials' properties. She devoted significant efforts to the development of the hot-pressing technique for the manufacture of B_4C, Si_3N_4, and SiC-based ceramics for armor and cutting-tools applications. Through her manufacturing work she has gained extensive experience not only in hot pressing but also in pressureless sintering of bulk ceramic materials, as well as in magnetron sputtering of the thin films. Recently she has also been working on spark plasma sintering to process B_4C, ZrB_2 and ReB_2 ceramics. One of Dr. Orlovskaya's major research interests is lightweight, hard, and tough boron-rich ceramic laminates. B_4C/B_4C–SiC laminates are designed such that the differences in the layers' compositions lead to the differences in the coefficients of thermal expansion and Young's moduli of the adjacent layers, thus facilitating the appearance of thermal residual stresses. If properly designed, the thermal residual stresses could bring a significant increase in the mechanical performance of laminates as compared to the traditional particulate B_4C–SiC composites. Another topic that Dr. Orlovskaya is currently pursuing is the mechanochemical synthesis of ReB_2, OsB_2, and IrB_2 powders. Additionally, stress- and temperature-altered vibrational properties of $Zr(Hf)B_2$–SiC ceramic composites are under intensive exploration. Dr. Orlovskaya's interest in materials availability and world production of lightweight materials led her to organize, as director of a NATO Advanced Research Workshop, a workshop entitled "Boron Rich Solids: Sensors for Chemical and Biological Detection, Ultra High Temperature Ceramics, Thermoelectrics and Armor," held at UCF in 2009. The workshop attracted attention from the international community interested in boron-rich solids, and scientists from the United States, France, Italy, Germany, Russia, Ukraine, Japan, Egypt, India, and South Africa presented their research results during the workshop.

Michael Ortiz, the Dotty and Dick Hayman Professor of Aeronautics and Mechanical Engineering, California Institute of Technology (Caltech), Department of Engineering and Applied Science, received a B.S. degree in civil engineering from the Polytechnic University of Madrid, Spain, and M.S. and Ph.D. degrees in civil engineering from the University of California, Berkeley. From 1984 to 1995 he held a faculty position in the Division of Engineering at Brown University, where he carried out research activities in the fields of the mechanics of materials and computational solid mechanics. He has been on the faculty at Caltech since 1995 and currently serves as the director of its Department of Energy/Predictive Science Academic Alliance Program Center on High-Energy Density Dynamics of Materials. Professor Ortiz has been a Fulbright Scholar, a Sherman Fairchild Distinguished Scholar at Caltech, a Midwest and Southwest Mechanics Seminar Series Distinguished Speaker, a fellow and an elected member-at-large of the U.S. Association for Computational Mechanics, and an elected fellow of the American Academy of Arts and Sciences. Professor Ortiz is the recipient of the Alexander von Humboldt Research Award for Senior U.S. Scientists, the International Computational Mechanics Awards for Research, the U.S. Association for Computational Mechanics Computational Structural Mechanics Award, the ISI Highly Cited Researcher Award, and the inaugural 2008 Rodney Hill Prize conferred every 4 years by the International Union of Theoretical and Applied Mechanics. Professor Ortiz has served on the Science and Technology Panel of the University of California's Office of the President and on the Los Alamos National Laboratory T-Division Review Committee. He currently serves on the Lawrence Livermore National Laboratory Predictive Science Panel; the Sandia National Laboratories Engineering Sciences External Review Panel; the Lawrence Livermore National Laboratory Chemistry, Materials, Earth and Life Sciences Directorate Review Committee; and the National Research Council's Panel for the Evaluation of Quantification of Methods and Uncertainty; he chairs the Lawrence Livermore National Laboratory Engineering Directorate Review Committee. He has been editor of the *Journal of Engineering Mechanics* of the American Society of Chemical Engineers and of the *Journal of Applied Mechanics* of the American Society of Mechanical Engineers, associate editor of the journal *Modeling and Simulation in Materials Science and Engineering* and of the *Journal for Computational Mechanics* and is currently associate editor of the *Journal of the Mechanics and Physics of Solids* and of the *Archive for Rational Mechanics and Analysis*.

Raúl A. Radovitzky, an associate professor of aeronautics and astronautics, Massachusetts Institute of Technology (MIT), is also the associate director, Institute for Soldier Nanotechnologies. Professor Radovitzky was born in Argentina and educated at the University of Buenos Aires, where he obtained his civil engineering degree in 1991. He received his S.M. in applied mathematics from Brown University in 1995 and his Ph.D. in aeronautical engineering from the California Institute of Technology in 1998. He joined MIT's Department of Aeronautics and Astronautics in 2001 as the Charles Stark Draper Assistant Professor. Professor Radovitzky's research interests are in the development of advanced concepts and material systems for blast protection. To this end, his research group develops theoretical and computational descriptions of the blast event and its effects on structures and humans, including advanced computational methods and algorithms for large-scale simulation. The resulting models help to improve the understanding of the various physical components of the problem and thus to design protective systems. Professor Radovitzky's educational interests include computational mechanics, continuum mechanics, aerospace structures, mechanics of materials, numerical methods, and high-performance computing. He is a member of the American Institute of Aeronautics and Astronautics, International Association of Computational Mechanics, American Academy of Mechanics, Materials Research Society, U.S. Association of Computational Mechanics, and American Society of Mechanical Engineers.

Kaliat T. Ramesh is the director of the Center for Advanced Metallic and Ceramic Systems (CAMCS) in the Department of Mechanical Engineering at the Johns Hopkins University. His degrees include a B.E. in mechanical engineering from Bangalore University (India) in 1982, an Sc.M. in engineering from Brown University in 1985, an Sc.M. in applied mathematics from Brown University in 1987, and a Ph.D. in engineering from Brown University in 1988. Dr. Ramesh was the chair of the Department of Mechanical Engineering at Johns Hopkins University in 1999-2002. He was appointed director of the CAMCS in 2001. His honors and awards include the following: M. Hetényi Award from the Society for Experimental Mechanics, 2006; elected fellow, American Society of Mechanical Engineers, 2001; William H. Huggins Award for Excellence in Teaching from Johns Hopkins University, 1995; elected member, Pi Tau Sigma, 1994; and best paper, ASME Tribology Division, 1987.

Donald A. Shockey, director of the SRI International Center for Fracture Physics, is an internationally recognized expert in the fracture of materials and structures and an authority on failure under impact and explosive loads. He joined SRI International in 1971 after earning a doctorate in materials science at Carnegie Mellon University and completing a 3-year postdoctoral appointment at the Ernst-Mach-Institut and the Institut für Werkstoffmechanik in Freiburg, Germany. In his 39 years at SRI, he has directed more than 350 research projects for government and industry, many of which involved ballistic testing, modeling, and post-test damage assessment of metals, ceramics, polymers, and fabrics. Inventor of engine fragment barriers for commercial aircraft, he is currently leading problem-solving efforts associated with developing new glass-based materials and new structural designs for more weight-efficient windows on military vehicles. He is also assessing transparent ceramics and novel structural designs for spacecraft windows that more effectively resist damage from hypervelocity impact of micrometeorites and orbital debris. Dr. Shockey's recent failure-related projects include the following: astronaut gloves—determining how high-strength fabric gets abraded and torn during space walks and what can be done to prevent glove damage; stents for peripheral arteries—devising mechanical tests that mimic loads imposed by blood vessels to enable the design of fracture-resistant stents; failure prognostics—developing and applying advanced fractographic methods to generate the ability to predict the future performance and remaining useful life of aircraft, bridges, and pipelines; and failure analysis—determining the root cause of and providing expert testimony with respect to equipment failures such as rotor hub cracking in a Chilean power plant. Dr. Shockey has written more than 150 technical articles, holds several patents, and serves on the NASA Panel of Materials Experts. He is a fellow of ASM International, the year 2000 recipient of the John S. Rinehart Award for pioneering work in the field of dynamic fracture, and the 2006 recipient of the Murray Medal for excellence in experimental mechanics.

Samuel Robert Skaggs, retired from Los Alamos National Laboratory (LANL), is a consultant for advanced armor design and evaluation. He has extensive experience in dynamic loading of materials under high strain rate. Dr. Skaggs was the LANL Armor Program manager from 1986 to 1993, assisting in the design of armors for Desert Storm and the Balkans conflict for both ground vehicles and aircraft. He is responsible for the add-on armor for the U.S. Marine Corps Light Armored Vehicle 25 (USMC LAV-25, now called the Stryker) and the cockpit armor for the C-141 Starlifter logistics aircraft flying into and out of Sarajevo. From 1982 to 1986 he served as program manager for the LANL Materials by Design program as well as the Fossil Energy program. In 1981-1982 he served at the Department of Energy, evaluating alternative-energy methods for clean coal, coal liquefaction, and coal gasification. Dr. Skaggs earned a Ph.D. in materials science and an M.S. degree in nuclear engineering at the University of New Mexico and a bachelor's degree in mechanical engineering at New Mexico College of Agriculture and Mechanic Arts (now New Mexico State University). Dr. Skaggs has written more than 60 journal articles and reports in classified and unclassified areas. He is a member of the American Association for the Advancement of Science, the Hypervelocity Impact Society, and the American Defense Preparedness Association. He is also a member and founding president of the NMSU Mechanical Engineering Academy and founder of the Ceramics Modeling Working Group (a joint working group of the Department of Energy, the Department of Defense, and university and nonprofit scientific research organizations) as well as a founding member of the Advisory Council to the Dean of the College of Engineering at NMSU, having served as secretary from its founding until 2010.

Steven G. Wax is a technology consultant specializing in defense research and development (R&D). He supports defense clients in strategic planning and technology innovation across a range of scientific and engineering disciplines, including the physical sciences, materials, biology, biomedical, and mathematics. Prior to holding executive-level positions at Strategic Analysis, Inc., and SRI, International, Dr. Wax spent 35 years working for the Department of Defense as a civilian and a military officer. During that period, he performed and managed government R&D across a broad spectrum of classified and unclassified technology areas. His last government position was as director of the Defense Advanced Research Projects Agency (DARPA), a $400-million-per-year office whose technology purview included the physical sciences, materials, mathematics, human effectiveness, and the biological sciences including biological warfare defense. As director, Dr. Wax was responsible for the office's investment strategy as well as the transition of the Defense Science Office's technologies to the military. His previous government positions also include deputy director of the Technology Reinvestment Project and an assignment to the National Reconnaissance Office. Dr. Wax is currently a member of the National Materials Advisory Board and past member of the Sandia National Laboratories' External Review Panel for Materials. He recently served as an external reviewer of the discovery and innovation portfolio of the Office of Naval Research. He is also a member of the Air Force Research Laboratory's Human Effectiveness Directorate's independent review team and has supported the Advanced Research Projects Agency of the Department of Energy in its white paper evaluations. He was the winner of the George Kimball Burgess Memorial Award in 2009. Dr. Wax's notable technical accomplishments include a major role in the development of the DARPA's strategic plans for both biology and materials science as well as the co-development of two material sciences program thrusts (Intelligent Processing of Materials and Accelerated Insertion of Materials) that have revolutionized materials processing and insertion. He has also supported work in such diverse areas as ceramics, ceramic composites and fibers, electroactive polymers, materials processing, space materials and systems, advanced batteries, and personnel armor. Dr. Wax holds a Ph.D. in ceramic engineering from Georgia Institute of Technology, an M.S. in chemical engineering from the University of Illinois, and a B.S. in chemical engineering from the University of Massachusetts. Dr. Wax is a retired Air Force officer.

Appendix C

Committee Meetings

FIRST MEETING
MARCH 9-11, 2010, WASHINGTON, D.C.

Armor Materials 101-501: Focus on Fundamental Issues Associated with Armor Ceramics: Kinetic Energy Passive Armor, *James W. McCauley*, Chief Scientist, Weapons and Materials Research Directorate (WMRD); fellow, Army Research Laboratory

Mesoscale Modeling and Experiments—Why and How? *Lalit Chhabildas*, Senior Scientist, Enhanced Energy Effects, Air Force Research Laboratory

Ceramics for Body Armor, *Don Bray*, Morgan Crucible Company PLC

Light Armor, 41 Years Later, *Carl Cline* (retired), Lawrence Livermore National Laboratory

Ceramic Behavior Under Ballistic Impact at High Pressure, *Michael Normandia*, Chief Scientist, Armor Development, Ceradyne, Inc.

Multiscale Modeling of Armor Materials, *Christopher Hoppel*, Chief, High Rate Mechanics and Failure Branch, Army Research Laboratory

Transparent Armor: Research Issues, *Stephan Bless*, Associate Director, Institute for Advanced Technology, University of Texas at Austin

Fiber Research for Soldier Protection, *Philip Cunniff*, U.S. Army Natick Soldier Systems Center

Open Discussion Among Participants, *Edwin (Ned) Thomas*, Chair; *Michael McGrath*, Vice Chair

Materials Initiatives and Needs for Lightening Ground Platforms, *Douglas W. Templeton*, Deputy for Ballistic Protection, Senior Technology Expert—Survivability (acting), U.S. Army Tank Automotive Research, Development and Engineering Center

Armor, *Timothy Holmquist*, Staff Engineer, Southwest Research Institute

Armor Material Considerations for Asymmetric Warfare, *Yellapu Murty*, Director of Research and Development, Cellular Materials International

The Origin and History of the ONR Transducer Materials Program, *Robert Pohanka*, Director, Materials Science and Technology Division, Office of Naval Research

SECOND MEETING
APRIL 28-29, 2010, WASHINGTON, D.C.

Development and Modeling of High Performance Fibers for Lightweight Protection, *Tucker Norton*, North American Technology Leader, Ballistics, DuPont Protection Technologies

Today's Lightweight Protective Equipment and the Challenges for the Next 10-20 Years, *Lt. Gen. George J. Flynn*, Deputy Commandant for Combat Development and Integration, Commanding General, Marine Corps Combat Development Command

Needed Materials Approach for Lightweight Protective Systems, *Lee Mastroianni*, Force Protection Thrust Lead, Office of Naval Research

Bi-modal to Tri-modal Aluminum, *Lawrence T. Kabacoff*, Program Officer, Materials Division, Office of Naval Research

Cellular Materials for Force Protection, *Steven Fishman*, Materials Consultant, Strategic Analysis, Inc.

ASAALT's View on Protection Materials, *Cary F. Chabalowski*, Acting Director, Research and Laboratory Management, Office of the Deputy Assistant Secretary of the Army for Research and Technology/Chief Scientist

Strain Rate Sensitive Polymers in Armor Applications—Science Issues, *Roshdy George S. Barsoum*, Manager, Explosion Resistant Coating, Joint Enhanced Explosion Resistance Coating Exploitation, Advanced Concept Technology Demonstration, Ships and Engineering Systems Division, Office of Naval Research.

From Fiber to Application: High Performance Polyethylene Fiber Research Needs, *Lori Wagner*, Technical Leader, Advanced Fibers and Composites, Honeywell

THIRD MEETING
JUNE 8-10, 2010, ABERDEEN, MARYLAND

DARPA's Role in Armor, *Judah M. Goldwasser*, Program Manager, Defense Advanced Research Projects Agency

A National Lab Perspective on Protection Materials, *Bruce Remington*, Lawrence Livermore National Laboratory

Army Applications of Magnesium Alloys: Past Lessons and Future Solutions *Suveen Mathaudhu*, Army Research Laboratory

Speakers on Transparent Armor, *J. Jay Zhang*, Program Manager, Corning; *Lee M. Goldman*, Vice President of Technology and Chief Technology Officer for the Optical Ceramics Division, Surmet; *Jeff Kutsch*, President, Technology Assessment and Transfer, Inc, *Kathie Leighton*, Schott; *Christopher Jones*, Saint-Gobain

Site Visit, Aberdeen Proving Ground, Welcome and WMRD, *J. Smith*

Summary of Towson Meeting 2008, *P. Plostins*

Aluminum History in the Army, *B. Cheeseman*

Multi Scale Modeling, *G. Gazonas*

Transparent Armor, *P. Patel*

High Strain Rate Lab Tour, *T. Weerasooriya*

Tours of Armor Test Facilities, *S. Schoenfeld*

Armor: A European Perspective, *Klaus Thoma*, Director, Fraunhofer Ernst-Mach-Institut

Hypervelocity Shields, *Eric Christiansen*, NASA

Ceramic-Metal Laminate (CML) Composites for Armor Applications, *Ken Kuang*, Torrey Hills Technologies LLC

FOURTH MEETING
JULY 22-23, 2010, WASHINGTON, D.C.

The Sagamore Conference, *K.T. Ramesh* and *Edwin (Ned) Thomas*

Army Research Laboratory on Protective Materials, *Ernest S.C. Chin*, Army Research Laboratory

FIFTH MEETING
AUGUST 24-26, 2010
WOODS HOLE, MASSACHUSETTS

Objective: Limited final information gathering; review of the report writing process thus far; walk through of the first full message draft; discussion of conclusions, recommendations, and next steps.

SIXTH MEETING
SEPTEMBER 28-30, 2010, BECKMAN CENTER, IRVINE, CALIFORNIA

Objective: Review of the draft report and determination of a path forward.

Appendix D

Improving Powder Production

For commercial-scale operations, SiC and B_4C powders are produced by the carbothermic reduction of a silicon oxide or boric oxide in contact with a carbon source. The resultant powder has large grains and must be comminuted to produce the micron- to submicron-sized particles required for ceramic processing. As a consequence, process-related impurities are introduced or process-induced changes occur within the particles, requiring extraordinary cleaning processes to remove impurities and a greater understanding of the changes that take place during processing.[1]

Aluminum nitride powder is primarily produced by carbothermal nitridation of alumina (Al_2O_3) in contact with carbon in a nitrogen atmosphere. Oxygen content can dramatically affect the structure of AlN, so large-scale Acheson-type furnaces cannot be employed. Typically, pusher-type furnaces are employed to provide improved control in the moving-bed furnace. Impurities condense near cold zones, which can lead to variable chemistry powders. Also, like SiC, AlN must be comminuted to achieve micron-sized powders, leading to process-related impurities that must be cleaned.[2,3]

Alumina is by far the most widely used ceramic powder, being a precursor to aluminum smelting. As a result, worldwide availability for commodity-grade Al_2O_3 has changed with the economic conditions in recent years. Across-the-board production cuts and future uncertainty have been prevalent. This has dramatically reduced the availability of low-soda, high-purity (>99.99 percent) Al_2O_3. Economic challenges brought many U.S. producers to the brink of bankruptcy. The impact for armor is that as production levels return, there may not be sufficient U.S. supplies to meet armor needs.[4]

Spinel and aluminum oxynitride (AlON) are specialty materials typically produced in very small volumes for transparent crystalline ceramics. AlON powder is not commercially available but is typically prepared by a vertically integrated ceramic producer. Common methods for forming AlON are either direct reaction of Al_2O_3 + AlN or reduction nitridation of Al_2O_3 + C + (Al or H_2) in nitrogen or ammonia. The latter process is the most widely utilized, although with this process it tends to be difficult to remove all residual carbon. As with AlN and SiC, this process results in powders that must be reduced in size by comminution. Consequently, these powders must be carefully milled to avoid particulate contaminations.[5]

Spinel powder is produced by direct reaction of magnesium and aluminum salts that are subsequently calcined to produce the powders. Spray pyrolysis has also been used for very high purity powders. There is one source, Baikowski International Corp. (France), of commodity spinel worldwide. As a result, the cost of spinel powder is high. Variability in chemistry, particle size, and degree of aggregation has led to challenges in producing transparent ceramics.[6] The current cost of spinel, at \$60/kg to \$80/kg, is much too high to expect widespread use for transparent armor. There is a need for research to be conducted to determine whether a more affordable, uniform, ceramic-grade powder can be produced.

---

[1]Guichelaar, P. 1977. Acheson process. Pp. 115-128 in Carbide, Nitride and Boride Materials Synthesis and Processing, A.W. Weimer, ed. London, U.K.: Chapman and Hall.

[2]Dunn, D., M. Paquette, H. Easter, and R. Pihlaja. Continuous carbothermal reactor. U.S. Patent 4,983,553, filed December 7, 1989, and issued January 8, 1991, to the Dow Chemical Company, Midland, Mich.

[3]Henley, J., G. Cochran, D. Dunn, G. Eisman, and A. Weimer. Moving bed process for carbothermally synthesizing nonoxide ceramic powders. U.S. Patent 5,370,854, filed January 8, 1993, and issued December 6, 1994, to the Dow Chemical Company, Midland, Mich.

[4]Moores, S. 2009. Economy crashes, alumina burns. Industrial Minerals 497: 30-37.

[5]Zheng, J., and B. Forslund. 1995. Carbothermal synthesis of aluminum oxynitride (AlON) powder: Influence of starting materials. Journal of the European Ceramic Society 15(11): 1087-1100.

[6]Bickmore, C., K. Waldner, D. Treadwell, and R. Laine. 1996. Ultrafine spinel powders by flame spray pyrolysis of a magnesium aluminum double alkoxid. Journal of the American Ceramic Society 79(5): 1419-1423.

SILICON CARBIDE

Silicon carbide (SiC) is not found in any appreciable quantities in nature but is one of the most widely used synthetic technical minerals. The market for SiC focuses on its hardness and refractoriness, but SiC is also used as a source of silicon in the metallurgical processing of iron. SiC's hardness and high-temperature stability make it as widely used as alumina as an abrasive grain. For higher-performance applications, the higher-purity (green) SiC powder is used, and for lesser requirements the lower-purity (black) SiC powder is used. For advanced ceramic applications such as armor, only the high-purity green materials are used. Other applications of high-purity SiC include space-based mirrors, semiconductor processing equipment, wire-impregnated saws for silicon wafer cutting, and automobile catalysts. These markets have driven the world supply of green SiC to more than 1 million tons per year. Armor ceramics make up less than 1 percent of the world market for high-purity SiC.[7]

There are many methods for producing SiC, including carbothermic reduction of silica, chemical vapor-phase reactions, and electrothermal techniques. The Acheson process, which dates from 1893, places electrodes into a graphite core laid within a mixture of reactant carbon, salt, and sand. The electric current resistively heats the graphite and in turn the surrounding reactants, resulting in the formation of a hollow cylinder of SiC and the evolution of carbon monoxide (CO) gas.[8] The chemical reaction that Acheson described for the manufacture of SiC from silica sand and carbon is as follows:

$$SiO_2 + 3C \rightarrow SiC + 2CO$$

Within the ceramic-grade zone, both green SiC (>99 percent SiC) and black SiC (95-98 percent SiC) can be found, with metallurgical SiC (80-94 percent SiC) making up the remainder of the reaction zone. The boundary between unreacted materials and the reaction zone is marked by a layer of condensed impurities. This layer is discarded, but the unreacted precursors can be used again.

The formation of SiC is the result of four subreactions, each of which provides vapor-phase mass transport:[9]

$$C + SiO_2 \rightarrow SiO(g) + CO(g)$$
$$SiO_2 + CO(g) \rightarrow SiO + CO_2(g)$$
$$C + CO_2(g) \rightarrow 2CO(g)$$
$$2C + SiO \rightarrow SiC + CO(g)$$

The exact kinetics of the reaction are highly dependent on carbon source, particle size, mixing uniformity, and packing of the silica and the carbon. During the heating of the graphite core, silica can react with carbon at temperatures as low as 1527°C to create β-SiC. At temperatures about 1900°C, the β-SiC converts to α-SiC. The various polytypes formed are dependent not only on temperature but also on the presence of impurities. For example, for α-SiC the 6H polytype is most prevalent. However, in the presence of aluminum, either intentionally or as an impurity, the 4H polytype becomes dominant. This change in polytype alters not only the shape of the resultant particles but also the microhardness, with the 4H being less hard.[10]

Today's Acheson furnaces are very large. The first commercial furnace was 2 meters long and had a power input rate of 58 kW; today the largest furnace has a 240-ton capacity and a power input rate of nearly 6 MW! Aside from SiC processing being a tremendous consumer of electricity, for every pound of SiC produced, 1.4 pounds of CO are emitted. Both electricity costs and environmental concerns shifted the manufacturing of powder offshore to the extent that today, the United States accounts for less than 5 percent of the world's production of SiC, whereas China accounts for more than 60 percent. However, that 5 percent produced in the United States supplies the abrasives and metallurgical markets, meaning that there was no supplier in 2010 providing SiC for advanced ceramics, including armor.

Work by Choi et al.[11] indicated that SiC sintered with AlN and oxide additives could have a marked effect on the mechanical properties of the resulting SiC. Zhou et al.[12] showed the strong influence of rare-earth additions and resulting intergranular properties on the mechanical properties of SiC. Thus a better understanding of the role of intergranular phases could be used to engineer high-performance armor materials.

BORON CARBIDE

Worldwide, 1,000 to 2,000 metric tons of boron carbide are produced annually. The boron carbide market is driven by the use of boron carbide based on selected properties, such as its hardness—for example, as an abrasive grit or powder; its neutron absorption capacity (for use as control rods and shielding in pressurized water nuclear reactors, among other applications); and its specific hardness—as an armor

[7]Moores, S. 2007. Energy prices prune SiC bloom. Industrial Minerals 475: 28-35.

[8]Guichelaar, P. 1977. Acheson process. Pp. 115-128 in Carbide, Nitride and Boride Materials Synthesis and Processing, A.W. Weimer, ed. London, U.K.: Chapman and Hall.

[9]Weimer, A., K. Nilsen, G. Cochran, and R. Roach. 1993. Kinetics of carbothermal reduction synthesis of beta silicon carbide. AIChE Journal 39(3): 493-503.

[10]Poch, W., and A. Dietzel. 1962. Formation of silicon carbide from silica and carbon. Berichte der Deutschen Keramischen Gesellschaft 39(8): 413-426 (in German).

[11]Choi, H-J., Y-W. Kim, M. Mitomo, T. Nishimura, J-H. Lee, and D-Y. Kim. 2004. Intergranular glassy phase free SiC ceramics retain strength at 1500°C. Scripta Materialia 50(9): 1203-1207.

[12]Zhou, Y., K. Hirao, M. Toriyama, Y. Yamauchi, and S. Kanzaki. 2001. Effects of intergranular phase chemistry on the microstructure and mechanical properties of silicon carbide ceramics densified with rare-earth oxide and alumina additions. Journal of the American Ceramic Society 84(7): 1642-1644.

ceramic, for example.[13,14,15] As mentioned in Chapter 5 of this report, boron carbide is a solid solution containing 10 percent to 20 percent carbon. The exact chemistry of boron carbide powders depends on the particular powder synthesis route. The carbothermic reduction processes provide the largest quantities of boron carbide powders produced.[16] Magnesiothermic reduction and vapor-phase reactions, while producing high-quality fine-grain powders, are very expensive (>$500/kg) and are not discussed here.

Carbothermic Reduction

Boron carbide, like silicon carbide, is most commonly produced by the reduction of boron oxide (or boric acid) with carbon. The reaction is commonly written as follows:

$$\text{Boric oxide: } 2B_2O_3 + 7C \rightarrow B_4C + 6CO$$

or

$$\text{Boric acid: } 4H_3BO_3 + 7C \rightarrow B_4C + 6CO + 6H_2O$$

This process occurs in two stages:

$$B_2O_3 + 3CO \rightarrow 2B + 3CO_2$$
$$4B + C \rightarrow B_4C$$

Carbothermic reduction of boron carbide utilizes a Higgins or an electric arc furnace. Here, a water-cooled crucible is insulated with a packed wall of the mixed boric oxide and carbon precursors. An electric arc is used to generate temperatures between approximately 2500°C and 2800°C. Mixed precursor powders are added where they slowly melt, near the highest temperature areas. Because the melt is highly viscous and evolved CO_2 must be allowed to escape, materials are gradually added and the electrode height is changed. When sufficient materials have been reacted, the electrodes are withdrawn and the melt is cooled. The result is an ingot that weighs between 25 kg and 1,000 kg. The outer edges of the ingot are covered with unreacted precursor powders, which must be manually removed and are typically recycled. The ingot then undergoes a series of crushing operations, and the powder grain is milled to size. Depending on the manufacturer, metallic impurities derived from the crushing and milling equipment can be eliminated through a series of acid leaching steps.[17]

The carbothermic method is a very high temperature operation having large temperature variations across the crucible, and the stoichiometry of the product boron carbide is typically rich in carbon, commonly $B_{4-x}C$. A few percent of essentially pure carbon is typically found in the powder, resulting from unreacted graphite, graphite originating from the electrode, decomposed B_4C, or vapor-phase condensates of CO/CO_2.

Direct carbothermic reduction has been demonstrated on a pilot scale, where boric oxide and carbon are reacted in a vertical tube furnace at between 1973°C and 2073°C. Although this method produces a fine-grained (0.5-5 µ) and very controlled stoichiometric boron carbide, its yield is lower than that of the arc-melted grain method and at present it is not considered a viable option.[18]

ALUMINA

In 1887, Bayer discovered that aluminum hydroxide precipitated from alkaline solution was crystalline and could be more easily filtered and washed than that precipitated from acid medium. The process was a key to the development of modern metallurgy, since aluminum hydroxide is the raw material for the electrolytic aluminum process that was invented in 1886. The process that Bayer invented has remained essentially the same and produces nearly all of the world's alumina as an intermediate in aluminum production. The Bayer process can be considered in three stages: (1) extraction, (2) precipitation, and (3) calcination.

The aluminum-bearing minerals in bauxite are dissolved in a solution of sodium hydroxide (caustic soda) to selectively extract them from the insoluble components (mostly oxides). Then the ore is milled to make the minerals more available for extraction and to reduce the particle size. It is then combined with the process liquor in a heated pressure digester. Temperature and pressure within the digester reflect the type of ore. Temperatures vary between 140°C and 240°C and pressures vary up to 35 atm. After the aluminum-containing components dissolve, the insoluble residue is separated from the liquor by settling.

Crystalline aluminium trihydroxide (ATH) is then precipitated from the digestion liquor:

$$Al(OH)_4 + Na^+ \rightarrow Al(OH)_3 + Na^+ + OH$$

The ATH crystals are then classified into size fractions and fed into a rotary kiln at temperatures greater than 1050°C

[13]Lipp, A., Pacific Northwest Laboratory, U.S. Atomic Energy Commission. 1970. Boron carbide: Production, properties, applications. Richland, Wash.: Battelle Northwest Laboratories.

[14]Thévenot, F. 1990. Boron carbide—A comprehensive review. Journal of the European Ceramic Society 6(4): 205-255.

[15]Schwetz, K. 2000. Boron-carbide, boron nitride, and metal borides. Ullmann's Encyclopedia of Industrial Chemistry. DOI: 10.1002/14356007.a04_295.

[16]Suri, A., C. Subramanian, J. Sonber, and T. Murthy. 2010. Synthesis and consolidation of boron carbide: A review. International Materials Review 55(1): 4-40.

[17]Scott, J. 1964. Arc furnace process for the production of boron carbide. U.S. Patent 3,161,471, filed February 25, 1958, and issued December 15, 1964, to Norton Company, Worcester, Mass.

[18]Rafaniello, W., and W. Moore. 1989. Producing boron carbide. U.S. Patent 4,804,525, filed July 14, 1987, and issued February 14, 1989, to the Dow Chemical Company, Midland, Mich.

for calcination. The ATH is calcined to form alumina, which can be directly used for aluminum processing or can be used for ceramic applications:

$$2Al(OH)_3 \rightarrow Al_2O_3 + 3H_2O$$

If the ATH is to be used for ceramics, it can undergo multiple washing steps to reduce the ionic sodium to less than 0.01 percent. The particle size of the calcined powder is reduced in size, depending on specifications determined by the end user.

Appendix E

Processing Techniques and Available Classes of Armor Ceramics

This appendix covers additional material and details relevant to Chapter 5 of this report. These pages address several topics related to processes used in the manufacturing of ceramics for armors and include discussions on potential armor materials such as functionally graded materials, biomimetics, foams, smart sensors, and phononic band gap materials.

Some of the key manufacturing processes, together with their advantages and disadvantages, are listed in Table 5-1 of Chapter 5. Table E-1 presents the relevant properties of materials listed in Table 5-1 of Chapter 5.

PRESSURELESS SINTERING

The pressureless sintering process offers cost reduction through net shape processing using innovative powders and processing methods to obtain full density without the application of pressure. The goal of densification is to create strong bonds in the material and eliminate porosity so that theoretical densities, along with homogeneous microstructures, can be achieved for the sintered bodies. Residual porosity, along with the shape and size distributions of the pores and grains, influences the characteristics of dynamic performance. The low-cost alumina armor plates manufactured by means of uniaxial pressing, slip casting, and sintering are used in vehicle armor applications in large volumes. The typical cost of these plates runs about \$2.50/lb to \$10/lb for the finished tile. Solid state sintering is achieved by heating the "green" compact to the temperature that is in the range between approximately 50 and 80 percent of the melting temperature.[1]

At these temperatures the powder does not melt, but fusing of adjacent powder particles and reduction in the overall porosity occur by atomic diffusion in the solid state. Solid state sintering is typically used for pure, single-phase polycrystalline materials, such as α–Al_2O_3. For many covalently bonded polycrystalline ceramics, the required dense microstructure is hard to achieve using solid state sintering; therefore, additives are used to form a small amount of liquid phase between the particles at the sintering temperature. The liquid phase provides a high-diffusivity path to transport matter into the pores and facilitates densification. For example, hard-to-sinter silicon carbide (SiC) is processed as a liquid-phase sintered ceramic. The price of SiC tiles manufactured by means of a pressureless sintered process is in the range of \$40/lb to \$50/lb.

Solid- and liquid-state sintering processes are widely employed to densify refractory ceramics, but at much higher temperatures than required by the hot-pressing (HP) technique. Sintering of SiC was first performed by Prochazka,[2] using boron and carbon as sintering aids to reduce the interfacial energy of the grains (boron),[3] and by reacting the carbon with residual silica (carbon) present on the SiC particle surface.[4,5] Sintering of β-SiC is more difficult than sintering of α–SiC because of the β to α phase transformation at 1900°C to 2000°C, which generates voids between grains owing to the difference in the growth morphology of β and α grains.[6] Several additives, such as Al–C, Al_2O_3–C, and Al_2O_3–Y_2O_3, have been tested as sintering aids for SiC powder to enhance the sintering rate and to reduce grain

[1]Rahaman, M. 2007. Sintering of Ceramics. Boca Raton, Fla.: Taylor and Francis Group.

[2]Prochazka, S. 1974. Hot pressed silicon carbide. U.S. Patent 3,853,566, filed December 21, 1972, and issued December 10, 1974, to General Electric Company. Schenectady, NY.

[3]Maddrell, E. 1987. Pressureless sintering of silicon carbide. Journal of Materials Science Letters 6(4): 486-488.

[4]van Rijswijk, W., and D. Shanefield. 1990. Effects of carbon as a sintering aid in silicon carbide. Journal of the American Ceramic Society 73(1): 148-149.

[5]Hamminger, R. 1989. Carbon inclusions in sintered silicon carbide. Journal of the American Ceramic Society 72(9): 1741-1744.

[6]Williams, R., B. Juterbock, C. Peters, T. Whalen, and A. Heuer. 1984. Forming and sintering behavior of B- and C-doped α- and β-SiC. Journal of the American Ceramics Society 67(4): C62-C64.

TABLE E-1 Summary of Properties of Various Ceramics for Personnel Armor Application

Material	Designation	Density ρ (g/cc)	Grain Size (μ)	Young's Modulus E (GPa)	Flex Strength Σ (MPa)	Fracture Toughness K ($MPa\text{-}m^{1/2}$)	Fracture Mode	Hardness (HK-Knoop hardness, HV-Vickers hardness)	HK 2 kg (kg/mm^2)	Areal Density
Al_2O_3	CAP-3	3.90	—	370	379	4-5	—	1,440 (HK 1 kg)	1,292	20.2
B_4C	Ceralloy-546 4E	2.50	10-15	460	410	2.5	TG	3,200 (HV 0.3 kg)	2,066	13.0
	Norbide	2.51	10-15	440	425	3.1	TG	2,800 (HK 0.1 kg)	1,997	13.0
SiC	SiC-N	3.22	2-5	453	486	4.0	IG, TG	—	1,905	16.7
	Ceralloy 146-3E	3.20	—	450	634	4.3	—	2,300 (HV 0.3 kg)	—	16.6
	Hexoloy	3.13	3-50	410	380	4.6	TG	2,800 (HK 0.1 kg)	1,924	16.2
	Purebide 5000	3.10	3-50	420	455	—	TG	—	1,922	16.1
	SC-DS	3.15	3-50	410	480	3-4	—	2,800 (HK 1 kg)	—	16.4
	MCT SSS	3.12	3-50	424	351	4.0	TG	—	1,969	16.2
	MCT LPS	3.24	1-3	425	372	5.7	IG	—	1,873	16.8
	Ekasic-T	3.25	1-3	453	612	6.4	IG	—	1,928	16.8
SiC (RB)	SSC-702	3.02	45	359	260	4.0	TG	1,757 (HK 0.5 kg)	—	15.7
	SSC-802	3.03	45	380	260	4.0	TG	—	1,332	15.7
	SSC-902	3.12	45	407	260	4.0	TG	—	1,536	16.2
SiC/B_4C (RB)	RBBC-751	2.56	45	390	271	5.0	TG +Ductile Si	—	1,626	13.3
TiB_2	Ceralloy 225	4.5	—	540	265	5.5	—	—	1,849	23.4

NOTE: Areal density in pounds per square foot (PSF): weight of a 12 × 12 × 1 in. panel in pounds; TG, transgranular fracture; IG, intergranular fracture.
SOURCES: CAP-3, SC DS: CoorsTek; Ceralloy, Ekasic-T: Ceradyne; Norbide, Hexoloy: Saint-Gobain; Purbide: Morgan AM&T; SiCN: Cercom (BAE); SSC, RBBC, BSC, SSS and LPS: M Cubed Technologies (MCT). Properties for other manufacturers' materials are from their respective Web sites except for 2 kg Knoop hardness, grain size, and fracture mode.

growth. Aluminum (Al)[7] and alumina[8] with carbon promote silicon carbide sintering by means of a solid state mechanism at a temperature over 2000°C, while alumina and yttria lead to a high-density sintered sample by means of a liquid-phase mechanism at temperatures below 2000°C.[9]

Boron carbide (B_4C) is mainly produced by the HP method. The cost of a B_4C tile is in the range of $75/lb to $85/lb. The pressureless sintering processes of B_4C and densification of B_4C by solid state sintering techniques[10] are slow, and it is difficult to reach high density due to low self-diffusion. Sintering aids such as SiC, Si, Al_2O_3, Mg, and Fe have been used to increase the density by means of liquid-phase sintering;[11] however, the mechanical performance of liquid-phase-sintered B_4C is inferior. It has been established that the presence of B_2O_3 coatings on B_4C particles inhibits densification and facilitates grain coarsening.[12] The boria can be removed by heat treatment in a hydrogen environment, which then permits direct contact between B_4C–B_4C grains, facilitating densification. As a result, the B_4C powders with a particle size of approximately 1 μ can then be sintered to 96 percent of theoretical density and with hardness values similar to hot-pressed samples. Methods used to produce pressureless sintered B_4C have been developed at the Georgia Institute of Technology and commercialized at Verco Materials,[13] as well as by larger armor producers such as Saint-Gobain. Armor-grade material of B_4C with a zero porosity state can be produced using pressureless sintering combined with hot isostatic pressing.

Both SiC and B_4C are harder materials with lower densities than alumina, yet alumina has been widely used in personnel and vehicle armor systems because of its lower

[7]Stutz, D., S. Prochazka, and J. Lorenz. 1985. Sintering and microstructure formation of β–silicon carbide. Journal of the American Ceramics Society 68(9): 479-482.

[8]Sakai, T., H. Watanabe, and T. Aikawa. 1987. Effects of carbon on phase transformation of β–SiC with Al_2O_3. Journal of Materials Science Letters 6(7): 865-866.

[9]Omori, M., and H. Takei. 1988. Preparation of pressureless-sintered SiC–Y_2O_3–Al2O_3. Journal of Materials Science 23(10): 3744-3749.

[10]Thévenot, F. 1990. Boron carbide—A comprehensive review. Journal of the European Ceramic Society 6(4): 205-225.

[11]H. Kim, H-W., Y-H. Koh, and H-E. Kim. 2000. Densification and mechanical properties of B_4C with Al_2O_3 as a sintering aid. Journal of the American Ceramic Society 83(11): 2863-2865.

[12]Cho, N., Z. Bao, and R. Speyer. 2005. Density and hardness-optimised pressureless sintered and post-hot isostatic pressed B_4C. Journal of Materials Research 20 (8): 2110-2116.

[13]Campbell, J., M. Klusewitz, J. LaSalvia, E. Chin, R. Speyer, N. Cho, N. Vanier, H. Cheng-Hung, E. Abbott, P. Votruba-Drzal, W. Coblenz, and T. Marcheaux. 2008. Novel processing of boron carbide (B_4C): Plasma synthesized nano powders and pressureless sintering forming of complex shapes. ADM002187. Proceedings of the Army Science Conference (26th). Accessed April 1, 2011.

raw material cost and ease of fabrication. Large-size alumina panels up to 400 mm × 550 mm are currently produced by means of pressureless sintering[14] for use in lightweight armor vehicles and police car door protection. Morgan is one is the main producers of armor-grade alumina made by the pressureless sintering technique.

HOT PRESSING

Hot pressing is often the procedure of choice for the manufacture of opaque ceramics, since it can produce fully dense ceramics at reasonably moderate temperatures and pressures. However, HP can only produce limited shapes such as flat plates or those with a small curvature. A current Army program is developing HP to fabricate SiC tiles at lower cost.[15]

Traditional hot pressing is a batch process in which the "green" powder compacts are formed by means of a suitable pressing method and then loaded into a hot-pressing die. Some armor manufacturers use tape casting or extrusion to build up green B_4C armor shapes for hot pressing. The die and powder are ramped up to the sintering temperature and pressure is applied to the die. To meet the required high sintering temperatures (>2000°C), heated dies made of graphite or other high-strength inert materials are used in special hot-pressing furnaces. As the ceramic part size increases, the load requirements increase, making the HP equipment large. HP is associated with small production volumes, and typically large billets are produced to be cut into individual armor tiles. Often, multiple parts, separated by spacers, are pressed together to increase production rate. Once the ceramic is hot-pressed, it is cooled and then machined to its final dimensions by diamond tools using slow machining steps and grinding rates.

Materials with different final densities and mechanical properties are produced by varying the nature of powder additives and hot-pressing conditions such as pressure, temperature, and time. Typically, in hot-pressing SiC armor material, powders of α–SiC are mixed with suitable sintering aids (boron and carbon, for example); additional carbon is added to remove the silica passivation layers from the SiC particles. Temperature steps are adjusted based on the specific mix of powders and additives, and during the last high-temperature cycle, pressure is applied to achieve maximum densification. Different final density and mechanical properties are reached by varying the hot-pressing conditions and the powder additives used. Additives and precise processing conditions are kept as proprietary information by different manufacturers.

Compared to lower-density, pressureless sintered products, the hot-pressing process is a slow batch process that typically yields a close-to-full-density product with superior ballistic properties. Hot-pressed SiC and B_4C provide superior ballistic properties but are manufactured at high cost and in limited volumes through a batch process. The U.S. Army Manufacturing Technology Program has developed an HP apparatus with multiple heating and cooling chambers and central hot pressing chamber to reduce the cost of hot pressed 4 in. × 4 in. tiles from \$135/lb to \$85/lb.[16] The goal is to increase production volumes, thus reducing the cost to \$35/lb, similar to pressureless sintered material.

Due to processing differences, hot-pressed and pressureless sintered materials have different microstructures; see, for example, Figure E-1. Hot-pressed material is typically fully dense with fine microstructures, whereas pressureless sintered materials have large grains with texture. However, recent dynamic magnetic compaction (DMC) work has shown the promise of obtaining fine grain structures similar to those of HP material by combining DMC and pressureless sintering. Such a process needs to be further developed.

CURRENT-ASSISTED SINTERING

Nano-grain-size ceramic powders are currently being explored by various laboratories, including the U.S. Army Research Laboratory, to obtain better mechanical properties through microstructural modifications. However, when ceramic powders are hot-pressed or sintered at very high temperatures for extended times, grain growth takes place. Processing methods such as dynamic compaction or spark plasma sintering (SPS) techniques can be used to retain the small grain size of nanograined powders.

One method of accelerating the sintering process of difficult-to-sinter armor ceramics involves the use of electrical current. The name most often used for such field-assisted sintering is SPS, but the process is also known as plasma pressure sintering, pulsed electric current, and electric-pulse-assisted consolidation. Significant advantages exist for using current-assisted sintering over that of hot-pressing, hot isostatic pressing, or pressureless sintering; the most important advantage is lower sintering temperature and reduced holding time, which results in marked comparative improvements in mechanical properties.[17] For example, attaining a heating rate of 600°C/min, typically used in SPS of ceramics, could help retain the homogeneous grain size distribution

[14]Medvedovski, E. 2010. Ballistic performance of armour ceramics: Influence of design and structure Part 2. Ceramics International 36(7): 2117-2127.

[15]Protection materials—Research to acquisition. Briefing by E. Chin, U.S. Army Research, Development and Engineering Command, to the committee, July 23, 2010.

[16]Campbell, J., J. LaSalvia, W. Roy, E. Chin, R. Palicka, and D. Ashkin. 2008. New Low-Cost Manufacturing Methods to Produce Silicon Carbide (SiC) for Lightweight Armor Systems. ADA504013. Proceedings of the Army Science Conference (26th). Accessed April 4, 2011.

[17]Munir, Z., U. Anselmi-Tamburini, and M. Ohyanagi. 2006. The effect of electric field and pressure on the synthesis and consolidation of materials: A review of the spark plasma sintering method. Journal of Materials Science 41(3): 763-777.

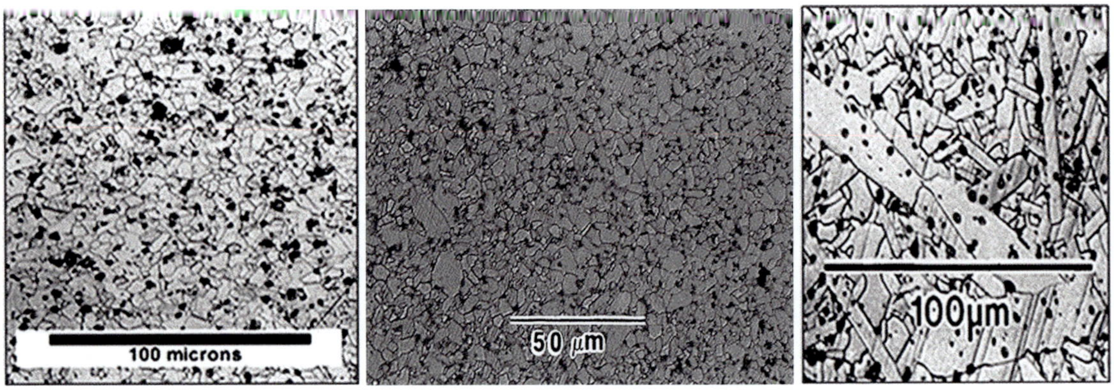

FIGURE E-1 Silicon carbide sample microstructures showing grains in (left) hot-pressing, (center) dynamic magnetic compaction followed by pressureless sintering, and (right) uniaxial pressing followed by pressureless sintering. SOURCE: Chelluri, B., and E.A. Knoth. 2008. SiC armor tiles via magnetic compaction and pressureless sintering. Presentation to the 32nd International Conference and Exposition on Advanced Ceramics and Composites, January 26-February 1, Daytona Beach, Florida.

along with the small average grain size of B_4C.[18,19] The spall strengths[20] of SPS-processed B_4C and SiC ceramics are improved over those from hot-pressing or pressureless sintering techniques. SPS is still under development, especially for sintering larger parts. Density variations are still observed in the case of difficult-to-sinter ceramics because of current flow along the highest-conductivity graphite die walls, although low or no current density is detected inside the part. As a result of such a temperature gradient, material density varies with its location relative to the die walls. The widespread adoption of SPS in the past decade was possible because of the availability of commercially built SPS systems. Currently the two major players are SPS SYNTEX Inc., Japan, and FCT Systeme GmbH, Germany. Recently, the U.S. firm Thermal Technology LLC also started selling field-assisted sintering furnaces. However, to the best of the committee's knowledge, there are no commercially produced ceramic armor tiles using SPS, suggesting a potential opportunity for improved processing of dense ceramics.

REACTION-BONDED CERAMICS

Reaction-bonded SiC and reaction-bonded boron carbide have been successfully used for armor applications over the past decade.[21,22,23,24] In reaction bonding, which uses silicon-based matrixes, the pressureless infiltration of a powder preform is achieved by good wetting and a highly exothermic reaction between liquid silicon and carbon.[25] The process is known variously as reaction bonding, reaction sintering, self-bonding, or melt infiltration.

M Cubed Technologies Inc., a developer of the reaction bonding process, uses a process that includes the following steps: (1) mixing of B_4C (or SiC) powder and a binder to make a slurry; (2) shaping of the slurry by various techniques, such as casting, injection molding, pressing, and others; (3) drying and carbonizing of the binder; (4) green machining; (5) infiltration (reaction bonding) with molten Si (or alloy) above 1410°C in an inert or vacuum atmosphere; and (6) solidification and cooling. During the infiltration step, carbon in the preform reacts with molten

[18]Hayun, S., S. Kalabukhov, V. Ezersky, M. Dariel, and N. Frage. 2010. Microstructural characterization of spark plasma sintered boron carbide ceramics. Ceramics International 36(2): 451-457.

[19]Hayun, S., V. Paris, M.P. Dariel, N. Frage, and E. Zaretzky. 2009. Static and dynamic mechanical properties of boron carbide processed by spark plasma sintering. Journal of the European Ceramics Society 29(16): 3395-3400.

[20]Paris, V., N. Frage, M., Dariel, and E. Zaretsky. 2010. The spall strength of silicon carbide and boron carbide ceramics processed by spark plasma sintering. International Journal of Impact Engineering 37(11): 1092-1099.

[21]Waggoner, W., B. Rossing, M. Richmond, M. Aghajanian, and A. McCormick. 2003. Silicon carbide composites and methods for making same. U.S. Patent 6,503,572, filed July 21, 2000, and issued January 7, 2003, to M Cubed Technologies, Monroe, Conn.

[22]Aghajanian, M., B. Morgan, J. Singh, J. Mears, and R. Wolffe. 2002. A new family of reaction bonded ceramics for armor applications. Ceramic Transactions 134: 527-539.

[23]Aghajanian, M., McCormick, B. Morgan, and A. Liszkiewicz, Jr. 2005. Boron carbide composite bodies, and methods for making same. U.S. Patent 6,862,970, filed November 20, 2001, and issued March 8, 2005, to M Cubed Technologies. Monroe, Conn.

[24]Karandikar, P., M. Aghajanian, and B. Morgan. 2003. Complex, NET-shape composite components for structural, lithography, mirror and armor applications. Pp. 561-566 in 27th Annual Cocoa Beach Conference on Advanced Ceramics and Composites: B. Ceramic Engineering and Science Proceedings 24(4). W. Kriven and H-T. Lin, eds. Hoboken, N.J.: John Wiley & Sons.

[25]Karandikar, P., S. Wong, G. Evans, and M. Aghajanian. 2010. Microstructural development and phase changes in reaction bonded boron carbide. Pp. 5-22 in Advances in Ceramic Armor VI: Ceramic Engineering and Science Proceedings 31(5). Swab, J., S. Mathur, and T. Ohji, eds. Hoboken, N.J.: John Wiley & Sons.

Si, forming SiC around the original ceramic particles and bonding them together—hence the term "reaction bonding." Typically the final product—that is, reaction-bonded boron carbide composite—consists of the original boron carbide particles, a newly formed ternary B–Si–C carbide, SiC, and some residual silicon. Because residual silicon adversely affects the mechanical properties of the composite,[26,27,28,29] its amount, which is related to the free carbon present in compacted preform and also to the initial porosity,[30] should be decreased. The porosity of the preform at the outset may be somewhat reduced by sintering[31] or by using multimodal powder mixtures.[32] Adding titanium or iron or compounds that react with the boron carbide and release more free carbon[33,34,35,36,37] or adding elements that react with silicon to form stable silicides[38] may also serve to reduce the amount of residual silicon.

FUNCTIONALLY GRADED MATERIALS

A functionally graded material (FGM) is a two-component composite system with a defined compositional gradient across its section; the system is structured in such a way as to preserve the inherent properties of each component. In a metal/ceramic FGM structure, for example, the gradual transition between an impact-resistant outer ceramic layer bonded to a tough metal backing can be advantageously applied in armor protection design.[39] FGMs such as titanium/titanium boride composites have the potential to reduce or eliminate the need for thermal protection in extreme environments such as those encountered by aerospace vehicles. They are also ideal for minimizing thermomechanical mismatch in metal/ceramic bonding.[40,41,42,43] FGMs are also of interest for other defense-related applications, including as heat shields for rockets, fusion energy devices, and thermal barrier coatings for turbine blades. For armor protection applications, FGMs with bulk dimensions on the order of millimeters to centimeters are needed, but commercially viable processes to make such structures are still in development. These processes are hindered by the high cost and specialized nature—they often involve segregation approaches employing sedimentation forming, slip casting, centrifugal casting, and thixotropic casting.[44,45,46]

Experimental and theoretical work recently revealed that controlled gradients in mechanical properties can guide the design of surfaces that are resistant to contact deformation and damage; such properties cannot be realized in conventional homogeneous materials. Wear-resistant, nano-

[26]Hayun, S., D. Rittel, N. Frage, M.P. Dariel. 2010. Static and dynamic mechanical properties of infiltrated B_4C-Si composites. Materials Science Engineering A 487(1-2): 405-409.

[27]Aghajanian, M., B. Morgan, J. Singh, J. Mears, and R. Wolffe. 2001. A new family of reaction bonded ceramics for armor applications. Pp. 527-555 in Ceramic Armor Materials by Design. Ceramic Transactions 134. J. McCauley, J., and A. Crowson, eds. Baltimore, Md.: American Ceramic Society.

[28]Chhillar P., M. Aghajanian, D. Marchant, R. Haber, and M. Sennett. 2009. The effect of Si content on the properties of B_4C-SiC-Si composites. Pp. 161-167 in Advances in Ceramic Armor III: Ceramic and Engineering Science Proceedings 28(5). Franks, L., J. Salem, and D. Zhu, eds. Hoboken, N.J.: John Wiley & Sons.

[29]Hayun, S., A. Weizmann, M. Dariel, and N. Frage. 2009. The effect of particle size distribution on the microstructure and the mechanical properties of boron carbide-based reaction-bonded composites. International Journal of Applied Ceramic Technology 6(4): 492-500.

[30]Hayun, S., N. Frage, M. Dariel, E. Zaretsky, and Y. Ashuah. 2006. Dynamic response of B_4C-SiC ceramic composites. Ceramic Transactions 178: 147-156.

[31]Hayun, S., N. Frage, and M. Dariel. 2006. The morphology of ceramic phases in B_xC-SiC-Si infiltrated composites. Journal of Solid State Chemistry 179(9): 2875-2879.

[32]Hayun, S., A. Weizmann, M. Dariel, and N. Frage. 2009. The effect of particle size distribution on the microstructure and the mechanical properties of boron carbide-based reaction-bonded composites. International Journal of Applied Ceramic Technology 6(4):492-500.

[33]Hayun, S., N. Frage, H. Dilman, V. Tourbabin, and M. Dariel. 2006. Synthesis of dense B_4C-SiC-TIB_2 composites. Ceramic Transactions 78: 37-44.

[34]Mizrahi, I., A. Raviv, H. Dilman, M. Aizenshtein, M. Dariel, and N. Frage. 2007. The effect of Fe addition on processing and mechanical properties of reaction infiltrated boron carbide-based composites. Journal of Materials Science 42(16): 6923-6928.

[35]Sigl, L., H. Thaler, and K-A. Schwetz. 1994. Elemental carbon-containing boron carbide-titanium diboride composites, and their manufacture and use. [Verbundwerkstoff auf Basis von Borcarbid, Titanborid und Elementarem Kihlenstoff sowie Verhahren zu ihrer Herstellung.] European Patent 628525, filed June 11, 1993, and issued June 9, 1994, to Elektroschmelzwerk Kempten GmbH, Munich, Germany.

[36]Sigl, L., H. Thaler, and K-A. Schwetz. 1996. Composite materials based on boron carbide, titanium diboride and elemental carbon and processes for preparation of same. U.S. Patent 5,543,337, filed May 24, 1994, and issued August 6, 1996, to Elektroschmelzwerk Kemptem GmbH. Munich, Germany.

[37]Sigl, L. 1998. Processing and mechanical properties of boron carbide sintered with TiC. Journal of the European Ceramic Society 18(1):1521-1529.

[38]Messner, R., and Y-M. Chiang. 2008. Processing of reaction-bonded silicon carbide without residual silicon phase. Pp. 1053-1059 in Proceedings of the 12th Annual Conference on Composites and Advanced Ceramic Materials, Part 1 of 2: Ceramic Engineering and Science Proceedings 9(7/8). J. Wachtman, ed. Hoboken, N.J.: John Wiley & Sons.

[39]Hirai, T. 1996. Functional gradient materials. Pp. 293-341 in Processing of Ceramics, Part 2. Cahn, R., and R. Brook, eds. New York, N.Y.: VCH.

[40]Ibid.

[41]Chin, E. 1999. Army focused research team on functionally graded armor composites. Materials Science and Engineering A259(2): 155-161.

[42]Suresh, S. 2001. Graded materials for resistance to contact deformation and damage. Science 292(5526): 2447-2451.

[43]Gooch, W., B. Chen, M. Burkins, R. Palicka, J. Rubin, and R. Ravichandran. 1999. Development and ballistic testing of a functionally gradient ceramic/metal appliqué. Materials Science Forum 308-311:614-621.

[44]Kleponis, D., A. Mihalcin, and G. Filbey, Jr. 2005. Material design paradigms for optimal functional gradient armors. Army Research Lab Weapons and Materials Directorate. Available at http://handle.dtic.mil/100.2/ADA436346. Accessed April 4, 2011.

[45]Clougherty, E. 1974. Graded impact resistant structure of titanium diboride in titanium. U.S. Patent 3,802,850, filed November 13, 1972, and issued April 9, 1974, to Man-Labs, Incorporated, Cambridge, Mass.

[46]Johnson, G., T. Holmquist, and S. Beissel. 2003. Response of aluminum nitride (including a phase change) to large strains, high strain rates, and high pressures. Journal of Applied Physics 94(3), 1639-1647.

crystalline surface coatings, with grain sizes as small as a few tens of nanometers, can be synthesized by means of electrodeposition, thermal spray, sputter deposition, metal-organic chemical vapor deposition, and eletrophoretic deposition.[47] Many of these processes can create surface layers wherein the grain sizes are smoothly graded from the surface to the bulk, bringing about controlled gradients in strength and fracture toughness. Similarly, improved resistance to contact damage can be achieved by tailoring gradients in porosity below the contact surfaces. Current materials synthesis and processing capabilities, engineered gradations in properties—from nanometer to macroscopic length scales—appear promising for the design of improved fracture-, damage-, and wear-resistant structures and surfaces and for armor protection applications.

BIOMIMETIC MATERIALS

Natural materials that are mechanically robust often have hierarchical designs. Abalone nacre, rat teeth, fish scales, wood, and spider silk exhibit highly complex hierarchical structures, multifunctionality, and even self-healing capabilities[48] and thus are appealing to mimic for use in advanced armor design. The remarkable mechanical performance of certain natural materials stems from their complex ordered microstructure, organized over several length scales, even though the materials are often made of relatively weak constituents. Nacre, which is found in a number of mollusk shells, combines stiffness and strength along with a high level of toughness. Mimicking of the abalone structure was first attempted in the 1980s, when a laminated structure of Al–B_4C was produced.[49] A significant increase in fracture toughness (up to 16 MPa-$m^{1/2}$) was achieved; however, Al_4C_3 was formed during the processing of the laminates, limiting the useful armor application of the produced laminates owing to low hardness and strength and the high brittleness of the Al_4C_3 phase. Another example of mimicking abalone nacre, but on a macroscale, was conducted a decade ago by Foster-Miller[50]—LAST (Light Applique Segmented Tile) armor plates were produced using Coors alumina and SiC or B_4C hexagonal tiles covered in a thermoset laminate of Kevlar and held together with a Velcro-type adhesive. A nacre-like armor structure such as this absorbs energy and serves to toughen by many of the same mechanisms that occur in natural nacre. The entire array is fastened onto the steel plates of tanks and similar vehicles using Velcro. The armor has been implemented onto various ground and air vehicles, including 1,000 High Mobility Multipurpose Wheeled Vehicles (Humvees) for the U.S. Marines.

Another example of a natural material for biologically inspired design is the skins of certain fish. Both the scales and dermis of fish skin are highly pliant, lightweight, and resist penetration—all of this in an ultrathin structure.[51] Although fish scales have received very little attention, it is recognized that bony scales are difficult to penetrate and dissipate energy quite well.[52] Scales form a physical barrier that deters attacks by predators; indeed, they were likely the inspiration for scale armor not only in ancient times but also in modern times.[53] While its hierarchical organization is important for the overall mechanical performance of fish skin,[54] the contributions at the different length scales are rarely investigated. It is not known, for instance, how adjacent scales interact to thwart penetration, but this mechanism should be understood if its performance in the new generation of ultralight pliant armor systems is to be replicated.

Biomorphic ceramics using natural products such as wood and cellulose-fiber paper and cardboard have also been of interest for their potential use as armor materials.[55] In particular, wood-based biomorphic SiC (bioSiC) is a promising material for armor. The fabrication of bioSiC entails the rapid mineralization of wood, during which the wood is carbonized and then infiltrated with either Si vapor or Si melt. The formed SiC replicates the wood microstructure, and the diversity of the wood texture results in a large and varied selection of bioSiC ceramics. Novel biomorphic SiC ceramics have been successfully developed at DLR, Germany's national research center for aeronautics and space.[56,57] At DLR, wood-based preforms are converted to SiSiC materials using the liquid silicon infiltration process. In this process, a green body or preform based on low-cost raw materials—for

[47]Suresh, S. 2001. Graded materials for resistance to contact deformation and damage. Science Magazine 292(5526):2447-2451. Available online at http://www.sciencemag.org/content/292/5526/2447.full. Accessed October, 11, 2010.

[48]Meyers, M., P-U. Chen, A. Lin, and Y. Seki. 2008. Biological materials: Structure and mechanical properties. Progress in Materials Science 53(1): 1-206.

[49]Sarikaya, M., and I.A. Aksay. 1992. Nacre of abalone shell: A natural multifunctional nanolaminated ceramic polymer composite material. Pp. 1-25 in Structure, Cellular Synthesis and Assembly of Biopolymers (Results and Problems in Cell Differentiation) 19(1). S. Case, ed. Amsterdam: Springer-Verlag.

[50]QinetiQ. www.foster-miller.com. Accessed October 13, 2010.

[51]Vernerey, F., and F. Barthelat. 2010. On the mechanics of fishscale structures. International Journal of Solids and Structures 47(17): 2268-2275.

[52]Bruet, B., J. Song, M. Boyce, and C. Ortiz. 2008. Materials design principles of ancient fish armour. Nature Materials 7: 748-756.

[53]Vernerey, F., and F. Barthelat. 2010. On the mechanics of fishscale structures. International Journal of Solids and Structures 47(17): 2268-2275.

[54]Fratzl, P., and R. Weinkamer. 2007. Nature's hierarchical materials. Progress in Materials Science 52(177): 1263-1334.

[55]Medvedovski, E. 2010. Ballistic performance of armour ceramics: Influence of design and structure. Part 1. Ceramics International 36(7): 21032115.

[56]Heidenreich, B., M. Gahr, E. Strassburger, and E. Lutz. 2010. Biomorphic SiSiC-materials for lightweight armour. Pp. 21-33 in Proceedings of 30th International Conference on Advanced Ceramics & Composites 2010. Hoboken, N.J.: John Wiley & Sons.

[57]Heidenreich, B., M. Crippa, H. Voggenreiter, H. Gedon, M. Nordmann, and E. Strassburger. 2010. Development of biomorphic SiSiC- and C/SiSiC- materials for lightweight armour. Advances in Ceramics Armor VI: Ceramic Engineering and Science Proceedings (31). Hoboken, N.J.: John Wiley & Sons.

example, wood fibers and phenolic resin—is manufactured by means of warm pressing. One preform used for manufacturing biomorphic SiSiC is medium-density fiberboard, which is widely used in the furniture industry. The preform is made by pressing fine fibers of needle wood with binders based on formaldehyde or phenolic resins in a mass production process, making very large panels—typically 1.22 m × 2.44 m (4 ft × 8 ft) up to 2.8 m × 6.5 m (9.2 ft × 21.3 ft)—at a cost of about $1.75/kg. After pyrolysis, the porous C-preform is siliconized in a vacuum at temperatures above 1450°C; next, capillary forces allow molten silicon to infiltrate the open pores of the C-preform. The resulting reaction with the carbon forms $SiSiC_x$.[58] The final composition—that is, the content of SiC, Si, and C—is heavily influenced by the porosity and microstructure of the C-preform and can be varied widely by using tailored green bodies. Because practically no change in geometry occurs during siliconization or in reproducible contraction rates during pyrolysis, even large and complicated shaped parts can be manufactured using a cost-effective, near-net-shape technique.

The processing technology for bioSiC makes the production of complex shapes relatively easy; production is much cheaper, because the bioSiC forms at much lower temperatures than those required for SiC sintering or hot-pressing techniques. Biomorphic SiC shows excellent thermomechanical performance along with structural stability over a wide range of temperatures.[59,60,61,62,63,64,65] BioSiC consists of β-SiC formed as a result of the interaction of Si and C, a significant quantity of residual Si (up to 30 vol percent), and a very small amount of carbon (up to 3 vol percent). The typical structure of bioSiC is more homogeneous than conventional reaction-bonded SiC and is defined by the particular wooden preform. The typical size of SiC grains is 5 to 20 μ, but owing to the relatively high content of residual Si, the density is about 2.8 g/cm^3. BioSiC ceramics, such as one manufactured using inexpensive preforms from medium-density fiberboard[66] can be produced inexpensively for armor systems consisting of large, single-piece components. Biomorphic siliconized silicon carbide (SiSiC) has demonstrated good potential for use in lightweight ceramic armor systems. Although manufacturing defects and excessive residual silicon in bioSiC reduce ballistic performance, especially in multi-hit situations, appropriate armor system design—that is, with the right selection of ceramic thickness and type and backing thickness—allows the materials to withstand even armor-piercing rounds. In order to increase resistance to multiple hits, novel materials based on the combination of biomorphic SiSiC and C/C-SiC ceramics have been developed, with carbon fibers integrated into biomorphic SiSiC to increase ductility and damage tolerance.

There is special interest in replicating dragline silk, the extremely strong silk that forms the framing threads of spider webs. The comparative properties of selected silk and manmade fibers are presented in Table E-2.

Mimicking the structure of materials found in nature might provide insight into the creation of armor materials with superior ballistic properties. However, the task of producing manmade materials with similar microstructures and performance is challenging precisely because the structures are so complex.

MACHINING, GRINDING, AND POLISHING CERAMICS

The machining, grinding, and polishing of ceramics are expensive processes. For example, the final shape of the armor product from a flat, hot-pressed part is created by grinding with diamond wheels. The pressureless sintered process, with its much larger shrinkage (20 to 30 percent), requires considerably more grinding to achieve final tolerances. Therefore, grinding and finishing costs make the final cost of hot-pressed parts higher than that of pressureless sintered parts. Part geometry and concentricity or parallelism also affect the final cost. For example, improving the tolerance of the outside diameter from 0.020 in. to 0.010 in. can double the cost of a piece. Typically, a ground part has a surface finish tolerance of 16 μin. or better. A better finish of 4 μin. can be obtained using lapping and honing, but will cost more.

[58]Gahr, M., J. Schmidt, W. Krenkel, A. Hofenauer, and O. Treusch. 2004. Dense SiSiC ceramics derived from different wood-based composites: processing, microstructure and properties. P. 425 in Proceedings of the 5th International Conference on High Temperature Ceramic Matrix Composites. Westerville, Ohio: The American Ceramic Society.

[59]Martinez-Fernández, J., F. Valera-Feria, and M. Singh. 2000. High temperature compressive mechanical behavior of biomorphic silicon carbide ceramics. Scripta Materialia 43(9): 813-818.

[60]Martínez-Fernández J., F. Valera-Feria, Rodríguez, A., and M. Singh. 2000. Microstructure and thermomechanical characterization of biomorphic silicon carbide-based ceramics. Pp. 733-740 in Environment Conscious Materials: Ecomaterials. 39th Annual Conference of Metallurgists. Ottawa, Canada: Canadian Institute of Mining.

[61]Singh, M., J. Martínez-Fernández, A., and de Arellano-López. 2003. Environmentally conscious ceramics (ecoceramics) from natural wood precursors. Current Opinion in Solid State and Materials Science 7(3): 247-254.

[62]de Arellano-López, A, J. Martínez-Fernández, P. González, C. Domínguez, V. Fernández-Quero, and M. Singh. 2004. Biomorphic SiC: A new engineering ceramic material. International Journal of Applied Ceramic Technology 1(1): 56-67.

[63]Varela-Feria, F., J. Martínez-Fernández, A. de Arellano-López, and M. Singh. 2002. Low density biomorphic silicon carbide: Microstructure and mechanical properties. Journal of the European Ceramic Society 22(14-15): 2719-2725.

[64]Varela-Feria, F., J. Ramírez-Rico, A. de Arellano-López, J. Martínez-Fernández, and M. Singh. 2008. Reaction-formation mechanisms and microstructure evolution of biomorphic SiC. Journal of Materials Science 43(3): 933-941.

[65]Bautista, M., A. de Arellano-López, J. Martínez-Fernández, A. Bravo-Léon, and J. López-Cepero. 2009. Optimization of the fabrication process for medium density fiberboard (MDF)-based biomimetic SiC. International Journal of Refractory Metals and Hard Materials 27(2): 431-437.

[66]Heidenreich, B., M. Gahr, and E. Medvedovski. 2005. Biomorphic reaction bonded silicon carbide ceramics for armor applications. Pp. 45-53 in Proceedings of the 107th Annual Meeting of the American Ceramic Society. Hoboken, N.J.: Wiley-Blackwell.

TABLE E-2 Tensile Mechanical Properties of Spider Silks and Other Materials

Material	Stiffness (GPa)	Strength (GPa)	Strain to Failure	Toughness ($MJ-m^{-3}$)
Natural fibers				
Araneus major Ampullate (MA) silk	10	1.1	0.27	160
A. viscid silk	0.003	0.5	2.7	150
Nephila clavipes silk	11-13	0.88-0.97	0.17-0.18	40-130[a]
N. edulis silk				208[b]
Bombyx mori cocoon silk	7	0.6	0.18	70
B. mori silk (w/sericin)	5-12	0.5	0.19	35-55[c]
B. mori silk (w/o sericin)	15-17	0.61-0.69	0.4-0.16	30-70[d]
Synthetic materials				
Nylon fiber	5	0.95	0.18	80
Kevlar 49 fiber	130	3.6	0.027	50
Carbon fiber	300	4	0.013	25
High-tensile steel	200	1.5	0.008	6

[a]Gosline, J., M. DeMont, and M. Denny. 1986. The structure and properties of spider silk. Endeavour 10(1): 37-43; Zemlin, J. 1968. A study of the mechanical behavior of spider silks. Technical Report 69-29-CM (AD 684333). U.S. Army Natick Laboratory, Natick, Mass.; Cunniff, P., S. Fossey, M. Auerbach, J. Song, D. Kaplan, W. Adams, R. Eby, D. Mahoney, and D. Vezie. 1994. Mechanical and thermal properties of dragline silk from the spider *Nephila clavipes*. Polymers for Advanced Technologies 5(8): 401-410.

[b]Vollrath, F., B. Madsen, and Z. Shao. 2001. The effect of spinning conditions on the mechanics of a spider's dragline silk. Proceedings of the Royal Society 268(1483): 2339-2346.

[c]Pérez-Rigueiro, J., C. Viney, J. Llorca, and M. Elices. 1998. Silkworm silk as an engineering material. Journal of Applied Polymer Science 70(12): 2439-2447.

[d]Pérez-Rigueiro, J., C. Viney, J., Llorca, and M. Elices. 2000. Mechanical properties of single-brin silkworm silk. Journal of Applied Polymer Science 75(10): 1270-1277.

FOAMS

Foam is a complex assemblage of dispersed voids or pores separated by a film. The reason for using foams to absorb shock wave energy is that as the shock wave passes into the foam, the individual cells collapse; it is through this deformation that energy is absorbed. Foams can be made from any number of materials and may be open- or closed-cell, but it is the metal foams, particularly aluminum, and the polymeric foams, particularly polyurethane, that are employed most frequently in shock wave research.[67,68,69] Two important directions for future research are (1) constructing foams from a wider variety of materials and (2) developing methods for greater control over foam microstructure. Such model foams will help computational efforts on porous structures.

The use of foams as impact barriers was demonstrated by Gama et al.,[70] who performed impact tests on layered composite armor systems with various foam positions and thicknesses, using 20 mm fragment simulating projectiles (FSP). A test performed with an impact velocity of 1,067 m/s on a baseline setup, followed by a test on the baseline with a 12.7-mm-thick piece of aluminum foam incorporated into the material stack, revealed that the rise time of the stress wave increased from 1 μs for the baseline sample to 2 μs for the sample with foam. The use of foam also delayed the time for the stress wave to reach the stress gauge by about 14.6 μs. The maximum stress reached in both cases was about 6.25 GPa. The air-filled cellular structure of the foam is not conducive to wave propagation because the waves are only transmitted along the cell walls, which, owing to their random orientations, tend to disperse the wave. For a foam thickness of 12.7 mm, a stress of 0.825 GPa was recorded in the ceramic tile, while a thickness of 30.48 mm completely eliminated the stress recorded in the tile. In this case, the foam was not fully compacted by the FSP and thus acted as an excellent wave barrier.

The location of the foam is also important, demonstrating the need not only to consider the inherent material properties in isolation but also to consider them as part of the overall armor system.

[67]Hanssen, A., L. Enstock, and M. Langseth. 2002. Close-range blast loading of aluminum foam panels. International Journal of Impact Engineering 27(6): 593-618.

[68]Ramachandra, S., P. Sudheer Kumar, and U. Ramamurty. 2003. Impact energy absorption in an Al foam at low velocities. Scripta Materialia 49(8): 741-745.

[69]Gama, B., T. Bogetti, B. Fink, C. Yu, T. Dennis Claar, H. Eifert, and J. Gillespie. 2001. Aluminum foam integral armor: A new dimension in armor design. Composite Structures 52(3-4): 381-395.

[70]Ibid.

TRANSPARENT CERAMICS AND EMBEDDED DAMAGE SENSORS

Transparent armor ceramics must provide good transparency in the visible (0.4-0.7 μ) and mid-infrared (1-5 μ) optical ranges and protection against fragmentation from ballistic impacts, with multi-hit capability and minimal distortion around the impacted regions.[71,72]

As mentioned in Chapter 5, transparent armor systems are typically constructed of multiple layers of armor-grade ceramic plates, separated by transparent polymer (for example, polycarbonate) interlayers, and are bonded together with a transparent adhesive. The polymer phase mitigates the stresses generated by thermal expansion mismatches and inhibits crack propagation from ceramic to polymer. Polymeric materials such as transparent nylons, polyurethanes, and acrylics have also been explored as separators, but they have not been widely used in armor protection owing to less-than-optimal optical and durability characteristics. Transparent alumina (Al_2O_3) and magnesia (MgO) are two commonly used transparent ceramic armor materials. The composite system formed by these two materials provides good protection against high-velocity ballistic projectiles.[73,74] Silicon nitride (Si_3N_4), a nonoxide ceramic, has also been employed for use in radomes because of its good transit of microwave energy and its superior mechanical strength.

These materials can be produced as transparent polycrystalline ceramic parts, often with complex geometries, by using standard ceramic-forming techniques such as pressing, (hot) isostatic pressing, and slip casting.[75]

Nanocomposite ceramic materials of yttria (Y_2O_3) and magnesia (MgO) have been explored for use in transparent armor protection. The materials exhibit an average grain size of approximately 200 nm, and near-theoretical transmission in the 3 to 5 μ infrared band range. These complex ceramic nanocomposites reportedly offer improved mechanical properties such as superplastic flow and metal-like machinability. However, mechanical failure modes and armor protection characteristics must still be fully evaluated for these nanocomposite materials systems.[76,77,78] These materials have potential for use in new dome construction requiring substantial durability and high transparency across the operational bandwidth for infrared-guided missile sensing.[79,80,81] Special glasses and glass-ceramics such as lithium disilicate or aluminum-lithium-based crystallized or partially crystallized structures also offer advantages as host materials for laser use, since their refractive indexes and strain-optical coefficients can be readily controlled through changes in chemical composition.[82]

Damage-Reactive Sensors for Armor

Combat vehicles could be outfitted with smart ceramic sensors built into the protective armor material. Such sensor could detect and report on structural damage in real time. Structural damage caused by a wide range of ballistic impacts can be expected to affect armor structures under battlefield conditions. Changes in the armor's structural condition can be detected by tiny piezoelectric transducers, or sensors, built into the protective armor plate material. Piezoelectric sensors are usually ferroelectric, perovskite structure materials—for example, lead zirconate/lead titanate, barium titanate, and others—that have been suitably doped and electrically poled to optimize their piezoelectric response characteristics.[83,84]

Given the above relationships, the piezoelectric transducers can be designed both to generate and to receive voltage responses when coupled with ultrasonic waves that are generated to pass through the material. To determine the best response characteristics for particular environmental conditions, the ultrasonic signals may vary over a wide frequency range (1.0 kHz to 200 kHz). A generated shock wave through the plate picks up the reflections of sound waves and converts them into electrical voltages from which, with suitable amplification, one can determine their spectra and whether the plate is cracked or damaged.

---

[71]Patel, P., G. Gilde, P. Dehmer, and J. McCauley. 2000. Transparent ceramics for armor and EM window applications. P. 1 in Proceedings of the International Society for Optics and Photonics 4102(1).

[72]Harris, D. 2009. Materials for infrared windows and domes: Properties and performance. Bellingham, Wash.: International Society of Optical Engineers.

[73]Villalobos, G., J. Sanghera, and I. Aggarwal. 2005. Transparent ceramics: Magnesium aluminate spinel. Naval Research Laboratory Optical Sciences Division.

[74]Hogan, P., R. Stefanik, C. Willingham, and R. Gentilman. 2004. Transparent yttria for IR windows and domes—Past and present. Raytheon Integrated Defense Systems.

[75]Ibid.

[76]Bisson, J-F., Lu Jianren, K. Takaichi, Yan Feng, M. Tokurakawa, A. Shirakawa, A. Kaminskii, H. Yagi, T. Yanagitani, and K-I. Ueda. 2004. Nanotechnology is stirring up solid-state laser fabrication technology. Recent Research Developments in Applied Physics 7(Part II): 475-469.

[77]Wen, L., X. Sun, S. Chen, and C-I. Tsai. 2003. Synthesis of nanocrystalline yttria powder and fabrication of transparent YAG ceramics. Journal of the European Ceramics Society 24(9): 2681-2688.

[78]Wen, L., X. Sun, Q. Lu, G. Xu, and X. Hu. 2006. Synthesis of yttria nanopowders for transparent yttria ceramics. Optical Materials 29(2-3): 239-245.

[79]Huang, Z., X. Sun, Z. Xiu, S. Chen, and C-T. Tsai. 2004. Precipitation synthesis and sintering of yttria nanopowders. Materials Letters 58(15): 2137-2142.

[80]Jeong, J., S. Park, D. Moon, and W. Kim. 2010. Synthesis of Y_2O_3 nanopowders by precipitation method using various precipitants and preparation of high stability dispersion for backlight unit (BLU). Journal of Industrial Engineering Chemistry 16(2): 243-250.

[81]Nihara, K., and T. Sekino. 1993. New nanocomposite structural ceramics. P. 405 in Materials Research Society Symposium Proceedings held at the Nanophase and Nanocomposite Materials Symposium. Warrendale, Pa.: Materials Research Society.

[82]Hartmann, P., R. Jedamzik, S. Reichel, and B. Schreder. 2010. Optical glass and glass ceramic historical aspects and recent developments: A Schott view. Applied Optics 49(16): D157-D176.

[83]Meitzler, T., G. Smith, M. Charbeneau, E. Sohn, M. Bienkowski, I. Wong, and A. Meitzler. 2008. Crack detection in armor plates using ultrasonic techniques. Materials Evaluation 66(6): 555-559.

[84]Song, J., and G. Washington. 2000. Plate vibration modes identification by using piezoelectric sensors. Pp. 867-878 in Smart Structures and Materials 2000. International Society for Optics and Photonics 3985.

Armor Damage Control Sensors

The piezoelectric transducers can perform other functions that contribute to armor and vehicle survivability. For example, they can be made to act as antennas; to monitor the temperature of armor structures, including that of body armor, and to detect and monitor projectile impacts on the armor surface. For the latter application, each projectile striking the armor will create an electrically generated shock wave and differing amounts of electricity; a smart sensor can integrate these effects to generate useful information. Complex mathematical algorithms can be used to analyze the amount of electricity generated by a bullet's impact to determine what kind of round was used, since a small-caliber projectile will generate less electricity than a large-caliber projectile.[85] This combination of detection and assessment of threat level in real time could be significant for developing armor survival strategies. The following conclusions may be drawn:

1. Infrared-transparent nanocomposite materials in the systems SiC/Al_2O_3, SiC/Si_3N_4, SiC/MgO, Al_2O_3/ZrO_2, and transparent Al_2O_3 offer greatly enhanced ballistic needs for transmission in the 3 to 5 μ range, significantly increased mechanical strength of (700 MPa), and fracture toughness and creep resistance of interest for next-generation armor use.
2. Piezoelectric transducers coupled with acoustic wave propagation and complex mathematical algorithms can be used to analyze impact damage to armor structures from ballistic projectiles.

PHONONIC BAND GAP CONCEPTS FOR PROTECTIVE MATERIALS

Shortly after research on photonic band gaps began to show promise of controlling the flow of photons, the idea was extended to mechanical waves in periodic elastic structures. Referred to as phononic crystals, such structures can create what are called phononic, or acoustic, band gaps. A phononic crystal prevents the propagation of elastic waves if the frequencies of the waves fall within a band gap. The normalized width of the band gap—the ratio of band-gap width to the central frequency of the gap—is a measure of the performance of the particular phononic crystal design. In addition to preventing the propagation of waves, phononic crystals dictate the nature of the modes that are allowed to propagate in the material; they can decrease the velocity of the waves and even force their negative refraction. This capability suggests numerous ideas that could someday be developed to influence material fabrication and enhance armor designs to take advantage of the new ability to control mechanical waves in armor materials.

Because there are three polarizations of elastic waves in solid materials—longitudinal, transverse (shear) in plane, and transverse out of plane—for a structure to possess a full band gap, it must prohibit the propagation of all types of waves in all directions. A phononic crystal can create an acoustic band gap through a combination of Bragg diffraction (destructive wave interference) and Mie resonances[86] as well as anticrossing of bands having the same mode symmetry.[87] Bragg scattering occurs when the wavelength of the phonon is approximately equal to the periodicity of the structure, and Mie resonances occur when the diameter of the scattering features is of the same order as the wavelength. Some of the earliest references to phononic crystals are from Sigalas and Economou[88,89] and Kushwaha et al.[90,91] A rather complete library of phononic crystal research may be found by consulting Vlasov and Dowling.[92,93]

The length scale of phononic crystals ranges from the macroscopic—meters for acoustic waves (kHz) to millimeters for the ultrasound typically used in medical imaging (MHz), and down to the nanoregime—approximately100 nm for waves in the gigahertz (GHz) regime. In general, design rules for creating a gap are based on Bragg scattering and the use of highly symmetric structures to minimize the irreducible Brillouin zone over which the gap occurs. The materials parameters that are important are density, elastic modulus, and Poisson ratio or, equivalently, density and the transverse and longitudinal speed of sound in the materials. Because the waves are scattered from interfaces that affect wave propagation implies that understanding the differences in mechanical impedance of the materials comprising the structure (impedance is the product of density and wave speed) is key to understanding how phononic crystals can control the propagation of mechanical waves. However, in

[85]Sands, J., C. Fountzoulas, G. Gilde, and P. Patel. 2009. Modelling transparent ceramics to improve military armour. Special Issue on Transparent Ceramics, Journal of the European Ceramics Society 29(2): 261-266.

[86]Mie theory, also called Lorenz-Mie theory, Lorenz-Mie-Debye theory, and Mie scattering, is an analytical solution of Maxwell's equations for the scattering of electromagnetic radiation by spherical particles.

[87]Kushwaha, M., A. Akjouj, B. Djafari-Rouhani, L. Dobrzynski, and J. Vasseur. 1998. Acoustic spectral gaps and discrete transmission in slender tubes. Solid State Communications 106(10): 659-663.

[88]Sigalas, M., and E. Economou. 1992. Elastic and acoustic wave band structure. Journal of Sound and Vibration 158(2): 377-382.

[89]Sigalas, M., and E. Economou. 1993. Band structure of elastic waves in two dimensional systems. Solid State Communications 86(3): 141-143.

[90]Kushwaha, M., P. Halevi, L. Dobrzynski, and B. Djafari-Rouhani. 1993. Acoustic band structure of periodic elastic composites. Physical Review Letters 71(13): 2022-2025.

[91]Kushwaha, M., P. Halevi, G. Martinez, L. Dobrzynski, and B. Djafari-Rouhani. 1994. Theory of acoustic band structure of periodic elastic composites. Physical Review B 49(4): 2313-2322.

[92]Yaslov, Y. Photonic band gap links. Available at http://www.pbglink.com.

[93]Dowling, J. 2008. Photonic and sonic band-gap bibliography. Available at http://phys.lsu.edu/~jdowling/pbgbib.html. Accessed on November 6, 2010.

solids, in addition to simple reflection at an interface, polarization conversion always occurs; making it difficult to form complete gaps, and a design strategy for optimal constructs is not yet available. The use of fluids (which support only longitudinal waves) makes it difficult to form gaps because of the conversion of shear modes into longitudinal modes at the solid-fluid interface.

The idea of using a structured material to influence the propagation of elastic waves is promising since in addition to creating a band gap, there is the possibility of creating a set of band gaps that would significantly block multiple frequencies. Moreover, phononic crystals permit the tailoring of the allowed modes and their wave speeds *inside* the material, such that the frequencies of various material loss regimes may be matched with the density of states and frequencies of the allowed modes to provide enhanced energy absorption.

Appendix F

High-Performance Fibers

ARAMID FIBERS (KEVLAR AND TECHNORA)

Poly(p-phenylene terephthalamide), Kevlar, was first synthesized by Kwolek at DuPont in the 1960s. Kevlar is processed from sulfuric acid, with the polymer concentration at about 20 wt percent. Surprisingly, there is a decrease in viscosity with increased polymer concentration due to local alignment of polymer molecules in the solution to form a nematic phase. Thus the solution becomes liquid crystalline, a feature that had earlier been predicted by Flory.[1] The solution is extruded through an air gap into an acid solvent, such as water, where it coagulates. Removal of the approximately 80 percent acid from solution during fiber drying and tension heat treatment (500°C) leads to the formation of a highly aligned, extended chain fiber. However, the coagulation process also creates undesirable defects. The number of defects can be estimated from the deviation of the actual fiber density from the theoretical crystal density of 100 percent (approximately 1.45 g/cm^3 versus 1.50 g/cm^3). Kevlar fiber was developed and commercialized at DuPont, originally for completely different applications than for body armor (for example, it was used for reinforcing tires). The potential of Kevlar for use in ballistic protection was realized only when the National Institute of Justice conducted ballistic testing on Kevlar fabric. Other polyaramids followed, including Technora, an aramid copolymer fiber that is produced in the Netherlands and Japan from terephthaloyl chloride and a mixture of p-phenylenediamine and 3,4'-diaminodiphenylether.

POLYETHYLENE (SPECTRA, DYNEEMA)

Unlike the extended rigid-rod molecular structure of Kevlar, polyethylene (PE) is one of the most flexible polymers. Since the 1930s, fibers and films have been manufactured from PE by melt processing. The morphology of these fibers and films is semicrystalline, consisting of 60 to 70 vol percent crystals; the remainder consists of amorphous, entangled polymer chains. Interestingly, melt-processed polyethylene contains chain-folded crystals with a modulus in the 1 GPa range. Trash bags and milk jugs, having typical molecular weights of 50,000 to 200,000 g/mole, are common examples of such polyethylene products. But if PE molecules could be extended into straight chains, the carbon-carbon backbone would give outstanding properties. Indeed, after nearly half a century of process development in the field of polyethylene, a new type of spinning was invented by Smith and Lemstra in the Netherlands in early 1980s.[2] Known as gel spinning, this process is able to extend the macromolecules to nearly their full length and results in a highly crystalline extended-chain polyethylene fiber exhibiting high strength and high modulus characteristics that show ballistic protection capability. Because the molecules are processed from a dilute solution, the molecular weight of the polyethylene used in gel spinning can be in excess of 3 million g/mole or higher, much higher than that in any other synthetic polymer. Fiber is processed from a decalin solution that typically contains less than 5 wt percent polymer. The polymer solution is extruded at between 130°C to 150°C or so into a cold coagulant such as water. This resulting gel-like fiber, which contains more than 95 percent solvent, is typically then drawn at between 90°C and 130°C to draw ratios of 50 to 100. The macromolecules become extended and form near-single-crystal fibers.

The theoretical density of polyethylene is 1.00 g/cm^3, while the density of Spectra and Dyneema fibers is about 0.97 g/cm^3. This underscores the fact that even today's highly extended-chain polyethylene fibers contain a significant number of defects and suggests an opportunity for even more significant gains in future development of this material.

[1] Flory, P. 1956. Phase equilibria in solutions of rod-like particles. Proceedings of the Royal Society of London Series A—Mathematical and Physical Sciences 234(1196): 73-89.

[2] Smith, P., and P. Lemstra. 1980. Ultra-high-strength polyethylene filaments by solution spinning/drawing. Journal of Materials Science 15(2): 505-514.

RIGID-ROD POLYMERS (ZYLON AND M5)

After the successful commercial development of Kevlar in the 1970s, significant research efforts were devoted to the development of other rigid-rod polymers. Rigid-rod polymers programs began in the 1960s at the U.S. Air Force Research Laboratory as well as in Russia. The U.S. program was accelerated in the 1970s, resulting in the development of poly-p-phenylene benzobisthiazole and polybenzoxazole (PBO) fibers.[3] PBO fiber was further developed initially at SRI International and later at Dow Chemical Company before being commercialized by Toyobo Company (Japan) in 1998 under the trade name Zylon. Among other applications, PBO fiber was also developed for use in fire-protective clothing as well as for ballistic protection. However, in the early 2000s it became clear that there were environmental stability issues with Zylon fiber causing decreased fiber strength over time and negatively affecting its ballistic performance. This is attributed to poor resistance to ultraviolet radiation as well as to poor hydrolytic stability.

In an attempt to improve intermolecular interactions in rigid-rod polymers with the intent of increasing the fiber compressive strength and torsional modulus, the Akzo Nobel firm in the Netherlands synthesized and processed polypyridobisimidazole (under the name M5) fiber during the 1990s.[4] The fiber was further developed by Magellan Systems International, and the technology now resides with DuPont, although the fiber has not yet been commercialized.

Similar to Kevlar, both the Zylon and M5 fibers are processed from a liquid crystalline polymer solution, except in this case the solution is one of polyphosphoric acid. Depending on the polymer molecular weight, for fiber spinning, polymer concentration in solution is again typically between 5 weight percent 15 weight percent. Like the process used to make Kevlar, the nematic solution is extruded through an air gap into an acid solvent such as water. The coagulated fiber is then heat-treated under tension up to about 500°C. Structure formation mechanism in the rigid-rod chains of Zylon and M5 fibers is very similar to the structure formation mechanism in Kevlar and is quite different from that of the flexible-chain gel-spun polyethylene (Dyneema and Spectra).

Intermolecular interactions in polyethylene are only van der Waals interactions, whereas in Kevlar there is hydrogen bonding in one dimension transverse to the fiber axis, and in M5 fibers there is hydrogen bonding in two transverse directions. Ranking fibers in from greatest to least, in terms of compressive and torsional properties, shows that M5 has highest compressive and torsional properties, followed by Kevlar, then Zylon, then Spectra and Dyneema, which are approximately equal.

[3] Chae, H., and S. Kumar. 2006. Rigid-rod polymeric fibers. Journal of Applied Polymer Science 100(1): 791-802.

[4] Sikkema, D. 1998. Design, synthesis and properties of a novel rigid rod polymer, PIPD or 'M5': High modulus and tenacity fibres with substantial compressive strength. Polymer 39(24): 5981-5986.

THERMOTROPIC LIQUID CRYSTALLINE POLYMERIC FIBERS

Thermotropic liquid crystalline polymeric fibers, developed in the 1970s, are melt processed (no solvent). These polymers exhibit liquid crystalline behavior in the melt state. Vectran, a copolyester and an example of a commercial fiber in this class, is spun at temperatures of 275°C or more. To further enhance mechanical properties, as-spun fiber may be further drawn and annealed below the polymer melting temperature. During this process, fiber may also undergo further solid state polymerization, resulting in a polymer of higher molecular weight. Unlike the liquid-crystalline-solution processing of rigid-rod polymers and the gel spinning of flexible-chain polyethylene—both of which are processed from polymer solutions containing 85 percent to 95 percent solvent (which must be removed during fiber processing)—there is no solvent to be removed in the processing of thermotropic liquid crystalline polymers. Compared to polyethylene, however, the molecular weights (and hence the chain length) of aramids, rigid-rod polymers, and thermotropic liquid crystalline polymers are much more limited. Vectran has more applications in injection-molded products than in fiber form.

CARBON FIBERS

The development of modern carbon fibers dates back to the 1960s with research by Shindo in Japan, Watt in England, and Bacon at Union Carbide in the United States. Early carbon fibers were made by pyrolyzing cellulose; today, carbon fibers are made starting from petroleum pitch or from polyacrylonitrile (PAN) copolymers. Pitch-based carbon fibers can have a very high tensile modulus and high electrical and thermal conductivities but exhibit relatively low tensile and compressive strength. By contrast, PAN-based carbon fibers have high tensile strength, good compressive strength, and intermediate modulus and electrical and thermal conductivities. High-purity mesophase pitch (a liquid crystalline pitch) is melted, extruded typically at about 400°C, and then carbonized in stages (Stage 1 at 600°C to 1000°C, Stage 2 at 1100°C to 1600°C, and Stage 3 at 2200°C to 2700°C) in an inert environment. Fibers carbonized at about 2700°C can exhibit up to 90 percent of the theoretical modulus. The theoretical modulus of graphite along graphene planes is 1,060 GPa, giving it a specific theoretical modulus of 469 N/tex,[5] which is equivalent to 469 GPa/(g/cm^3).

PAN fibers are either wet spun or dry-jet wet spun from solutions in sodium thiocyanate and water, dimethyl acetate,

[5] "Tex" is the mass of a 1,000-meter length of fiber in grams.

dimethylsulfoxide, or zinc chloride and water.[6] Depending on the molecular weight, solvent, and the copolymer composition, the polymer concentration in solution is typically 5 to 25 wt percent. After spinning, fibers are successively drawn at several different temperatures (typically between room temperature and 175°C). Drawn fibers are oxidized under tension typically between 200°C and 350°C for approximately 2 hours. Oxidized fibers are then carbonized under tension in stages, similar to the carbonization of pitch-based fiber. Fibers with the highest tensile strength are typically obtained at about 1300°C to 1500°C.

CARBON NANOTUBE FIBERS

Carbon nanotube (CNT) fibers to date have been processed primarily by one of the following two techniques: (1) CNT smoke drawn directly from the chemical vapor deposition reactor in the form of aerogel fibers[7] and (2) fiber processed from aqueous[8] or acidic[9] dispersions of CNTs. In both cases, it is important that the CNTs be as long as possible and as perfect as possible, and they should be free of catalyst and other foreign impurities, including amorphous carbon. The tube-to-tube diameter variation should be minimized and the diameter should be relatively small. Nanotube orientation also plays a critical role with respect to mechanical properties.[10] Multiwall CNTs tend to undergo telescoping, with the individual tubular shells slipping past one another, whereas single-wall CNTs are essentially the ultimate for a high-strength polymer molecule, having a theoretical strength as high as 150 GPa and modulus values as high as 1,050 GPa, respectively. The theoretical modulus of carbon nanotubes is dependent on their diameter since their central portion is empty; however, their specific theoretical modulus is 469 N/tex irrespective of the diameter.

[6]Gupta, V., and V. Kothari. 1997. Manufactured Fibre Technology. New York, N.Y.: Chapman and Hall.

[7]Koziol, K., J. Vilatela, A. Moisala, M. Motta, P. Cunniff, M. Sennett, and A. Windle. 2007. High-performance carbon nanotube fiber. Science 318(5858): 1892-1895.

[8]Vigolo, B., A. Penicaud, C. Coulon, C. Sauder, R. Pailler, C. Journet, P. Bernier, and P. Poulin. 2000. Macroscopic fibers and ribbons of oriented carbon nanotubes. Science 290(5495): 1331-1334.

[9]Ericson, L., H. Fan, H. Peng, V. Davis, W. Zhou, J. Sulpizio, Y. Wang, R. Booker, J. Vavro, C. Guthy, A. Parra-Vasquez, M. Kim, S. Ramesh, R. Saini, C. Kittrell, G. Lavin, H. Schmidt, W. Adams, W. Billups, M. Pasquali, W-F. Hwang, R. Hauge, J. Fisher, and R. Smalley. 2004. Macroscopic, neat, single-walled carbon nanotube fibers. Science 305(5689): 1447-1450.

[10]Liu, T., and S. Kumar. 2003. Effect of orientation on the modulus of SWNT films and fibers. Nano Letters 3 (5): 647-650.

ALUMINA, BORON, SILICON CARBIDE, GLASS, AND ALUMINA BOROSILICATE CERAMIC FIBERS

Boron fiber is processed using chemical vapor deposition on substrates such as tungsten or carbon, whereas silicon carbide fibers can be processed either by chemical vapor deposition or by a precursor method similar to the processing of carbon fibers. Alumina and alumina borosilicate fibers are typically processed using a sol-gel precursor followed by sintering. Nextel fibers (from 3M Company) are ceramic oxide fibers that belong to the category of alumina-boro-silicate. Compared to polymeric and carbon fibers, these fibers retain their mechanical properties to much higher temperatures. Although the tensile strength of these fibers is not quite as high as that of some of the polymeric fibers, their compressive strength can be comparable to or higher than that of carbon fiber having the best compressive strength. Owing to ionic-covalent bonds in all directions, these fibers are much more isotropic than are carbon and polymer fibers, which exhibit a very high degree of anisotropy.[11,12,13]

Glass is melt-extruded and drawn into fibers typically at 1000°C to 1200°C. Fiber tensile strength is limited by defects, residual stresses, and structural inhomogeneities in the fibers.

[11]Chawla, K. 1998. Fibrous Materials. Cambridge, U.K.: Cambridge University Press.

[12]Elices, M., and J. Llorca. 2002. Fiber Fracture. Oxford, U.K.: Elsevier Science.

[13]Watt, W., and B. Perov, eds. 1986. Strong Fibers. North-Holland: Elsevier Science.

Appendix G

Failure Mechanisms of Ballistic Fabrics and Concepts for Improvement

FAILURE MECHANISMS

Breakage of Fiber Bonds and Yarns

As in all materials, when a force is applied to the fiber or yarn or fabric, a set of competing deformation processes can take place, depending on the loading rate, stress state, temperature, and other factors. Polymer fibers are normally highly crystalline and highly anisotropic due to the high molecular orientation and the covalent bonds along the fiber axis versus van der Waals or hydrogen bonding in the transverse directions. However, glass and ceramic fibers can be essentially isotropic due to their multidirectional ionic-covalent bonds. The assembly of fibers into yarns and yarns into a fabric with a given architecture or geometry leads to different overall symmetries for the actual armor.

When a molecular bond is excited beyond its activation energy, bond breakage occurs. The activation energies for shear and interchain slip are lower than for covalent bond rupture and are strongly affected by ambient temperature, pressure, and the polymer's intrinsic glass transition temperature. When a projectile hits the fabric, the fiber is stretched along the axial direction owing to the longitudinal stress wave. Also, penetration of the projectile leads to shearing across the direction of the fiber thickness. Normally in the contact area of projectile and fabrics, if induced strain is larger than the failure strain of the fibers, the fiber will break. For polymer regions that are in a rubbery state (the noncrystalline component of which may be above its T_g), shear yielding is expected to occur before fracture. However, under a very high strain rate, as is the case for ballistic impact, the time interval that a stressed bond spends at a certain stress level is shortened and there is a lower probability for bond breaking at that level; thus, strength increases with the increase of strain rate. Termonia et al.[1] calculated the strain-rate dependence of strength of perfectly ordered polyethylene (PE) and found that the maximum strength may increase from 1.5 GPa to 21 GPa for PE with a molecular weight of 2.2×10^4 g/mol when strain rate increases from 10^{-1} min^{-1} to 10^5 min^{-1}. Also, at low strain rate, before bond breakage, molecular slippage occurs and plastic deformation is observed. By comparison, at the higher strain rates observed in ballistic impact, bond breakage and molecular slippage may occur simultaneously, or the primary bond breakage may even become predominant.[2] Although the tensile properties of fibers such as aramid and carbon fibers are relatively less sensitive to the strain rate, fibers such as Spectra are sensitive to strain rate, and their failure strain and mechanism at high strain rate may be distinctly different from that at low strain rate. There are relatively few studies of the strain-rate dependence of tensile behavior, and more efforts are needed to fully characterize the strain-rate dependence. Gu[3] observed that strength/modulus increased from 2.4 GPa and 62 GPa to 2.75 GPa and 72 GPa for Twaron [poly(paraphenylene terephthalamide)] and from 1.19 GPa and 20.3 GPa to 1.85 GPa and 51.2 GPa for Kuralon (a polyvinyl alcohol), when the strain rate increased from 10^{-2} s^{-1} to 10^3 s^{-1}. Wang and Xia[4] tested Kevlar in the strain-rate range from 10^{-4} s^{-1} to 10^3 s^{-1} and observed that the strength of Kevlar 49 increased from 2.34 GPa to 3.08 GPa and its modulus from 97 GPa to 125 GPa. Zhou et al.[5] studied the strain-rate dependence of mechanical properties of T-300 and M40J carbon fibers in the range 10^{-3} s^{-1} to 1.3×10^3 s^{-1} and observed that these

[1]Termonia, Y., P. Meakin, and P. Smith. 1986. Theoretical study of the influence of strain rate and temperature on the maximum strength of perfectly ordered and oriented polyethylene. Macromolecules 19(1): 154-159.

[2]Shim, V., C. Lim, and K. Foo. 2001. Dynamic mechanical properties of fabric armour. International Journal of Impact Engineering 25(1): 1-15.

[3]Gu, B. 2003. Analytical modeling for the ballistic perforation of planar plain-woven fabric target by projectile. Composites Part B: Engineering 34(4): 361-371.

[4]Wang, Y., and Y. Xia. 1998. The effects of strain rate on the mechanical behaviour of kevlar fibre bundles: an experimental and theoretical study. Composites Part A: Applied Science and Manufacturing 29(11): 1411-1415.

[5]Zhou, Y., D. Jiang, and Y. Xia. 2001. Tensile mechanical behavior of T300 and M40J fiber bundles at different strain rate. Journal of Materials Science 36(4): 919-922.

fibers were strain-rate-insensitive materials. Wang and Xia[6] observed that for Kevlar 49 fiber, at a fixed strain rate, the initial tensile modulus decreased and elongation at break increased with the increase in test temperature.

Yarn Pullout

If yarn is not well gripped at its ends, the ends may be pulled out from the fabric mesh. In this case, yarn pullout may occur and none of the fibers inside this portion of the yarn break. The pullout force is dependent on interyarn friction and pre-tension. The interyarn friction is related to friction efficiency and interyarn contact area. Yarn pullout may be the major energy dissipation path only when fabric is ungripped or not well gripped.

Remote Yarn Failure

Yarn failure may happen away from the impact area but between the impact point and the gripping boundary. Shockey et al.[7] observed remote yarn failure during Zylon tensile testing. The remote yarn failure occurs in tests of both transverse load (perpendicular to the yarn direction) and cylindrical load (along the yarn direction). The remote yarn failure may be hard to detect, as broken fibers may be buried inside the fabric mesh. Remote yarn failure will not affect the load on the projectile until friction force on the yarns decreases to a value that cannot sustain additional remote yarn failure. Since remote yarn failure involves yarns in a large area of fabric target, it may significantly increase the energy absorbance. Remote yarn failure has been observed in penetration by a blunt projectile in both two-edge-gripped and four-edge-gripped fabric targets.

Wedge-Through Phenomenon

The wedge-through phenomenon occurs when the formed hole is smaller than the diameter of the projectile. The phenomenon is more predominant in the back side of a multi-ply system. When a projectile hits the fabric, the transverse movement of the yarns locally expands the mesh and increases the space between woven yarns. For a projectile with a small cross-section and a fabric with only a few layers, the projectile may push the yarns aside and slip through the hole. There is a greater possibility of a wedge-through projectile phenomenon in loosely woven fabric than in tightly woven fabric, as has been observed by many researchers.[8,9] The wedge-through phenomenon is affected by projectile geometry, fabric structure, and mobility of yarns, which is correlated to frictional behavior of the yarns.

Fibrillation

Anisotropic fibers are subject to splitting along their axial direction.[10] High-strength fibers with highly oriented and extended polymer chains may fail in compression at very low strains, normally less than 1 percent; kinking and microbuckling are major failure responses.[11] When polymer chains are highly aligned in a fiber, the tensile modulus along the fiber axis is very high, whereas the shear modulus is relatively low. Fibrillation can occur during compression and results in high energy absorption during failure, which will be useful for the ballistic performance.[12] Fibrillation was found in para-aramid fibers[13] after ballistic impact, and its level was found to increase at low impact energy as compared to high impact energy. Fibrillation is caused by the abrasion of a projectile with yarns in the lateral direction to the fiber axis. Flat head projectiles with less possibility of penetration do not promote much fibrillation.[14,15]

Other Damage Forms

During impact, the friction between projectile, fabric, yarns, and filaments may cause heat generation and lead to temperature increase. This is more of an issue for thermoplastic polymer fibers such as PE and nylons than for aromatic heterocyclic backbone fibers such as Kevlar due to the vastly higher melting points of the latter type of fiber. Carr[16] observed the melting of fibers after the high energy impact

---

[6]Wang, Y., and Y. Xia. 1999. Experimental and theoretical study on the strain rate and temperature dependence of mechanical behaviour of Kevlar fibre. Composites Part A: Applied Science and Manufacturing 30(11): 1251-1257.

[7]Shockey, D., J. Simons, and D. Elrich. 2001. Improved barriers to turbine engine fragments: interim report III. May, 2001. Available online http://oai.dtic.mil/oai/oai?verb=getRecord&metadataPrefix=html&identifier=ADA392533. Accessed April 5, 2011.

[8]Montgomery, T., P Grady, and C. Tomasino. 1982. The effects of projectile geometry on the performance of ballistic fabrics. Textile Research Journal 52(7): 442-450.

[9]Kirkland, K., T. Tam, and G. Weedon. 1991. New third-generation protective clothing from high-performance polyethylene fiber: From knives to bullets. Pp. 214-237 in High-Tech Fibrous Materials, ACS Symposium Series. American Chemical Society.

[10]Carr, D. 1999. Failure mechanisms of yarns subjected to ballistic impact. Journal of Materials Science Letters 18(7): 585-588.

[11]Kozey,V. H. Jiang, V. Mehta,and S. Kumar. 1995. Compressive behavior of materials: Part 2. high-performance fibers. Journal of Materials Research 10)4): 1044-1061.

[12]Chawla, K. 2002. Fiber fracture: An introduction. Pp. 3-26 in Fiber Fracture. M. Elices and J. Llorca, eds. Oxford, U.K.: Elsevier Science.

[13]Carr, D. 1999. Failure mechanisms of yarns subjected to ballistic impact. Journal of Materials Science Letters 18(7): 585-588.

[14]Tan, V., C. Lim, and C. Cheong. 2003. Perforation of high-strength fabric by projectiles of different geometry. International Journal of Impact Engineering 28(2): 207-222.

[15]Lim, C., V. Tan, and C. Cheong. 2002. Perforation of high-strength double-ply fabric system by varying shaped projectiles. International Journal of Impact Engineering 27(6): 577-591.

[16]Carr, D. 1999. Failure mechanisms of yarns subjected to ballistic impact. Journal of Materials Science Letters 18(7): 585-588.

of Spectra fabrics. Prosser et al.[17] observed a temperature increase on the back surface of a ballistic panel containing 40 layers of nylon fabrics to as high as 76.6°C after perforation by a .22 caliber projectile.

CONCEPTS FOR ENHANCING BALLISTIC PERFORMANCE OF FABRICS

There is an opportunity to develop new fibers, coming up with entirely new methods of processing fibers that eliminate defects, and to make fibers from other desirable materials. Magnesium, with a density of only 1.7 g/cm^3, is an example of such a desirable material. The tensile strength of most magnesium alloys is in the range 200 MPa to 400 MPa.[18] Alumina fiber, with a tensile strength of 1.7 GPa, is the high-performance fiber with the lowest tensile strength. Thus the development of even 1 GPa tensile strength magnesium fiber that could be used to replace bulk magnesium alloy in helmets with a magnesium alloy and Spectra fiber construction could be significant.

In carbon-nanotube-reinforced composites, polymers such as poly(paraphenylene terephthalamide), poly(benzobisoxazole), poly(diimidazo pyridinylene [dihydroxy]phenylene), ultrahigh-molecular-weight PE, polyurethane, and so on can be used as a matrix system, with the carbon nanotube as the reinforcing entity. Similarly, carbon-nanotube-reinforced fibers can also be made from metals, ceramics, and glasses, wherein during high-temperature processing there exists the probability of compound formation and new types of interfacial bonds.

[17]Prosser, R., S. Cohen, and S. Cohen. 2000. Heat as a factor in the penetration of cloth ballistic panels by 0.22 caliber projectiles. Textile Research Journal 70(8): 709-722.

[18]Mathaudhu, S., and E. Nyberg. 2010. Magnesium alloys in army applications: Past, current and future solutions in magnesium technology. Pp. 27-33 in Magnesium Technology 2010: Proceedings of a Symposium Sponsored by the Magnesium Committee of the Light Metals Division of TMS, 2010. Warrendale, Pa.: Minerals, Metals, and Materials Society.

Appendix H

Metals as Lightweight Protection Materials

TITANIUM AND TITANIUM ALLOYS

Titanium is a hexagonally close-packed metal with a density of 4,950 kg/m^3; it can have a specific strength (the ratio of the yield strength to the density) that is greater than that of some (but not all) steels. Commercially pure titanium has a yield strength of about 400 MPa, with strong strain hardening and substantial rate sensitivity at high strain rates.[1] Strengths of this magnitude are not sufficient to provide significant benefit in comparison to that of rolled homogeneous armor (RHA) for protection material applications, given the density of titanium. However, titanium alloys can have much greater strengths, and in particular the Ti-6Al-4V alloy has a strength approaching 1 GPa in the solution treated and aged condition. As a consequence, there is at least one specification of Ti-6Al-4V for armor applications,[2] and there are several specific components of military vehicles in which this titanium alloy has been substituted for steel, with significant weight savings.[3] Titanium alloys have good corrosion resistance, offer good ballistic protection with some weight savings, and can be welded.

The primary obstacles to the expanded use of titanium as protection materials are twofold. First, and most important, is cost: the extraction, processing, and forming of titanium all result in a final component that is significantly more expensive than a component made of steel. Second, titanium alloys, like many hexagonally close-packed metals, have a relatively high susceptibility to adiabatic shear localization. These factors have resulted in the greater use of aluminum and aluminum alloys as substitutes for steels.

[1]Meyers, M., G. Subhash, B. Kad, and L. Prasad. 1994. Evolution of microstructure and shear-band formation in α-hcp titanium. Mechanics of Materials 17(2-3): 175-193.

[2]MIL-T-9046J.

[3]Montgomery, J., M. Wells, B. Roopchand, and J. Ogilvy. 1997. Low-cost titanium armors for combat vehicles. Journal of the Minerals, Metals and Materials Society 49(5): 45-47.

ALUMINUM AND ALUMINUM ALLOYS

Aluminum and aluminum alloys were developed early in the twentieth century, and beginning around the time of World War II, they were pressed into service to reduce the weight of protective materials (beginning with armor for aircraft). The introduction in the late 1950s of the T113 (later M113) personnel carrier using an aluminum alloy structure resulted in the deployment of a significant amount of aluminum alloys to the armored fleet. Whereas pure aluminum is very soft, conventional aluminum alloys can have yield strengths that easily compete with the simpler steels. Specific approaches such as solid solution strengthening and age hardening have been developed to strengthen aluminum alloys. Note that the range of strengths attainable with steels is very large, and there are no conventional aluminum alloys that can compete with the highest-strength steels in terms of yield strength. However, when one considers the specific strength (that is, the strength per unit weight, or σ_y/ρ), some of the commercial aluminum alloys can be very competitive.

Figure H-1 shows the typical specific strengths and specific stiffnesses of many metals and ceramics—the specific stiffnesses are of interest when deflection-limited design is important, as with some ceramic tiles, whereas specific strength is important for some strength-limited applications. Ceramics generally have higher specific strengths than metals and metal alloys, and ceramics indeed have a major role to play in protection material systems. The figure shows, among the metals, the relative locations of RHA and one aluminum alloy (Al 5083, which is 4.4 wt percent Mg, 0.7 wt percent Mn, and 0.15 wt percent Cr; the balance is Al). This alloy is commonly used in military vehicles such as personnel carriers.

A critical question for metals that meet both structural and armor roles in vehicles involves weldability, since this has a large impact on both production cost and maintenance. The welding of steels is a finely developed technology, but the weldability of aluminum alloys is much more variable.

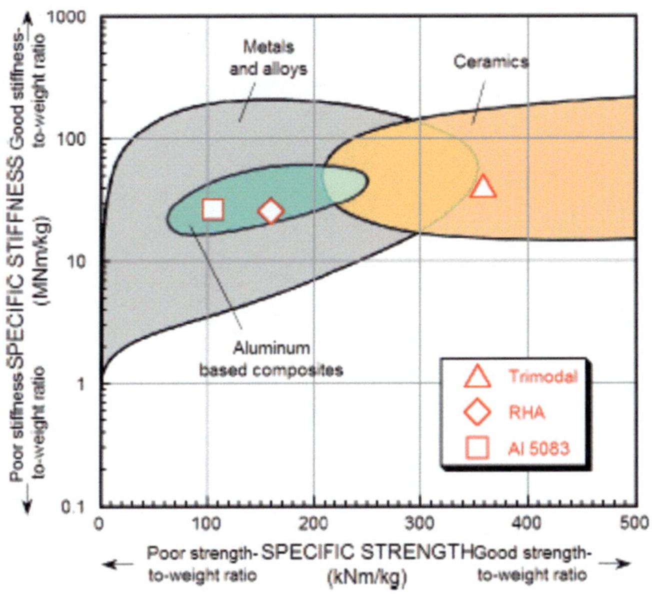

FIGURE H-1 Specific stiffness versus specific strength of various materials, including metals and ceramics. The position occupied by rolled homogeneous armor is identified, as is the conventional aluminum alloy 5083. Note the substantially greater specific strength that can be obtained by using aluminum-based nanocrystalline matrix composites such as the so-called trimodal aluminum materials. SOURCE: Zhang, H., J. Ye, S. Joshi, J. Schoenung, E. Chin, G. Gazonas, and K. Ramesh: Superlightweight nanoengineered aluminum for strength under impact. Advanced Engineering Materials. 2007. 9. 335-423. Copyright Wiley-VCH Verlag GmbH & Co. KGaA. Reproduced with permission.

Those aluminum alloys that are easily weldable are therefore preferred in these applications, even if some penalty is paid in terms of strength and ballistic performance. The trade-offs between weight, structural performance, ballistic performance, ease of production, and ease of maintenance (including resistance to corrosion) play a very significant role in the choice of alloy for vehicular applications. Because most of these alloys are used as rolled plate, work-hardening alloys such as the 5000 series (Al 5083 being the prime example) have some advantages. Aluminum alloys used as armor in Army vehicles also include Al 2024, Al 2519, Al 5059, Al 6061, Al 7039, and Al 7075. Promising new commercial alloys include Al 2139, which is a commercial alloy with significant strength (around 600 MPa at high strain rates) and reasonable ductility.

There is significant potential for the development of novel aluminum-based materials with very high strengths through alloying approaches, the development of nanostructured systems, and the development of aluminum-based composites. The nanostructured aluminum approach is exemplified by the so-called trimodal aluminum material developed by Li and Zhao and their coworkers.[4] This aluminum-based material exhibits a very high strength (950-1,000 MPa) when loaded at high strain rates, although the ductility (as of 2009) is relatively low. The material achieves dramatic mechanical properties at impact rates of deformation through a combination of three microstructural approaches: strengthening through a nanocrystalline core architecture; additional strengthening through length-scale-dependent reinforcement with micron-size ceramic particles; and enhanced ductility through the incorporation of a certain volume fraction of micron-scale grains. The resulting trimodal aluminum-based material achieves high specific strengths under very high rates of deformation and shows promise as a protective material, although the ductility remains a major concern. The material is produced by cryomilling Al 5083 aluminum powders with boron carbide ceramic particulates. This composite powder is then degassed and blended with microscale Al 5083. This trimodal composite powder is then consolidated with conventional powder metallurgy techniques such as cold isostatic pressing plus extrusion to generate a bulk trimodal aluminum-based composite.

Figure H-2 presents stress versus strain curves obtained on a trimodal aluminum alloy at strain rates of 3,200 s^{-1} and 11,000 s^{-1} using a compression Kolsky bar. Strength levels of this magnitude are remarkable for an aluminum-based material. The mechanical response of the most common current armor steel (RHA) measured at similar strain rates is also shown in Figure H-2—note that this steel is nearly three times as dense as the aluminum alloy. The specific strength of the trimodal material is also shown in Figure H-2.

Mechanical milling, temperature and consolidation lead to a peculiar microstructure for this material; as a result its strength is derived from, in addition to the normal load transfer characteristics of the composite, four strengthening mechanisms. They are (1) grain boundary strengthening, via the refinement of grain size, (2) particle-size strengthening through ceramic reinforcement, (3) dispersoid strengthening, and (4) work-hardening owing to prior plastic work from extrusion and cryomilling. This material can be considered to be a sophisticated alloy, a nanostructured material, or a specific metal-matrix composite—the value is in the use of all of the associated strengthening mechanisms.

Advanced aluminum-based materials of this type, including wrought alloys such as Al 2139 and aluminum-based metal-matrix composites, discussed below, show promise of dramatic improvements as protection materials in terms of mass efficiency. The key research questions in terms of the utility of such advanced materials are those concerning the failure processes within the material: ductility, resistance

[4]Li, Y., Y.H. Zhao, V. Ortalan, W. Liu, Z.H. Zhang, R.G. Vogt, N.D. Browning, E.J. Lavernia, and J.M. Schoenung. 2009. Investigation of aluminum-based nanocomposites with ultra-high strength. Materials Science and Engineering: A 527(1-2): 305-316.

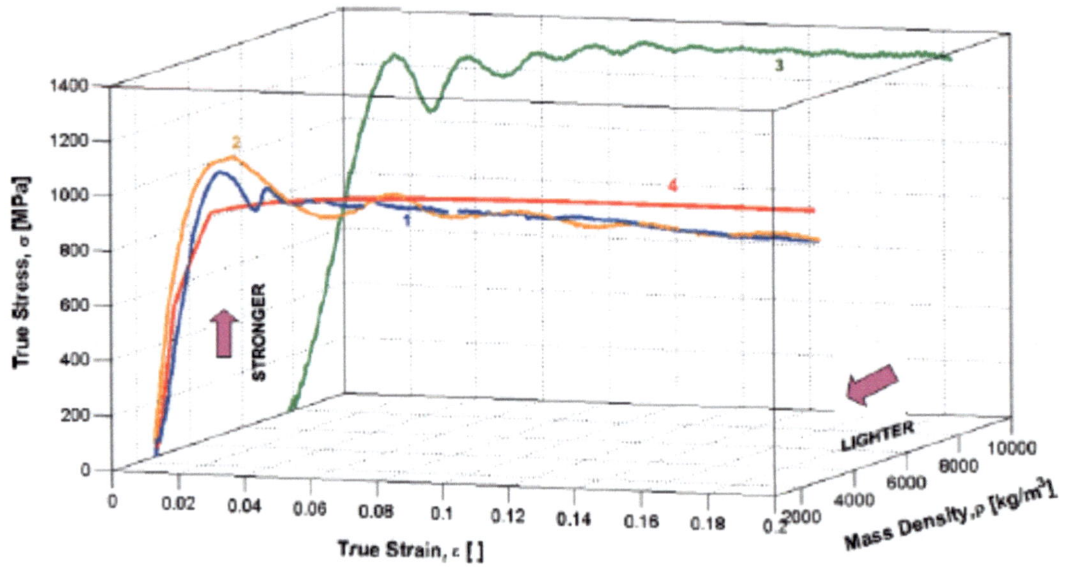

FIGURE H-2 High-strain-rate compressive response of a trimodal aluminum alloy, in comparison with that of rolled homogeneous armor at similar strain rates (10^3 s^{-1}). Curve 4 represents not experimental data but the prediction of a model based on composite micromechanics. SOURCE: Zhang, H., J. Ye, S. Joshi, J. Schoenung, E. Chin, G. Gazonas, and K. Ramesh: Superlightweight nanoengineered aluminum for strength under impact. Advanced Engineering Materials. 2007. 9. 335-423. Copyright Wiley-VCH Verlag GmbH & Co. KGaA. Reproduced with permission.

to crack growth, resistance to spall, and resistance to shear band development.

MAGNESIUM AND MAGNESIUM ALLOYS

Magnesium has a remarkably low density of 1,700 kg/m^3 (in comparison, the density of Al is 2,800 kg/m^3, that of Ti is 4,950 kg/m^3 and those of steels are 7,800 kg/m^3). The density of magnesium approaches that of polymers. Magnesium and magnesium alloys, which are among the lightest structural metals, are becoming increasingly important in the automotive and hand-tool industries. The rapid growth in the commercial use of magnesium is intimately tied to the increasing cost of energy. The low density makes these materials very attractive for defense applications, but magnesium alloys historically have had relatively low strengths (in the range 250-300 MPa) in comparison to aluminum alloys. There has also been lingering (and somewhat exaggerated) concern about the flammability of magnesium and about the relative ease with which these alloys can be corroded in severe environments. However, these potential problems are relatively easily mitigated by proper design and the appropriate protocols for maintenance.

A substantial effort was begun over the past decade to generate high-strength magnesium alloys using a variety of approaches, including solid solution strengthening and precipitation strengthening. Commercial magnesium alloys that can substitute for some aluminum alloys include AZ31[5] and ZK60, and several alloys containing rare earths show promise. Most of the innovation in this area is currently occurring outside this country, particularly in China and Japan, which may present a long-term risk for the United States. A recent workshop at the Johns Hopkins University on the potential of magnesium and magnesium alloys as protection materials highlighted a variety of opportunities. One of the more promising strengthening approaches appears to be the development of ultra-fine-grained or nanostructured magnesium alloys through severe plastic deformation. A major research effort in developing a fundamental understanding of strengthening mechanisms in magnesium alloys promises to be fruitful, and the opportunities presented by low-density alloys should not be missed.

Since magnesium is a hexagonally close-packed material, the plastic deformation of this metal is much more complex than that of cubic metals like aluminums and steels. Two features of the plastic deformation are particularly important: the development of deformation twins and the development of strong textures. Both topics require careful investigation in order to increase the utility of magnesium-based materials as components of protection material systems.

[5]Mukai, T., M. Yamanoi, H. Watanabe, and K. Higashi. 2001. Ductility enhancement in AZ31 magnesium alloy by controlling its grain structure. Scripta Materialia 45(1): 89-94.

CERMETS

The term "cermet" describes a structure that is a composite mixture of a metal phase and a ceramic phase. The combination of ceramic and metal in cermets works synergistically to improve the toughness of the composite material: The ceramic phase is a strengthening (a "hard" material) phase with the function of breaking or eroding the penetrator, and the ductile metal phase inhibits failure. The metals usually used are aluminum, magnesium, and titanium. Because of the synergism between the two materials, which in concert can defeat an incoming kinetic energy penetrator, cermets have a significant potential for expanded use in lightweight armor development. Cermets can be divided into two subgroups: ceramic-matrix composites and metal-matrix composites (MMC), depending on whether the ceramic is in continuous or matrix phase. Figure H-3 shows a micrograph of an MMC with a dispersed SiC phase in an aluminum matrix.[6]

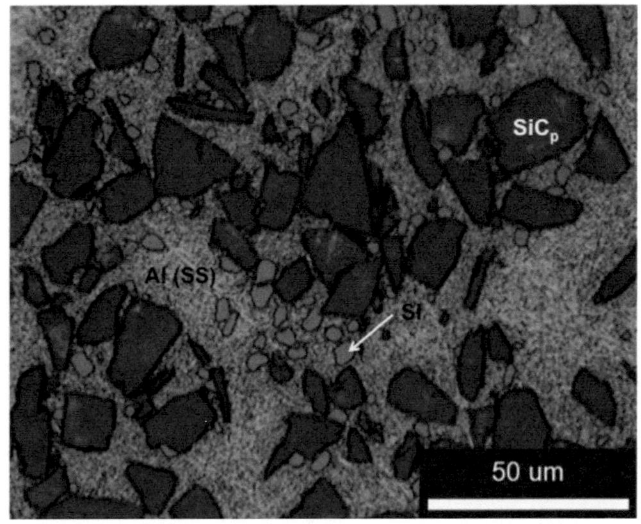

FIGURE H-3 Optical micrograph of Al-SiC cermet. Aluminum is the light-gray matrix, with discrete silicon carbide particles. SOURCE: Unpublished research. Permission granted by K.T. Ramesh.

A number of metal-matrix composites show potential for protective material applications. Typically these materials consist of ceramic particulate or ceramic fiber reinforcements within a ductile metal matrix, with the volume fractions of the reinforcements ranging from 5 to 50 percent. The typical result of incorporating a ceramic reinforcement into a metallic matrix is enhanced strength and some loss of ductility. Most of the MMCs used commercially are aluminum-based and ceramic-reinforced,[7] and these have been investigated thoroughly. However, there is also potential for magnesium-based systems and steel-based systems. Such MMCs could also lead to the development of functionally graded materials that have microstructures graded to provide optimum resistance to a specific threat. The high-strain-rate mechanical properties and dynamic failure processes in MMCs (see, for example, Li and Ramesh, 1998,[8] and Li et al., 2000[9]) have not been investigated in detail, and further work in this area is likely to be very useful in the development of armor packages in which the MMC may be used as a backing for a ceramic material.

The conventional method for fabrication of MMCs is to compress a porous compact of ceramic powder to approximately 65 percent of its theoretical density, leaving an open and continuous pore phase, which can be readily infiltrated with molten metal, usually aluminum. Finally, the compact undergoes a heat-treatment process at a somewhat more elevated temperature, causing a reaction between the aluminum metal and the ceramic, forming a strong interphase bond. In situ processes for making cermets—such as Lanxide's PRIMEX process, Martin Marietta's XD process, self-propagating high temperature, and reactive gas injections—have also been developed.[10,11,12,13] The PRIMEX process involves cermet fabrication under a pressureless condition, in which a spontaneous infiltration of molten aluminum into a porous ceramic preform in the presence of magnesium and nitrogen occurs without using vacuum or externally applied pressure.[14] A cermet material in the form of silicon carbide-aluminum was produced by Lanxide Armor Products and was employed to protect against artillery fragments and small arms. It has largely been replaced, however, by an improved material developed by M Cubed Technologies, in which the SiC+C and B_4C+SiC are infiltrated with molten silicon to form a tough SiC bonding phase that provides superior performance as a cermet armor protection material. A lightweight cermet material was also developed at Lawrence Livermore National Laboratory, using boron carbide for the ceramic compact, backfilled with aluminum metal, and sub-

[6]Uribe, Y., and H. Sohn, unpublished research.

[7]Lloyd, D. 1994. Particle reinforced aluminum and magnesium matrix composites. International Materials Reviews 39(1):1-23.

[8]Li, Y., and K. Ramesh. 1998. Influence of particle volume fraction, shape, and aspect ratio on the behavior of particle-reinforced metal–matrix composites at high rates of strain. Acta Materialia 46(16): 5633-5646.

[9]Li, Y., K. Ramesh, and E. Chin. 2000. The compressive viscoplastic response of an A359/SiC$_p$ metal-matrix composite and of the A359 aluminum alloy matrix. International Journal of Solids and Structures 37(51): 7547-7562.

[10]Mortensen, A., and I. Jin. 1992. Solidification processing of metal matrix composites. International Materials Review 37: 101-128.

[11]Ibrahim, A., F. Mohamed, and E. Lavernia. 1991. Particulate reinforced metal matrix composites—A review. Journal of Materials Science 26(5): 1137-1156.

[12]Koczak, M., and M. Premkumar. 1993. Emerging technologies for the in situ production of MMC's. The Journal of the Minerals, Metals, and Materials Society 45(1): 44-48.

[13]Asthana, R. 1998. Reinforced cast metals: part I solidification microstructure. Journal of Materials Science 33(7): 1679-1698.

[14]Aghajanian, M., A. Rocazella, J. Burke, and S. Keck. 1991. The fabrication of metal matrix composites by a pressureless infiltration technique. Journal of Materials Science 26(2): 447-454.

sequently heat-treated to form a delta phase chemical bond between the ceramic and the metal. The processing of B_4C and Al composites, especially when the B_4C content is high (above 55 vol percent), faces the problem of poor wettability of the aluminum on B_4C at β temperatures, especially near the melting point of aluminum (660°C). Aluminum begins to wet the B_4C surface at temperatures just above 1000°C, which results in an increase in the driving force of chemical reactions. The high temperatures (1000°C to 1200°C) used for improved infiltration increase the wettability of the materials, but at the same time, chemical reactions between Al and B_4C can result in the formation of intermediate phases, such as binary AlB_2, β-AlB_{12}, AlB_{10}, borides, and ternary Al-borocarbides $AlB_{24}C_4$, $Al_3B_{48}C_2$, and Al_3BC.[15] Al_3C_4 is also formed. It has been reported that about 30 vol percent of new phases are formed from initially 38 vol percent aluminum and 62 vol percent B_4C.[16] Al_4C_3 is the most undesirable phase because of its hygroscopic nature and pure mechanical properties. Some products of the interfacial reactions are not desirable and can cause premature failure and poor ballistic performance, while other interphases are desired and even required to form a good interfacial bond and bring significant strengthening and high tensile strength of the cermet.

It is understood, however, that for an armor cermet material to be of high quality, a clean metallurgical interface between the ceramic reinforcement and metal matrix is highly desirable, since it allows a more effective strengthening from the reinforcement.[17] To avoid formation of intermediate interphases, low-temperature cryomilling was developed to synthesize a composite powder with clean metallurgical interfaces and without voids.[18] In addition, to increase the ductility, which is always sacrificed when strength is increased, a trimodal Al-B_4C cermet was developed, in which coarse-grained aluminum was introduced into the nanocrystalline Al reinforced with B_4C particles.[19] A trimodal composition with 10 wt percent B_4C, 50 wt percent coarse-grained Al 5083, and the remainder nanocrystalline Al 5083 exhibited 1,065 MPa yield strength under compressive loading while still showing 0.04 true strain deformation.

As noted by Chin,[20] in addition to particulate-reinforced cermets with excellent work-hardening characteristics under dynamic loading, the functionally graded armor composites (FGACs) were developed. In FGACs, ballistic space and mass efficiency of cermets were enhanced by tailoring the through-thickness incorporation and distribution of various reinforcement morphologies, sizes, and chemistries to mitigate shock damage. The idea of improving FGAC performance is to disrupt the shock wave in order to minimize collateral damage during a ballistic event. The FGAC structure is composed of a series—a hard (ceramic) layer interspersed with a high strain-to-failure material such as aluminum. The hard outer surface is usually designed to be the ballistic impact layer, and behind this layer is a thin-bonded layer of the ductile material. The design feature is such that in successive layers going toward the back surface, the volume fraction of the ductile material is increased and the volume fraction of the hard layer is decreased. Thus, the strain-to-failure ratio is increased as the depth of the penetration increases. The perturbations will be tailored throughout the microstructural design, which prolongs projectile-through-target-material dwell time. The extended dwell time promotes the breaking up of the projectile prior to the occurrence of complete penetration or unacceptable collateral damage of the armor material.[21] There is a clear realization of the importance of and need for a better understanding of the character of the interfaces in FGAC because of the softening of the material due to interfacial and particle damage from high-rate loading.

The self-propagating high-temperature synthesis methodology is another important technique used to produce metal-matrix composites where dissimilar phases (metal and ceramics) are integrated through a self-propagating exothermic reaction.[22] The development of nanoscale, multilayer, self-propagating exothermic reaction foils, which can be ignited by a simple electrical spark, is important for joining FGACs to a wide range of structural surfaces as well as for modular armor repair.

In summary, cermet materials exhibit light weight and excellent ballistic properties suitable for personnel armor use. However, cermets have not been extensively utilized in armor protection applications, in part due to high fabrication costs but also because the optimal composite properties have not always been fully realized, owing to poorly understood interfacial bonding and properties. The field of armor cermets is, therefore, ripe for exploitation using combinations of the common refractory ceramic materials (alumina, silicon carbide, boron carbide) and light metals such as magnesium, titanium, and aluminum. Cermets have been successfully

[15]Lee, K., B, Sim, S. Cho, and H. Kwon. 1991. Reaction products of Al-Mg/B_4C composite fabricated by pressureless infiltration technique. Journal of Materials Science and Engineering A 302(2): 227-234.

[16]Beidler, C., W. Hauth, and A. Goel, 1992. Development of a B_4C/Al cermet for use as an improved structural neutron absorber. Journal of Testing and Evaluation 20(1): 57-60.

[17]Lloyd, D. 1992. Particle reinforced aluminium and magnesium matrix composites. International Materials Reviews 39(1): 1-23.

[18]Schoenung, J., J. Ye, J. He, F. Tang, and D. Witkin. 2005. B_4C reinforced nanocrystalline aluminum composites: Synthesis, characterization, and cost analysis. Pp. 123-128 in Materials Forum Volume 29. J.F. Nie and M. Barnett, eds. Institute of Materials Engineering Australia Ltd.

[19]Ye, J., B. Han, Z. Lee, B. Ahn, S. Nutt, and J. Schoenung. 2005. A trimodal aluminum based composite with super-high strength. Scripta Materialia 53(5): 481-486.

[20]Chin, E. 1999. Army focused research team on functionally graded armor composites. Materials Science and Engineering A 259(2): 155-161.

[21]Ibid.

[22]Michaelsen, C., K. Barmak, and T.Weihs. 1997. Investigating the thermodynamics and kinetics of thin film reactions by differential scanning calorimetry. Journal of Physics D: Applied Physics 30(23): 3167-3186.

used in armor protection applications because of their ruggedness and ability to withstand impact, but the best properties of each component phase are often not fully realized in the composite structure. Cermets can be fabricated in a relatively straightforward manner and in a wide variety of forms, but most MMCs, like aluminum, have been developed with relatively low-melting metal phases. For higher-temperature components, special fabrication techniques are needed. Mechanistic research on high-temperature ceramic-metal bonding in cermets, the fabrication of these structures, and their relationship to projectile defeat and armor performance can productively be researched.

Appendix I

Nondestructive Evaluation for Armor

Various nondestructive methods have historically been used to rapidly locate and identify anomalous internal flaws within dense armor materials; these methods have included resonant ultrasound spectroscopy, high-frequency ultrasound C scans, infrared thermography, and microfocus x-ray computed tomography (XCT). Testing before the materials have been used in their particular applications can be further subdivided into tests on individual armor materials and tests on arrays of tiles or body armor plates assembled with other confining materials.

Resonant ultrasound spectroscopy has recently been shown to demonstrate excellent potential for rapid go/no-go testing of armor materials.[1] In this technique, a tile of armor material is held at the corners and struck to create a set of vibrations at the tile's harmonic frequencies. Each peak in the spectrum is determined by the material's geometry, elastic properties, and microstructure. Shifts in expected peak positions can identify the presence of internal flaws such as cracks, anomalous inclusions, and large porosity. Spectra are used to identify quickly whether the component is suitable for armor applications. Since a single spectrum is measured for the entire sample, determination of the location within the material where flaws exist is not currently possible.

High-frequency ultrasound has been successfully demonstrated for quickly evaluating armor material homogeneity and measuring properties of interest.[2] Ultrasound testing can be performed at individual points to measure acoustic energy loss, elastic properties, and surface roughness. These measurements can be extended over the entire material in a rastered full area or C scan[3] to generate mapped images of sample properties. Brennan et al.[4] illustrate this technique's ability to determine how properties vary as a function of distance from the sample edges. Elastic property maps serve as a visual representation of density variations throughout a material.

Ultrasound C scans of acoustic energy loss can map changes in sample composition.[5] This nondestructive evaluation (NDE) technique is founded on an understanding of how a material's microstructure attenuates an acoustic wave as the wave interacts with grains, inclusions, and porosity. This technique can identify anomalous defects as well as more subtle compositional variations throughout a SiC tile. Interpretation of maps of acoustic energy loss result in an understanding of how mean grain size and inclusion concentration vary, aiding in an assessment of the material's suitability for armor applications. Acoustic spectroscopy, the analysis of the frequency dependency of acoustic loss, can be used to estimate distributions of bulk inclusions and mean grain size.

Although ultrasound C scans provide additional information regarding sample homogeneity, this information comes at the price of increased testing time. Conventional ultrasound testing requires approximately 10 to 20 minutes to characterize a 4-in. × 4-in. tile. Through use of ultrasound phased arrays, however, the time requirement can be reduced by an order of magnitude. Phased-array probes contain an assembly of several dozen ultrasound transducers, allowing for digital beam steering, focusing, and rastering, all of which increase the rapidity of testing. Phased-array probes

[1]Ashkin, D., R. Brennan, J. Campbell, S. Klann, R. Palicka, and R. Sisneros. Resonant ultrasound testing of hot pressed silicon carbide. Proc. 2010 International Conference and Exposition on Advanced Ceramics and Composites.

[2]Brennan, R. 2007. Ultrasonic Nondestructive Evaluation of Armor Ceramic. Ph.D. Dissertation, Publication Number AAI3319593. New Brunswick, N.J.: Rutgers University.

[3]A C scan is a nondestructive technique that uses ultrasound to inspect materials.

[4]Brennan, R., R. Haber, D. Niesz, G. Sigel, and J. McCauley. 2009. Elastic property mapping using ultrasonic imaging. Advances in Ceramic Armor III: Ceramic Engineering and Science Proceedings 28(5): 213-222.

[5]Portune, A., and R. Haber. 2010. Microstructural study of sintered SiC via high frequency ultrasound spectroscopy. Pp. 159-170 in Advances in Ceramic Armor V. J. Swab, ed. Hoboken, N.J.: John Wiley & Sons.

can thus characterize specific material layers within armor assemblies. Whereas conventional ultrasound can effectively test materials before their inclusion in final pieces, phased-array techniques can evaluate materials both before and after assembly.[6] Although phased-array instruments have advanced capabilities, they currently exhibit significant hardware limitations and increased costs.

XCT has proven to be a powerful tool for evaluating armor integrity and visualizing compositional variations in three dimensions. Layer, or X-ray slice, data are generated by an x-ray source rotating around an object; x-ray sensors are placed on the other side of the circle from the x-ray source. Testing is then repeated until the entire material has been characterized. By assembling these layers with a computer, three-dimensional images are created. XCT is used to evaluate samples prior to assembly to map variations in sample density and to locate anomalous flaws or microcracks.

One benefit of XCT is its capability for rapidly assessing sample homogeneity in armor assemblies. Devices have been created that can quickly examine armor in the field prior to engagements.[7] Inspection devices for use in the field can be optimized toward a single expected part geometry, increasing the speed by which crucial parts of the armor composite can be identified and characterized. An example is a device to characterize a small-arms protective insert plate and identify an internal crack.[8] Rapid characterization is necessary in the field because flaws in armor that were not present after production or assembly may be introduced during handling.

Nondestructive tests are also used to characterize damage incurred by armor materials after destructive testing. NDE is an excellent tool for this purpose as it does not introduce further damage to the material or change the damage state that already exists. To date, XCT has proven most efficient at this task because it can provide three-dimensional images of damage zones.

XCT has also been applied to the characterization of damage in confined armor materials.[9,10] The XCT reconstruction can be used as a damage diagnostic for understanding crack-propagation behavior and the extent of damage spread. XCT can be performed on an armor piece assembled from multiple tiles and used to illustrate how this configuration minimizes the spread of damage to surrounding areas. Additionally, since testing can be performed without changing the sample state, it is possible to visualize residual projectile fragments.

Each NDE technique acquires different kinds of information about the armor material. No single technique has been shown to be sufficient for full sample characterization. XCT provides excellent visualizations of damage incurred by materials and can map large compositional variations, but it cannot provide the level of microstructural information possible through ultrasound spectroscopic analysis. Ultrasound C scan testing provides excellent maps of fine microstructural variations in a material, but it requires more time than other techniques do and may be unsuitable for the rapid testing of full sample lots. Resonant ultrasound spectroscopy provides rapid go/no-go tests, but it cannot identify where flaws exist in a material, as only a single curve is measured for the entire sample. A separation therefore exists between using NDE for studied characterization and using it for rapid identification of a material's suitability for use.

Many challenges exist for the future development of NDE for armor. Ideally NDE would be employed in production lines for all armor materials. However, the assessment of individual components requires the standardization of test techniques and the integration of testing equipment. The characterization of armor material microstructures through NDE could be improved through the study of defined standards. The use of standard sample sets that could be used across industry, in governmental institutions, and in research facilities would benefit this process. It is clear that there is room for improvements: The characterization of damage and defects can still be made faster and more robust, as many defects beneath a critical size currently go undetected. Finally, any future improvements in test equipment and software need to decrease the time required to perform analyses, increasing the feasibility of the use of such analyses outside dedicated laboratories.

---

[6]Steckenrider, S., W. Ellingson, E. Koehl, and T.J. Meitzler. 2010. Inspecting composite ceramic armor using advanced signal processing together with phased array ultrasound. Advances in Ceramic Armor VI: Ceramic Engineering and Science Proceedings 31. J. Swab, S. Mathur, and T. Ohji, eds. Hoboken, N.J.: John Wiley & Sons.

[7]Haynes, N., K. Masters, C. Perritt, D. Simmons, J. Zheng, and J. Youngberg. 2009. Automated non-destructive evaluation system for hard armor protective inserts of body armor in Advances in Ceramic Armor IV: Ceramic Engineering and Science Proceedings 29(6). L. Franks, ed. Hoboken, N.J.: John Wiley & Sons.

[8]Ibid.

[9]Wells, J., and N. Rupert. 2009. Ballistic damage assessment of a thin compound curved B_4C ceramic plate using XCT. Advances in Ceramic Armor IV: Ceramic Engineering and Science Proceedings 29(6). L. Franks, ed. Hoboken, N.J.: John Wiley & Sons.

[10]Wells, J., N. Rupert, and M. Neal. 2010. Impact damage analysis in a Level III flexible body armor vest using XCT diagnostics. Advances in Ceramic Armor V. J. Swab, D. Singh, and J. Salem, eds. Hoboken, N.J.: John Wiley & Sons.

Appendix J

Fiber-Reinforced Polymer Matrix Composites

Polymer matrix composites (PMCs) consist of a polymer resin reinforced with fibers, an example of which is the combat helmet. PMCs can be subdivided into two categories, based on whether the fiber reinforcement is continuous or discontinuous. PMCs with discontinuous fibers (less than 100 mm long) are made with thermoplastic or thermosetting resins, whereas PMCs with continuous fibers usually employ thermosetting resins. This appendix primarily addresses PMCs containing continuous fibers. The most common design for PMCs is a laminate structure made of woven fabrics held together by a polymer resin. Fabrics are incorporated in order to take advantage of their high strength and stiffness and to improve energy absorption and distribute the kinetic energy laterally. Owing to their highly engineered structures, PMCs are lightweight with high specific strength and high specific stiffness.

Commonly used reinforcement materials include carbon, glass, aramid, and polyethylene fibers. PMCs can be manufactured by wet and hand lay-up; molding (compression, injection, and transfer); vacuum bag molding; infusion molding; vacuum-assisted resin transfer molding; prepreg[1] molding; and other common fabrication techniques. Unlike common structural composites, which typically contain up to about 60 vol percent fibers, ballistic PMCs contain a higher volume fraction of fibers or fabrics (up to about 80 vol percent). The effect of this variation in structure on the ballistic protection properties of PMCs has not been thoroughly investigated.

PMCs respond to ballistic impact in ways that depend on their particular structure and thus are different from other protective materials. Unlike fabrics, with PMCs only the material in the neighborhood of the impact position shows a response; thus the response is completely governed by the local behavior of the material and unaffected by boundary conditions. Additionally, the penetration mechanism is dependent on the thickness of the composite. For thin composites the deformation across the thickness direction does not vary with depth, whereas for thick composites it does.[2] Ballistic performance initially increases linearly with the increased thickness; however, as the composite becomes thicker the marginal protective gain incurred by increasing the thickness becomes smaller,[3,4] although the rate at which the weight increases is maintained.

[1]Semifinished fiber products preimpregnated with epoxy resin (prepregs).

DEFORMATION AND FAILURE MECHANISMS

When a PMC is subjected to high-velocity impact, the kinetic energy is transferred from the projectile to the PMC. The existence of two components, the fabric and the matrix, and their interface, makes the energy absorption mechanism more complex than that of ballistic fabrics. The commonly recognized energy absorption and failure mechanisms are discussed here.

Cone Formation on the Back Face

As with ballistic fabrics, the mode of impact response known as cone formation has also been observed in PMCs. Guoqi et al.[5] observed the formation of a cone-shaped $\sigma_f(\varepsilon,\dot{\varepsilon},T)$ deformation zone in the back surface of Kevlar/polyester laminates during the ballistic impact of a blunt projectile; using high speed photography, Morye et al.[6] documented the temporal evolution of this response for the ballistic behavior of nylon fabric preimpregnated with

[2]Naik, N., and A. Doshi. 2008. Ballistic impact behaviour of thick composites: Parametric studies. Composite Structures 82(3): 447-464.

[3]Ibid.

[4]Faur-Csukat, G. 2006. A study on the ballistic performance of composites. Macromolecular Symposia 239 (1): 217-226.

[5]Guoqi, Z., W. Goldsmith, and C.K.H. Dharan. 1992. Penetration of laminated Kevlar by projectiles—I. Experimental investigation. International Journal of Solids and Structures 29(4): 399-420.

[6]Morye, S., P. Hine, R. Duckett, D. Carr, and I. Ward. 2000. Modelling of the energy absorption by polymer composites upon ballistic impact. Composites Science and Technology 60(14): 2631-2642.

a matrix of a 50:50 mixture of phenol formaldehyde resin and polyvinyl butyral resin. Figure J-1 shows the scheme of cone formation in two-dimensional woven fabric composites during projectile impact. The yarns that the bullet directly contacts are called primary yarns; these yarns resist penetration and undergo deformation due to cone formation. The longitudinal compressive stress wave generated upon impact propagates outward along the yarn direction, forming a quasi-circular shape. The conical portion moves backward and stores kinetic energy by its motion.

Deformation of Yarns and Failure

When a PMC undergoes ballistic impact, the primary yarns deform and resist projectile penetration. The other yarns (called orthogonal yarns) also deform, but to a lesser extent due to primary yarn deformation; this process stores kinetic energy. During cone formation, strain is highest along the middle primary yarns in each layer of the composite. The highest overall strain is at the point of impact, and the strain falls off along the radial direction. After the cone forms, the top layers of the PMC are compressed, leading to an increase in the tensile strain of the yarns there. A linear relation between strain and depth along the thickness direction can be assumed; see Figure J-1. Once the strain is beyond the failure strain, sequential breakage will occur beginning at the top layer. This yarn failure absorbs additional kinetic energy.

Delamination and Matrix Cracks

During ballistic impact, transverse and longitudinal waves are formed. The geometry of the deformation influences the terminology used to describe the deformation: The waves that move out in the lateral direction (having both longitudinal and transverse polarization) from the point of impact are called transverse, and the waves propagating along the direction of the incident projectile are called longitudinal. A cone of deformation, quasi-lemniscate in shape, is formed due to transverse waves.[7] As the longitudinal waves propagate along the yarns, attenuation occurs, leading to strain variations radially from the impact site in the target. The matrix has mechanical properties different from those of the yarns, but it must carry the same deformation lest delamination or slippage occur due to weak adhesion between the yarn and the matrix; there may be damage if the yarn strain is higher than the strain at failure in the matrix. As the material deforms, cracking and delamination will continue until total perforation occurs.[8] Research has shown[9] that initiation and propagation of delamination occur more frequently along the warp and fill directions than along other directions. Compared to conventional materials, composite materials contain numerous interfaces between the matrix and the fibers, providing multiple locations for cracking to occur. Energy absorption occurs through a combination of cracking, delamination, and shear banding (the latter is dependent on the plasticity of the matrix and possibly of the fibers). Typical shapes of delaminated regions after impact are shown in Figure J-2;[10] the noncircular shape is attributed to the anisotropic nature of these materials (different paths of the stress waves, hence different distances that the stress information must travel).

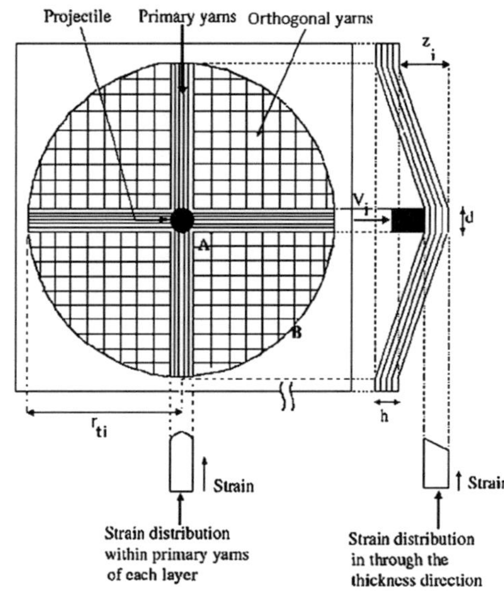

FIGURE J-1 Cone formation during ballistic impact on the back face of the composite target. SOURCE: Naik, N.K. 2005. Ballistic impact behaviour of woven fabric composites: Parametric studies. Materials, Science and Engineering: A 412(1-2): 104-116.

Shear Plugs

During impact experiments on conventional carbon-fiber-reinforced plastic laminates, it was observed[11] that a small area of the laminate was sheared off by the projectile

[7]Wu, E., and L.-C. Chang. 1995. Woven glass/epoxy laminates subject to projectile impact. International Journal of Impact Engineering 16(4): 607-619.

[8]Naik, N., and K. Reddy. 2002. Delaminated woven fabric composite plates under transverse quasi-static loading: experimental studies. Journal of Reinforced Plastics and Composites 21(10): 869-877.

[9]Wu, E., and L.-C. Chang. 1995. Woven glass/epoxy laminates subject to projectile impact. International Journal of Impact Engineering 16(4): 607-619.

[10]Naik, N. 2006. Ballistic impact behaviour of woven fabric composites: Formulation. International Journal of Impact Engineering 32(9): 1521-1552.

[11]Cantwell, W., and J. Morton. 1990. Impact perforation of carbon fibre reinforced plastic. Composites Science and Technology 38(2): 119-141.

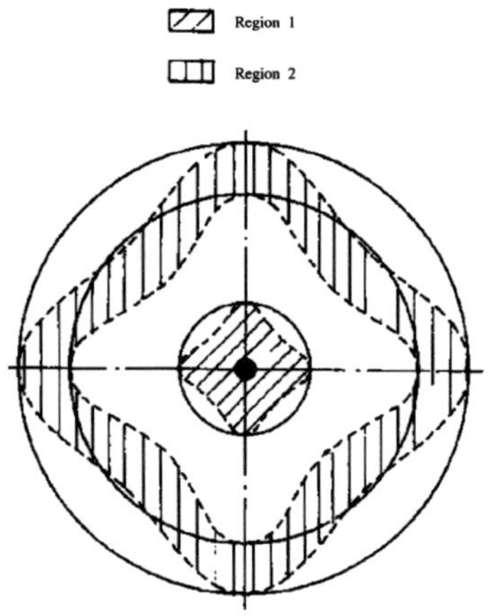

FIGURE J-2 Schematic shape of delaminated regions observed in impact experiments. Region 1: area damage in the first time interval after impact; Region 2: area damaged in the (i + 1) time interval. SOURCE: Naik, N. 2006. Ballistic impact behaviour of woven fabric composites: Formulation. International Journal of Impact Engineering 32(9): 1521-1552.

during impact and that a distinct conical-shaped zone was formed. The schematic is shown in Figure J-3. The shear plug phenomenon has never been observed in glass-fiber-reinforced composites, which may be due to the much higher failure strain of glass fibers compared to that of carbon fibers at high strain rates.

Friction and Hole Enlargement

In contrast to the complex frictional forces present in neat fabrics (including friction between yarns, between the projectile and the yarn, and between the individual fibers), the only friction present in PMCs during impact occurs between the projectile and the laminate. After the yarns and the fabrics fail, friction between the damaged laminates dissipates some of the kinetic energy from the projectile. Goldsmith et al.[12] calculated the frictional work by using the friction efficiency between projectile and laminate measured by the quasi-static method. They found that the friction resistance depends on the shape of the projectile and that it increases with increasing composite thickness. Additionally, they calculated the energy dissipated when the projectile enlarges the hole and found that this process also contributes to energy dissipation. Although the energy absorbed due to friction is much larger than that due to hole enlargement, neither of these modes is the major energy absorption mechanism.

[12]Goldsmith, W., C.K.H. Dharan, and H. Chang. 1995. Quasi-static and ballistic perforation of carbon fiber laminates. International Journal of Solids and Structures 32(1): 89-103.

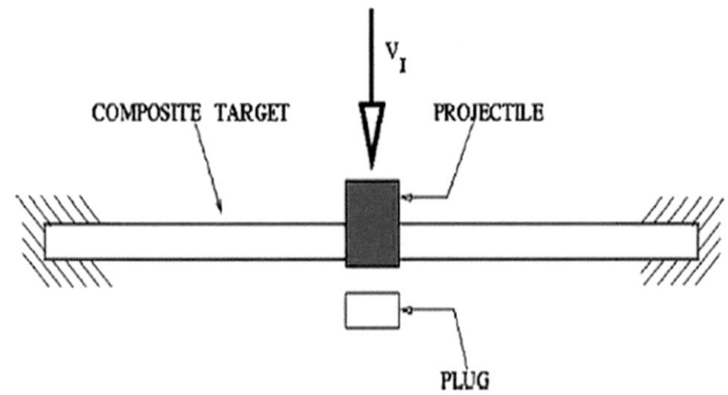

FIGURE J-3 Schematic showing plug formation. SOURCE: Naik, N. 2004. Composite structures under ballistic impact. Composite Structures 66(1-4): 579-590

The Contribution of Different Types of Energy Absorption Paths

Naik and Shrirao[13] analyzed the ballistic impact behavior of woven fabric composites under a flat head projectile using wave theory and presented an analytical formulation for each energy absorption mechanism. The calculation is based on the material properties at high strain rate, and analytical prediction shows a good match with experimental results. During the ballistic impact, the moving area of the cone increases, leading to an increase in the kinetic energy of the cone even though the speed of the projectile is reduced. Next, as the moving speed decreases significantly, the kinetic energy of the cone decreases and becomes zero when the projectile's speed reaches zero. The kinetic energy of the cone is the major energy absorption factor, followed by deformation of the orthogonal yarns and tensile breakage of primary yarns; delamination and cracking provide only a small fraction of the energy absorption. The calculations assume a relatively thin and flexible PMC system; for thicker systems, the variation of deformation as a function of thickness changes the relevant material behavior and requires a consideration of friction.

CURRENT ISSUES AND RELATED STUDIES

As noted above, the ballistic performance of laminated PMCs depends on the properties of the polymer matrix and of the reinforcement material, on the stacking sequence, on the fiber architecture, on the qualities of the interface and the interphase, on the environmental conditions, and on the characteristics of the projectile. To date, however, the experimental studies have only focused on certain types

[13]Naik, N., and P. Shrirao. 2004. Composite structures under ballistic impact. Composite Structures 66(1-4): 579-590.

of composites and ballistic conditions. Thus, the full map of ballistic performance of this class of composites is still unknown, and more analytical experiments and simulations are needed to improve the understanding of the ballistic performance of PMCs.

Material Properties

The properties of the fabrics, the surrounding matrix, and the interfaces affect the overall performance of laminates. Although no thorough map of the effects of the properties of fabrics and polymer matrix has been drawn, an examination of the experimental literature allows for some preliminary conclusions. Faur-Csukat[14] prepared fabric composites with a fabric wt percent of approximately 55 by hand lay-up followed by compression molding. The ballistic performance of carbon-, glass-, aramid-, and polyethylene-fabric-reinforced composites showed that the efficacy of reinforcing fibers was as follows: Glass is better than aramid, which is better than or equal to ultrahigh-molecular-weight polyethylene (UHMWPE), which is better than carbon fibers. Among the different PMCs studied, carbon-fiber-reinforced composites exhibited the worst ballistic performance owing to their low strain to failure. Roughly, fibers with high strain at high strain rate are better energy absorbers than high-strength fibers with low strain to failure. This conclusion is the same as that of Naik.[15] The fiber-matrix interface and interphase also play a critical role in ballistic performance. It was observed that weaker interfacial interaction resulted in higher energy absorption.[16,17] In composites, fiber-matrix debonding, cracks, and friction slippage improve energy absorption; this is different from the behavior of noncomposite materials. However, excessively low interaction and interfacial strength will lead to pre-ballistic failure problems. For a full understanding of the effects of material properties, more analytical experiments as well as further modeling and simulation are needed.

Fabric Structure

Weave architecture also influences the ballistic performance of composites. It was shown that (under the conditions investigated), the performance of basket-weave fabrics was better by about 10 percent than that of plain-weave fabrics.[18] Satin and twill weaves also tended to absorb more energy than the plain weaves,[19] possibly due to a decrease in the crimp angle. It was also found that the architecture of the fabric is more important in thicker composites than in thinner composites, as the decreased crimp angle decreases stress concentration.

Improved ballistic performance can be obtained by using three-dimensional woven fabrics instead of two-dimensional woven fabrics.[20] Walter et al.[21] quantitatively analyzed results from three-dimensional woven glass-fiber-reinforced composites and observed that delamination along the weak layer is the most severe shortcoming in current three-dimensional woven composites at high load and high loading rates. In general, Z-stitching increased the resistance to damage, and it restricted damage to a smaller total area than that in unstitched samples. However, in one study[22] a decrease in ballistic limit was observed in Z-stitched targets, although no explanation of this decrease was provided.

Cohen et al.[23] used Spectra 1000 yarns to reinforce a UHMWPE matrix with a fiber content of up to 85 percent. The shear strength (20-25 MPa) and tensile strength (longitudinal, 1.3-1.5 GPa; transversal, 21-25 MPa) of the prepared composite are better than those of composites like UHMWPE fiber/epoxy matrix composites and UHMWPE fiber/high-density polyethylene (PE) matrix composites. Furthermore, the tensile strength of the prepared composite is similar to that of Kevlar fiber/epoxy matrix composites. These improvements are attributed to the good self-adhesion and strong bonding of PE fibers to the PE matrix. The ballistic response of PE/PE composites under the impact of bullets (9 mm in diameter, weighing 8 g, velocity approximately 400 m/s) shot from an Uzi submachine gun has been investigated.[24] High-density PE was used as the matrix, and UHMWPE fibers such as Spectra and Dyneema were used as the

[14]Faur-Csukat, G. 2006. A study on the ballistic performance of composites. Macromolecular Symposia 239 (1): 217-226.

[15]Naik, N. 2004. Composite structures under ballistic impact. Composite Structures 66(1-4): 579-590.

[16]Park, R., and J. Jang. 1998. A study of the impact properties of composites consisting of surface-modified glass fibers in vinyl ester resin. Composites Science and Technology 58(6): 979-985.

[17]Tanoglu, M., S. McKnight, G. Palmese, and J. Gillespie Jr. 2001. The effects of glass-fiber sizings on the strength and energy absorption of the fiber/matrix interphase under high loading rates. Composites Science and Technology 61(2): 205-220.

[18]Faur-Csukat, G. 2006. A study on the ballistic performance of composites. Macromolecular Symposia 239 (1): 217-226.

[19]Hosur, M., U. Vaidya, C. Ulven, and S. Jeelani. 2004. Performance of stitched/unstitched woven carbon/epoxy composites under high velocity impact loading. Composite Structures 64(3-4), 455-466.

[20]Shukla, A., J. Grogan, S. Tekalur, A. Bogdanovich, and R. Coffelt. 2005. Ballistic resistance of 2D & 3D woven sandwich composites. Pp. 625-634 in Sandwich Structures 7: Advancing with Sandwich Structures and Materials: Proceedings of the 7th International Conference on Sandwich Structures, O. Thomsen, E. Bozhevolnaya, and A. Lyckegaard, eds. New York, N.Y.: Springer.

[21]Walter, T., G. Subhash, B. Sankar, and C. Yen. 2009. Damage modes in 3D glass fiber epoxy woven composites under high rate of impact loading. Composites Part B: Engineering 40(6): 584-589.

[22]Hosur, M., U. Vaidya, C. Ulven, and S. Jeelani. 2004. Performance of stitched/unstitched woven carbon/epoxy composites under high velocity impact loading. Composite Structures 64(3-4), 455-466.

[23]Cohen, Y., D. Rein, and L. Vaykhansky. 1997. A novel composite based on ultra-high-molecular-weight polyethylene. Composites Science and Technology 57(8): 1149-1154.

[24]Harel, H., G. Marom, and S. Kenig. 2002. Delamination controlled ballistic resistance of polyethylene/polyethylene composite materials. Applied Composite Materials 9 (1): 33-42.

reinforcement phase. The material was created by winding fibers in a unidirectional pattern on large-diameter mandrels which were then flattened into film; these films were stacked on top of one another, with each layer rotated 90° from the one below it to achieve a 0°/90° laminate.

PERSPECTIVE: NEW TYPES OF FIBERS

Nanocomposites

When incorporated into composite materials,[25] nanosized fillers have been shown to provide superior reinforcement due to their outstanding mechanical properties.[26] Thus, the ballistic resistance dynamics and capacity of carbon nanotubes (CNTs) were simulated, and their potential use in armor was discussed.[27] Simulations found that CNTs with the highest ballistic resistance could be resilient to a projectile at speeds of 200 m/s to 1,400 m/s if one end is fixed.[28] Additionally, CNT hybrid composites and CNT-reinforced fibers all have potential for improving ballistic performance.

Polymer Laminates

In matrix composites, the reinforcing fibers have mechanical properties that are much higher than those of the matrix. Because this mismatch can cause delamination and cracking, which do not absorb as much kinetic energy as other modes of failure, or for other reasons relevant to the intended use of the product, polymer laminates that contain two or more kinds of polymer have also been investigated. For example, polycarbonate (PC) is widely used in transparent ballistic applications, but its susceptibility to chemicals, scratching, and other possible service conditions limit applications. Two possible solutions have been investigated: (1) blending a second polymer with PC and (2) applying a hard surface coating to the PC. Blending another transparent, chemically resistant polymer such as polymethyl methacrylate (PMMA) with PC can improve the chemical sensitivity, but it can also reduce ballistic performance. Similarly, a hard coating may provide abrasion and chemical protection, but it also reduces impact resistance.[29] PC and PMMA are not normally miscible, so blending can only be achieved by solvent casting, which may trap a nonequilibrium structure during solvent evaporation as the solution goes through its glass transition concentration.[30] Thus, further phase separation can occur when the temperature is higher than the glass transition temperatures (PMMA T_g = 100°C; PC T_g = 150°C, depending on the component polymer molecular weights). This further phase separation results in strong optical scattering from the larger domains and loss of transparency. Component immiscibility causes opaque materials for melt processing of PMMA and PC blends. Additionally, solvent-induced crystallization of PC decreases optical clarity. Another strategy for addressing the transparency problem is to produce multinanolayer polymer laminates by co-extrusion of PMMA and PC; this results in laminates containing individual layers as thin as 100 nm and an overall structure that has good optical clarity. This method was originally developed at Dow Chemical Company in the 1960s and further refined at the 3M Company and at Case Western Reserve University.[31] A system with two extruders and a co-extrusion block is used to extrude two layers that are first sliced vertically, then spread horizontally, and finally recombined. This step can be repeated n times and generate $2^{(n+1)}$ polymer layers while the thickness of the layers is decreased in proportion to their increased number. The thickness of the PMMA layers plays a critical role in the ballistic performance of PC/PMMA polymer laminates.[32] The adhesion between PMMA and PC is strong enough to overcome delamination.[33] In this case, the mode of failure depends strongly on the thickness of the individual component layers. For laminates containing the thickest layers (greater than 0.5 μ thick), the composite film is brittle, and the laminate fails in brittle mode. For intermediate layer thicknesses (between 150 nm and 0.5 μ), several different failure mechanisms are present, with microcracking in the PMMA layers appearing to be the dominant one. Materials with layer thicknesses less than 150 nm behave in a ductile manner and fail with a large amount of plastic flow, resulting in increased ballistic impact energy. The ballistic performance of polymer laminates of PC with PMMA as well as with poly(styrene-co-acrylonitrile) (SAN) processed with varying layer thicknesses has also been reported.[34] The adhesion between PC and PMMA is 10 times higher than that

[25] Njuguna, J., K. Pielichowski, and S. Desai. 2008. Nanofiller-reinforced polymer nanocomposites. Polymers for Advanced Technologies 19(8): 947-959.

[26] Koziol, K., J. Vilatela, A. Moisala, M. Motta, P. Cunniff, M. Sennett, and A. Windle. 2007. High-performance carbon nanotube fiber. Science 318(5858): 1892-1895.

[27] Mylvaganam, K., and L. Zhang. 2007. Ballistic resistance capacity of carbon nanotubes. Nanotechnology 18(47).

[28] Mylvaganam, K., and L. Zhang. 2006. Energy absorption capacity of carbon nanotubes under ballistic impact. Applied Physics Letters 89(12).

[29] Hsieh, A., and J. Song. 2001. Measurements of ballistic impact response of novel coextruded PC/PMMA multilayered-composites. Journal of Reinforced Plastics and Composites 20(3): 239-254.

[30] Kyu, T., and J. Saldanha. 1998. Miscible blends of polycarbonate and polymethyl methacrylate. Journal of Polymer Science Part C: Polymer Letters 26(1): 33-40.

[31] Mueller, C., S. Nazarenko, T. Ebeling, T. Schuman, A. Hiltner, and E. Baer. 1997. Novel structures by microlayer coextrusion–talc-filled PP, PC/SAN, and HDPE/LLDPE. Polymer Engineering & Science 37(2): 355-362.

[32] Hsieh, A., and J. Song. 2001. Measurements of ballistic impact response of novel coextruded PC/PMMA multilayered-composites. Journal of Reinforced Plastics and Composites 20(3): 239-254.

[33] Ibid.

[34] Kerns, J., A. Hsieh, A. Hiltner, and E. Baer. 2000. Comparison of irreversible deformation and yielding in microlayers of polycarbonate with poly(methylmethacrylate) and poly(styrene-co-acrylonitrile). Journal of Applied Polymer Science 77(7): 1545-1557.

between PC and SAN as measured by the T-peel method.[35] However, the difference in adhesion has almost no effect on the deformation mechanisms. The ductility of thin layers of laminates was attributed to the cooperative yielding of both components, and both PC/SAN and PC/PMMA laminates with thin layers exhibited superior ballistic performance to that of laminates with thicker layers. Further decreases in the thickness of the PMMA layer should produce better ballistic performance.

[35]The T-peel method is a way to measure the peel resistance of adhesives.